Look at the Facts
Climate Change Denial

Contents

Chapter 1

Climate Change Denial

1.1 Climate change denial

This article is about views which undermine public confidence in scientific consensus on climate change. For the public debate over scientific conclusions, see global warming controversy.

Climate change denial, or **global warming denial**, involves denial, dismissal, or unwarranted doubt about the scientific consensus on the rate and extent of global warming, the extent to which it is caused by humans, its impacts on nature and human society, or the potential for human actions to reduce these impacts.[1][2] **Climate change skepticism** and *climate change denial* form an overlapping range of views, and generally have the same characteristics; both reject to a greater or lesser extent current scientific opinion on climate change.[3][4] Climate change denial can also be implicit, when individuals or social groups accept the science but divert their attention to less difficult topics rather than take action.[5] Several social science studies have analyzed these positions as forms of denialism.[6][7][8]

In the global warming controversy, campaigning to undermine public trust in climate science has been described as a "denial machine" of industrial, political and ideological interests, supported by conservative media and skeptical bloggers in manufacturing uncertainty about global warming.[9][10] In the public debate, phrases such as *climate skepticism* have frequently been used with the same meaning as *climate denialism*.[11] The labels are contested: those actively challenging climate science commonly describe themselves as "skeptics", but many do not comply with scientific skepticism and, regardless of evidence, continue to deny the validity of human caused global warming.[3]

Although there is a scientific consensus that human activity is the primary driver of climate change,[12][13] the politics of global warming has been impacted by climate change denial, hindering efforts to prevent climate change and adapt to the warming climate.[14][15][16] Typically, public debate on climate change denial may have the appearance of legitimate scientific discourse, but does not conform to scientific principles.[17][18]

Organised campaigning to undermine public trust in climate science is associated with conservative economic policies and backed by industrial interests opposed to the regulation of CO_2 emissions.[19] Climate change denial has been associated with the fossil fuels lobby, the Koch brothers, industry advocates and libertarian think tanks, often in the United States.[15][20][21][22] Between 2002 and 2010, nearly $120 million (£77 million) was anonymously donated, some by conservative billionaires via the Donors Trust and Donors Capital Fund, to more than 100 organizations seeking to undermine the public perception of the science on climate change.[23] In 2013 the Center for Media and Democracy reported that the State Policy Network (SPN), an umbrella group of 64 U.S. think tanks, had been lobbying on behalf of major corporations and conservative donors to oppose climate change regulation.[24]

1.1.1 Terminology

"Climate change skepticism" and "climate change denial" refer to denial, dismissal or unwarranted doubt of the scientific consensus on the rate and extent of global warming, its significance, or its connection to human behavior, in whole or in part.[25][26] Though there is a distinction between skepticism which indicates doubting the truth of an assertion and outright denial of the truth of an assertion, in the public debate phrases such as "climate scepticism" have frequently been used with the same meaning as climate denialism or contrarianism.[11][27]

The terminology emerged in the 1990s. Even though all scientists adhere to scientific skepticism as an inherent part of the process, by mid November 1995 the word "skeptic" was being used specifically for the minority who publicised views contrary to the scientific consensus. This small group of scientists presented their views in public statements and the media, rather than to the scientific community.[28][29] This usage continued.[30] In his December 1995 article *The Heat is On: The warming of the world's climate sparks a blaze of denial* , Ross Gelbspan said industry had engaged "a small band of skeptics" to confuse public opinion in a "persistent and well-funded campaign of denial".[31] His 1997 book *The Heat is On* may have been the first to concentrate specifically on the topic.[32] In it, Gelbspan discussed a "pervasive denial of global warming" in a "persistent campaign of denial and suppression" involving "undisclosed funding of these 'greenhouse skeptics' " with "the climate skeptics" confusing the public and influencing decision makers.[33] A November 2006 CBC Television documentary on the campaign was titled "The Denial Machine".[34][35] In 2007 journalist Sharon Begley reported on the "denial machine",[36] a phrase subsequently used by academics.[9][35]

In addition to *explicit denial*, social groups have shown *implicit denial* by accepting the scientific consensus, but failing to come to terms with its implications or take action to reduce the problem.[5] This was exemplified in Kari Norgaard's study of a village in Norway affected by climate change, where residents diverted their attention to other issues.[37]

The terminology is debated: most of those actively rejecting the scientific consensus use the terms *skeptic* and *climate change skepticism*, and only a few have expressed preference for being described as deniers,[26][38] but the word "skepticism" is incorrectly used, as scientific skepticism is an intrinsic part of scientific methodology.[39][40][41] The term *contrarian* is more specific, but used less frequently. In academic literature and journalism, the terms *climate change denial* and *climate change deniers* have well established usage as descriptive terms without any pejorative intent. Both the National Center for Science Education and historian Spencer R. Weart recognise that either option is problematic, but have decided to use "climate change denial" rather than "skepticism".[42][43]

Terms related to *denialism* have been criticised for introducing a moralistic tone, and potentially implying a link with Holocaust denial.[39][44] There have been claims that this link is intentional, which academics have strongly disputed.[45] The usage of "denial" long predates the Holocaust, and is commonly applied in other areas such as HIV/AIDS denialism: the claim is described by John Timmer of *Ars Technica* as itself being a form of denial.[46]

In December 2014, an open letter from the Committee for Skeptical Inquiry called on the media to stop using the term "skepticism" when referring to climate change denial. They contrasted scientific skepticism–which is "foundational to the scientific method"–with denial–"the a priori rejection of ideas without objective consideration", and the behavior of those involved in political attempts to undermine climate science. They said "Not all individuals who call themselves climate change skeptics are deniers. But virtually all deniers have falsely branded themselves as skeptics. By perpetrating this misnomer, journalists have granted undeserved credibility to those who reject science and scientific inquiry."[45][47] The letter was taken up by the advocacy group Face the Facts as the basis for an online petition to news media.[45][48] In June 2015 Media Matters for America were told by the *New York Times* Public Editor that the newspaper was increasingly tending to use "denier" when "someone is challenging established science", but assessing this on an individual basis with no fixed policy, and would not use the term when someone was "kind of wishy-washy on the subject or in the middle." The executive director of the Society of Environmental Journalists said that while there was reasonable skepticism about specific issues, she felt that denier was "the most accurate term when someone claims there is no such thing as global warming, or agrees that it exists but denies that it has any cause we could understand or any impact that could be measured."[49]

1.1.2   History

Further information: History of climate change science

Research on the effect of CO_2 on the climate began in the 19th century; Joseph Fourier discovered the atmospheric "greenhouse effect" in 1824, and in 1860 John Tyndall quantified the effect of each gas. This potential explanation of ice ages was investigated by Svante Arrhenius, who published research in 1896 showing that a geometric increase in CO_2 would cause an arithmetical increase in temperatures. He suggested that coal burning could cause the effect, and in a 1938 article Guy Stewart Callendar presented evidence that it was already happening. Both viewed this as a benign

Joseph Fourier is credited with first discovering the greenhouse effect in 1824, beginning scientific research into the effects of increased greenhouse gasses in the atmosphere.

possibility.[50][51]

Military services in the 1940s and 1950s supported scientific research into the environment, primarily for operational data and potential for warfare, but open to scientific discoveries. For example, Gilbert Plass worked on radiation transmission through the atmosphere for weapons systems, and "in the evening" wrote papers giving new impetus to greenhouse effect theory. Oceanographer Roger Revelle played a key role; his 1957 paper co-authored with Hans Suess overturned the presumption that oceans would quickly absorb increased CO_2, and has been described as "the opening shot in the global warming debates". Revelle was quick to inform both the public and government officials of the risks, spreading his theme that "In consuming our fossil fuels at a prodigious rate, our civilization is conducting a grandiose scientific experiment."[52][53] In the following decades, public and scientific concerns about this and other environmental issues increased. More research was needed, and was taken on by new agencies including NASA and NOAA, but funding became sporadic. The 1979 Charney Report reviewed the state of climate research, concluding that substantial warming was already on the way, and "the ocean, the great and ponderous flywheel of the global climate system, may be expected to slow the course of observable climatic change. A wait-and-see policy may mean waiting until it is too late."[54]

A conservative reaction built up, denying environmental concerns which could lead to government regulation. With the 1981 Presidency of Ronald Reagan, global warming became a political issue, with immediate plans to cut spending on environmental research, particularly climate related, and stop funding for CO_2 monitoring. Reagan appointed as Secretary of Energy James B. Edwards, who said that there was no real global warming problem. Congressman Al Gore had studied under Revelle and was aware of the developing science: he joined others in arranging congressional hearings from 1981 onwards, with testimony by scientists including Revelle, Stephen Schneider and Wallace Smith Broecker. The hearings gained enough public attention to reduce the cuts in atmospheric research.[55] A polarized party-political debate developed. In 1982 Sherwood B. Idso published his book *Carbon Dioxide: Friend or Foe?* which said increases in CO_2 would not warm the planet, but would fertilize crops and were "something to be encouraged and not suppressed", while complaining that his theories had been rejected by the "scientific establishment". An Environmental Protection Agency (EPA) report in 1983 said global warming was "not a theoretical problem but a threat whose effects will be felt within a few years", and could potentially be "catastrophic". The Reagan administration reacted by calling the report "alarmist", and the dispute got wide news coverage. Public attention turned to other issues, then the 1985 finding of a polar ozone hole brought a swift international response. To the public, this was related to climate change and the possibility of effective action, but news interest faded.[56]

Public attention was renewed amidst summer droughts and heat waves when James Hansen testified to a Congressional hearing on 23 June 1988, stating with high confidence that long term warming was under way with severe warming likely within the next 50 years, and warning of likely storms and floods. There was increasing media attention: the scientific community had reached a broad consensus that the climate was warming, human activity was very likely the primary cause, and there would be significant consequences if the warming trend was not curbed.[57] These facts encouraged discussion about new laws concerning environmental regulation, which was opposed by the fossil fuel industry.[58]

From 1989 onwards industry funded organisations including the Global Climate Coalition and the George C. Marshall Institute sought to spread doubt among the public, in a strategy already developed by the tobacco industry.[59][60][61] A small group of scientists opposed to the consensus on global warming became politically involved, and with support from conservative political interests, began publishing in books and the press rather than in scientific journals.[62] Spencer Weart identifies this period as the point where legitimate skepticism about basic aspects of climate science was no longer justified, and those spreading mistrust about these issues became deniers.[63] As their arguments were increasingly refuted by the scientific community and new data, deniers turned to political arguments, making personal attacks on the reputation of scientists, and promoting ideas of a global warming conspiracy.[64]

With the 1989 fall of communism and the environmental movement's international reach at the 1992 Rio Earth Summit, the attention of U.S. conservative think tanks, which had been organised in the 1970s as an intellectual counter-movement to socialism, turned from the "red scare" to the "green scare" which they saw as a threat to their aims of private property, free trade market economies and global capitalism. As a counter-movement, they used environmental skepticism to promote denial of the reality of problems such as loss of biodiversity and climate change.[65]

In 1992, an EPA report linked second-hand smoke with lung cancer. The tobacco industry engaged the APCO Worldwide public relations company, which set out a strategy of astroturfing campaigns to cast doubt on the science by linking smoking anxieties with other issues, including global warming, in order to turn public opinion against calls for government intervention. The campaign depicted public concerns as "unfounded fears" supposedly based only on "junk science" in contrast to their "sound science", and operated through front groups, primarily the Advancement of Sound Science Center (TASSC) and its Junk Science website, run by Steven Milloy. A tobacco company memo commented "Doubt is our product since it is the best means of competing with the 'body of fact' that exists in the mind of the general public. It is also the means of establishing a controversy." During the 1990s, the tobacco campaign died away, and TASSC began taking funding from oil companies including Exxon. Its website became central in distributing "almost every kind of climate-change denial that has found its way into the popular press."[66]

In the 1990s, the Marshall Institute began campaigning against increased regulations on environmental issues such as acid rain, ozone depletion, second-hand smoke, and the dangers of DDT.[60][66][67] In each case their argument was that the science was too uncertain to justify any government intervention, a strategy it borrowed from earlier efforts to downplay the health effects of tobacco in the 1980s.[59][61] This campaign would continue for the next two decades.[68]

These efforts succeeded in influencing public perception of climate science.[69] Between 1988 and the 1990s, public discourse shifted from the science and data of climate change to discussion of politics and surrounding controversy.[70]

The campaign to spread doubt continued into the 1990s, including an advertising campaign funded by coal industry advocates intended to "reposition global warming as theory rather than fact,"[71][72] and a 1998 proposal written by the American Petroleum Institute intending to recruit scientists to convince politicians, the media and the public that climate science was too uncertain to warrant environmental regulation.[73] The proposal included a US$ 5,000,000 multi-point strategy to "maximize the impact of scientific views consistent with ours on Congress, the media and other key audiences", with a goal of "raising questions about and undercutting the 'prevailing scientific wisdom'".[74]

In 1998, Gelbspan noted that his fellow journalists accepted that global warming was occurring, but said they were in "'stage-two' denial of the climate crisis", unable to accept the feasibility of answers to the problem.[75] A subsequent book by Milburn and Conrad on *The Politics of Denial* described "economic and psychological forces" producing denial of the consensus on global warming issues.[76]

These efforts by climate change denial groups were recognized as an organized campaign beginning in the 2000s.[77] Riley Dunlap and Aaron McCright played a significant role in this shift when they published an article in 2000 exploring the connection between conservative think tanks and climate change denial.[78]

Gelbspan's *Boiling Point*, published in 2004, detailed the fossil-fuel industry's campaign to deny climate change and undermine public confidence in climate science.[79] In *Newsweek*'s August 2007 cover story "The Truth About Denial", Sharon Begley reported that "the denial machine is running at full throttle", and said that this "well-coordinated, well-funded campaign" by contrarian scientists, free-market think tanks, and industry had "created a paralyzing fog of doubt around climate change."[36]

Referencing work of sociologists Robert Antonio and Robert Brulle, Wayne A. White has written that climate change denial has become the top priority in a broader agenda against environmental regulation being pursued by neoliberals.[80] Today, climate change skepticism is most prominently seen in the United States, where the media disproportionately features views of the climate change denial community.[81] In addition to the media, the contrarian movement has also been sustained by the growth of the internet, having gained some of its support from internet bloggers and amateur radio.[82]

1.1.3 Arguments and positions on global warming

Some climate change denial groups allege that CO_2 is only a trace gas in the atmosphere, and has little effect on the climate.[84] The scientific consensus, as summarized by the IPCC's fourth assessment report, the U.S. Geological Survey, and other reports, is that human activity is the leading cause of climate change. The burning of fossil fuels accounts for around 30 billion tons of CO_2 each year, which is 130 times the amount produced by volcanoes.[85] Some groups allege that water vapor is a more significant greenhouse gas, and is left out of many climate models.[84] However, water vapor has been incorporated into these models since the inception of climatology in the 1800s, and while it is also a greenhouse gas, CO_2 remains the primary driver of increasing temperatures.[86]

Climate denial groups may also argue that global warming stopped recently, a global warming hiatus, or that global temperatures are actually decreasing, leading to global cooling.[87]

These groups often point to natural variability, such as sunspots and cosmic rays, to explain the warming trend.[88] According to these groups, there is natural variability that will abate over time, and human influences have little to do with it. These factors are already taken into account when developing climate models, and the scientific consensus is that they cannot explain the recent warming trend.[89]

Global warming conspiracy theories have been posited which allege that the scientific consensus is illusory, or that clima-

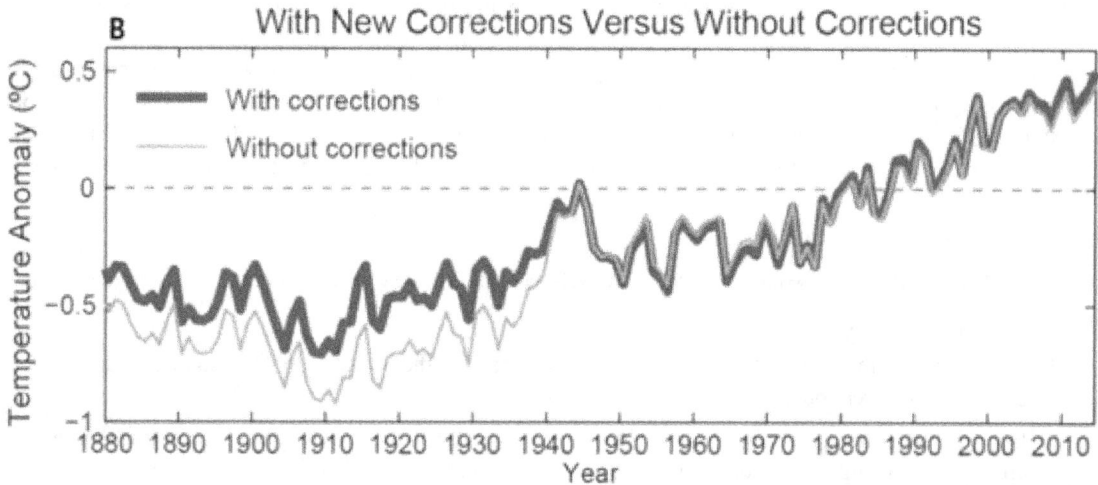

One argument is that global warming has recently stopped. However, temperature anomalies in an updated NOAA dataset show no evidence of a recent hiatus.[83]

tologists are acting on their own financial interests by causing undue alarm about a changing climate.[90][91] Despite leaked emails during climategate, as well as multinational, independent research on the topic, no evidence of such a conspiracy has been presented, and strong consensus exists among scientists from a multitude of political, social, organizational and national backgrounds about the extent and cause of climate change.[92][93] Several researchers have concluded that around 97% of climate scientists agree with this consensus.[94] As well, much of the data used in climate science is publicly available to be viewed and interpreted by competing researchers as well as the public.[95]

In 2012, research by Stephan Lewandowsky (then of the University of Western Australia) concluded that belief in other conspiracy theories, such as that the FBI was responsible for the assassination of Martin Luther King, Jr., was associated with being more likely to endorse climate change denial.[96]

Climate change denial literature often features the suggestion that we should wait for better technologies before addressing climate change, when they will be more affordable and effective.[97]

Taxonomy of climate change denial

In 2004 Stefan Rahmstorf described how the media give the misleading impression that climate change was still disputed within the scientific community, attributing this impression to PR efforts of climate change skeptics. He identified different positions argued by climate skeptics, which he used as a taxonomy of climate change skepticism:[98]

> 1. Trend sceptics (who deny there is global warming), [and] argue that no significant climate warming is taking place at all, claiming that the warming trend measured by weather stations is an artefact due to urbanisation around those stations ("urban heat island effect").
> 2. Attribution sceptics (who accept the global warming trend but see natural causes for this), [and] doubt that human activities are responsible for the observed trends. A few of them even deny that the rise in the atmospheric CO_2 content is anthropogenic [while others argue that] additional CO_2 does not lead to discernible warming [and] that there must be other – natural – causes for warming.
> 3. Impact sceptics (who think global warming is harmless or even beneficial).
> — [98][numbering added]

This taxonomy has been used in social science for analysis of publications, and to categorize climate change skepticism and climate change denial.[99][100]

The National Center for Science Education describes climate change denial as disputing differing points in the scientific consensus, a sequential range of arguments from denying the occurrence of climate change, accepting that but denying any significant human contribution, accepting these but denying scientific findings on how this would affect nature and human society, to accepting all these but denying that humans can mitigate or reduce the problems.[1] James L. Powell provides a more extended list,[2] as does climatologist Michael E. Mann in "six stages of denial", a ladder in which deniers have over time conceded acceptance of points, while retreating to a position which still rejects the mainstream consensus:[101]

1. CO_2 is not actually increasing.

2. Even if it is, the increase has no impact on the climate since there is no convincing evidence of warming.

3. Even if there is warming, it is due to natural causes.

4. Even if the warming cannot be explained by natural causes, the human impact is small, and the impact of continued greenhouse gas emissions will be minor.

5. Even if the current and future projected human effects on Earth's climate are not negligible, the changes are generally going to be good for us.

6. Whether or not the changes are going to be good for us, humans are very adept at adapting to changes; besides, it's too late to do anything about it , and/or a technological fix is bound to come along when we really need it.
   — [101]

Denialism in this context has been defined by Chris and Mark Hoofnagle as the use of rhetorical devices "to give the appearance of legitimate debate where there is none, an approach that has the ultimate goal of rejecting a proposition on which a scientific consensus exists." This process characteristically uses one or more of the following tactics:[18][102][103]

1. Allegations that scientific consensus involves conspiring to fake data or suppress the truth: a global warming conspiracy theory.

2. Fake experts, or individuals with views at odds with established knowledge, at the same time marginalising or denigrating published topic experts. Like the manufactured doubt over smoking and health, a few contrarian scientists oppose the climate consensus, some of them the same individuals.

3. Selectivity, such as cherry picking atypical or even obsolete papers, in the same way that the MMR vaccine controversy was based on one paper: examples include discredited ideas of the medieval warm period.

4. Unworkable demands of research, claiming that any uncertainty invalidates the field or exaggerating uncertainty while rejecting probabilities and mathematical models.

5. Logical fallacies, such as claiming that findings supporting environmental controls are a threat to the democratic way of life.

1.1.4 Pseudoscience

Various groups, including the National Center for Science Education, have described climate change denial as be a form of pseudoscience.[104][105][106] Climate change skepticism, while in some cases professing to do research on climate change, has focused instead on influencing the opinion of the public, legislators and the media, in contrast to legitimate science.[107]

In a review of the book *The Pseudoscience Wars: Immanuel Velikovsky and the Birth of the Modern Fringe* by Michael D. Gordin, David Morrison wrote:

"In his final chapter, Gordin turns to the new phase of pseudoscience, practiced by a few rogue scientists themselves. Climate change denialism is the prime example, where a handful of scientists, allied with an effective PR machine, are publicly challenging the scientific consensus that global warming is real and is due primarily to human consumption of fossil fuels. Scientists have watched in disbelief that as the evidence for global warming has become ever more solid, the deniers have been increasingly successful in the public and political arena.... Today pseudoscience is still with us, and is as dangerous a challenge to science as it ever was in the past.[108]

Journalists and newspaper columnists including George Monbiot[109][110][111] and Ellen Goodman,[110] among others,[7][8] have described climate change denial as a form of denialism.[6]

1.1.5 Public opinion

Main article: Public opinion on climate change

Public opinion on climate change is significantly impacted by media coverage of climate change, and the effects of climate change denial campaigns. Campaigns to undermine public confidence in climate science have decreased public belief in climate change, which in turn have impacted legislative efforts to curb CO_2 emissions.[112]

The popular media in the U.S. gives greater attention to climate change skeptics than the scientific community as a whole, and the level of agreement within the scientific community has not been accurately communicated.[113][114][115] In some cases, news outlets have allowed climate change skeptics to explain the science of climate change instead of experts in climatology.[116] US and UK media coverage differ from that presented in other countries, where reporting is more consistent with the scientific literature.[117][118] Some journalists attribute the difference to climate change denial being propagated, mainly in the US, by business-centered organizations employing tactics worked out previously by the US tobacco lobby.[59][119][120] In France, the US and the UK, the opinions of climate change skeptics appear much more frequently in conservative news outlets than other news, and in many cases those opinions are left uncontested.[121]

The efforts of Al Gore and other environmental campaigns have focused on the effects of global warming and have managed to increase awareness and concern, but despite these efforts, the number of Americans believing humans are the cause of global warming was holding steady at 61% in 2007, and those believing the popular media was understating the issue remained about 35%.[122]

A study assessed the public perception and actions to climate change, on grounds of belief systems, and identified seven psychological barriers affecting the behavior that otherwise would facilitate mitigation, adaptation, and environmental stewardship. The author found the following barriers: cognition, ideological world views, comparisons to key people, costs and momentum, discredence toward experts and authorities, perceived risks of change, and inadequate behavioral changes.[123]

1.1.6 Lobbying

Efforts to lobby against environmental regulation have included campaigns to manufacture doubt about the science behind climate change, and to obscure the scientific consensus and data.[124] These efforts have undermined public confidence in climate science, and impacted climate change lobbying.[15][112]

The political advocacy organizations FreedomWorks and Americans for Prosperity, funded by brothers David and Charles Koch of Koch Industries, were important in supporting the Tea Party movement and in encouraging the movement to focus on climate change.[125] Other conservative organizations such as the Heritage Foundation, Marshall Institute, Cato Institute and the American Enterprise Institute were significant participants in these lobbying attempts, seeking to halt or eliminate environmental regulations.[126]

This approach to downplay the significance of climate change were copied from tobacco lobbyists; in the face of scientific evidence linking tobacco to lung cancer, to prevent or delay the introduction of regulation. Lobbyists attempted to discredit the scientific research by creating doubt and manipulating debate. They worked to discredit the scientists involved, to dispute their findings, and to create and maintain an apparent controversy by promoting claims that contradicted scientific research. ""Doubt is our product," boasted a now infamous 1969 industry memo. Doubt would shield the tobacco industry from litigation and regulation for decades to come."[127] In 2006, George Monbiot wrote in *The Guardian* about similarities between the methods of groups funded by Exxon, and those of the tobacco giant Philip Morris, including direct attacks on peer-reviewed science, and attempts to create public controversy and doubt.[109]

Former National Academy of Sciences president Frederick Seitz, who, according to an article by Mark Hertsgaard in *Vanity Fair*, earned about US$585,000 in the 1970s and 1980s as a consultant to R.J. Reynolds Tobacco Company,[128] went on to chair groups such as the Science and Environmental Policy Project and the George C. Marshall Institute alleged to have made efforts to "downplay" global warming. Seitz stated in the 1980s that "Global warming is far more a matter of politics than of climate." Seitz authored the Oregon Petition, a document published jointly by the Marshall Institute and Oregon Institute of Science and Medicine in opposition to the Kyoto protocol. The petition and accompanying "Research Review of Global Warming Evidence" claimed:

> The proposed limits on greenhouse gases would harm the environment, hinder the advance of science and technology, and damage the health and welfare of mankind. There is no convincing scientific evidence that human release of carbon dioxide, methane, or other greenhouse gases is causing or will, in the foreseeable future, cause catastrophic heating of the Earth's atmosphere and disruption of the Earth's climate. ... We are living in an increasingly lush environment of plants and animals as a result of the carbon dioxide increase. Our children will enjoy an Earth with far more plant and animal life than that with which we now are blessed. This is a wonderful and unexpected gift from the Industrial Revolution.[109]

George Monbiot wrote in *The Guardian* that this petition, which he criticizes as misleading and tied to industry funding, "has been cited by almost every journalist who claims that climate change is a myth." Efforts by climate change denial groups played a significant role in the eventual rejection of the Kyoto protocol in the US.[129]

Monbiot has written about another group founded by the tobacco lobby, The Advancement of Sound Science Coalition (TASSC), that now campaigns against measures to combat global warming. In again trying to manufacture the appearance of a grass-roots movement against "unfounded fear" and "over-regulation," Monbiot states that TASSC "has done more damage to the campaign to halt [climate change] than any other body."[109]

Drexel University environmental sociologist Robert Brulle analysed the funding of 91 organizations opposed to restrictions on carbon emissions, which he termed the "climate change counter-movement." Between 2003 and 2013, the donor-advised funds Donors Trust and Donors Capital Fund, combined, were the largest funders, accounting for about one quarter of the total funds, and the American Enterprise Institute was the largest recipient, 16% of the total funds. The study also found that the amount of money donated to these organizations by means of foundations whose funding sources cannot be traced had risen.[130][131][132][133][134]

Private sector

See also: Business action on climate change

Several large corporations within the fossil fuel industry provide significant funding to the climate change denial movement.[135]

After the IPCC released its February 2007 report, the American Enterprise Institute offered British, American and other scientists $10,000, plus travel expenses to publish articles critical of the assessment. The institute had received more than $US 1.6 million from Exxon, and its vice-chairman of trustees was former head of Exxon Lee Raymond. Raymond sent letters that alleged the IPCC report was not "supported by the analytical work." More than 20 AEI employees worked as consultants to the George W. Bush administration.[136] Despite her initial conviction that climate change denial would abate with time, Senator Barbara Boxer said that when she learned of the AEI's offer, she "realized there was a movement behind this that just wasn't giving up."[137]

The Royal Society conducted a survey that found ExxonMobil had given US$ 2.9 million to American groups that "misinformed the public about climate change," 39 of which "misrepresented the science of climate change by outright denial of the evidence".[138][139] In 2006, the Royal Society issued a demand that ExxonMobil withdraw funding for climate change denial. The letter drew criticism, notably from Timothy Ball who argued the society attempted to "politicize the private funding of science and to censor scientific debate."[140]

ExxonMobil denied that it has been trying to mislead the public about global warming. A spokesman, Gantt Walton, said that ExxonMobil's funding of research does not mean that it acts to influence the research, and that ExxonMobil supports taking action to curb the output of greenhouse gasses.[141] Research conducted at a Exxon archival collection at the University of Texas and interviews with former employees by journalists indicate the scientific opinion within the company and their public posture towards climate change was contradictory.[142]

Between 1989 and 2002 the Global Climate Coalition, a group of mainly United States businesses, used aggressive lobbying and public relations tactics to oppose action to reduce greenhouse gas emissions and fight the Kyoto Protocol. The coalition was financed by large corporations and trade groups from the oil, coal and auto industries. The *New York Times* reported that "even as the coalition worked to sway opinion [towards skepticism], its own scientific and technical experts were advising that the science backing the role of greenhouse gases in global warming could not be refuted."[143] In 2000, Ford Motor Company was the first company to leave the coalition as a result of pressure from environmentalists,[144] followed by Daimler-Chrysler, Texaco, the Southern Company and General Motors subsequently left to GCC.[145] The organization closed in 2002.

In early 2015, several media reports emerged saying that Willie Soon, a popular scientist in climate denialist circles, had failed to disclose conflicts of interest in at least 11 scientific papers published since 2008.[146] They reported that he received a total of $1.25m from ExxonMobil, Southern Company, the American Petroleum Institute and a foundation run by the Koch brothers.[147] Charles R. Alcock, director of the Harvard–Smithsonian Center for Astrophysics, where Soon was based, said that allowing funders of Dr. Soon's work to prohibit disclosure of funding sources was a mistake, which will not be permitted in future grant agreements.[148]

Public sector

In 1994, according to a leaked memo, the Republican strategist Frank Luntz advised members of the Republican Party, with regard to climate change, that "you need to continue to make the lack of scientific certainty a primary issue" and "challenge the science" by "recruiting experts who are sympathetic to your view."[137] In 2006, Luntz stated that he still believes "back [in] '97, '98, the science was uncertain", but he now agrees with the scientific consensus.[149]

In 2005, the *New York Times* reported that Philip Cooney, former fossil fuel lobbyist and "climate team leader" at the American Petroleum Institute and President George W. Bush's chief of staff of the Council on Environmental Quality, had "repeatedly edited government climate reports in ways that play down links between such emissions and global warming, according to internal documents."[150] Sharon Begley reported in *Newsweek* that Cooney "edited a 2002 report on climate science by sprinkling it with phrases such as 'lack of understanding' and 'considerable uncertainty.'" Cooney reportedly removed an entire section on climate in one report, whereupon another lobbyist sent him a fax saying "You are doing a great job."[137] Cooney announced his resignation two days after the story of his tampering with scientific reports broke,[151] but a few days later it was announced that Cooney would take up a position with ExxonMobil.[152]

In 2015, environmentalist Bill McKibben accused President Obama of "Catastrophic Climate-Change Denial", for his approval of oil-drilling permits in offshore Alaska. According to McKibben, the President has also "opened huge swaths of the Powder River basin to new coal mining." McKibben calls this "climate denial of the status quo sort", where the President denies "the meaning of the science, which is that we must keep carbon in the ground." [153]

Schools

According to documents leaked in February, 2012, The Heartland Institute is developing a curriculum for use in schools which frames climate change as a scientific controversy.[154][155][156]

1.1.7 Effect

Manufactured uncertainty over climate change, the fundamental strategy of climate change denial, has been very effective, particularly in the US. It has contributed to low levels of public concern and to government inaction worldwide.[16][157] An Angus Reid poll released in 2010 indicates that global warming skepticism in the United States, Canada, and the United Kingdom has been rising.[158][159] There may be multiple causes of this trend, including a focus on economic rather than environmental issues, and a negative perception of the United Nations and its role in discussing climate change.[160] Another cause may be weariness from overexposure to the topic: secondary polls suggest that the public may have been discouraged by extremism when discussing the topic,[158] while other polls show 54% of U.S. voters believe that "the news media make global warming appear worse than it really is."[161] A poll in 2009 regarding the issue of whether "some scientists have falsified research data to support their own theories and beliefs about global warming" showed that 59% of Americans believed it "at least somewhat likely", with 35% believing it was "very likely".[160]

According to Tim Wirth, "They patterned what they did after the tobacco industry. [...] Both figured, sow enough doubt, call the science uncertain and in dispute. That's had a huge impact on both the public and Congress."[59] This approach has been propagated by the US media, presenting a false balance between climate science and climate skeptics.[162] *Newsweek* reports that the majority of Europe and Japan accept the consensus on scientific climate change, but only one third of Americans considered human activity to play a major role in climate change in 2006; 64% believed that scientists disagreed about it "a lot."[163] A 2007 *Newsweek* poll found these numbers were declining, although majorities of Americans still believed that scientists were uncertain about climate change and its causes.[164] Rush Holt wrote a piece for *Science*, which appeared in *Newsweek*:

> "...for more than two decades scientists have been issuing warnings that the release of greenhouse gases, principally carbon dioxide (CO_2), is probably altering Earth's climate in ways that will be expensive and even deadly. The American public yawned and bought bigger cars. Statements by the American Association for the Advancement of Science, American Geophysical Union, American Meteorological Society, Intergovernmental Panel on Climate Change, and others underscored the warnings and called for new government policies to deal with climate change. Politicians, presented with noisy statistics, shrugged, said there is too much doubt among scientists, and did nothing."[165]

Deliberate attempts by the Western Fuels Association "to confuse the public" have succeeded in their objectives. This has been "exacerbated by media treatment of the climate issue". According to a Pew poll in 2012, 57% of the US public are unaware of, or outright reject, the scientific consensus on climate change.[166] Some organizations promoting climate change denial have asserted that scientists are increasingly rejecting climate change, but this notion is contradicted by research showing that 97% of published papers endorse the scientific consensus, and that percentage is increasing with time.[166]

1.1.8 See also

- Agnotology

- Anti-environmentalism

- Information Council on the Environment

- International Conference on Climate Change

- Renewable energy commercialization#Non-technical barriers to acceptance

- Semmelweis reflex

1.1.9 References

[1] National Center for Science Education 2010: "The first pillar of climate change denial — that climate change is bad science — attacks various aspects of the scientific consensus about climate change... there are climate change deniers:

- who deny that significant climate change is occurring

- who... deny that human activity is significantly responsible

- who... deny the scientific evidence about its significant effects on the world and our society...

- who... deny that humans can take significant actions to reduce or mitigate its impact.

Of these varieties of climate change denial, the most visible are the first and the second."

[2] Powell 2012, pp. 170–173: "Anatomy of Denial – Global warming deniers.... throw up a succession of claims, and fall back from one line of defense to the next as scientists refute each one in turn. Then they start over:
'The earth is not warming.'
'All right, it is warming but the Sun is the cause.'
'Well then, humans are the cause, but it doesn't matter, because it warming will do no harm. More carbon dioxide will actually be beneficial. More crops will grow.'
'Admittedly, global warming could turn out to be harmful, but we can do nothing about it.'
'Sure, we could do something about global warming, but the cost would be too great. We have more pressing problems here and now, like AIDS and poverty.'
'We might be able to afford to do something to address global warming some-day, but we need to wait for sound science, new technologies, and geoengineering.'
'The earth is not warming. Global warming ended in 1998; it was never a crisis.'

[3] Dunlap 2013, pp. 691–698: "There is debate over which term is most appropriate... Those involved in challenging climate science label themselves "skeptics"... Yet skepticism is...a common characteristic of scientists, making it inappropriate to allow those who deny AGW to don the mantle of skeptics...It seems best to think of skepticism-denial as a continuum, with some individuals (and interest groups) holding a skeptical view of AGW...and others in complete denial"

[4] Timmer 2014

[5] National Center for Science Education 2012: "Climate change denial is most conspicuous when it is explicit, as it is in controversies over climate education. The idea of implicit (or "implicatory") denial, however, is increasingly discussed among those who study the controversies over climate change. Implicit denial occurs when people who accept the scientific community's consensus on the answers to the central questions of climate change on the intellectual level fail to come to terms with it or to translate their acceptance into action. Such people are in denial, so to speak, about climate change."

[6] "Timeline, Climate Change and its Naysayers". Newsweek. 13 August 2007.

[7] Christoff, Peter (July 9, 2007). "Climate change is another grim tale to be treated with respect - Opinion". Melbourne: Theage.com.au. Retrieved 2010-03-19.

[8] Connelly, Joel (2007-07-10). "Deniers of global warming harm us". Seattle Post-Intelligencer. Retrieved 2009-12-25.

[9] Dunlap 2013, pp. 691–698: "From the outset, there has been an organized "disinformation" campaign... to "manufacture uncertainty" over AGW ... especially by attacking climate science and scientists ... waged by a loose coalition of industrial (especially fossil fuels) interests and conservative foundations and think tanks ... often assisted by a small number of 'contrarian scientists. ... greatly aided by conservative media and politicians ... and more recently by a bevy of skeptical bloggers. This 'denial machine' has played a crucial role in generating skepticism toward AGW among laypeople and policy makers "

[10] Begley 2007: "ICE and the Global Climate Coalition lobbied hard against a global treaty to curb greenhouse gases, and were joined by a central cog in the denial machine: the George C. Marshall Institute, a conservative think tank. .... the denial machine—think tanks linking up with like-minded, contrarian researchers"

[11] Nerlich 2010, pp. 419, 437: "Climate scepticism in the sense of climate denialism or contrarianism is not a new phenomenon, but it has recently been very much in the media spotlight. .... Such disagreements are not new but the emails provided climate sceptics, in the sense of deniers or contrarians, with a golden opportunity to mount a sustained effort aimed at demonstrating the legitimacy of their views. This allowed them to question climate science and climate policies based on it and to promote political inaction and inertia. .... footnote 1. I shall use 'climate sceptics' here in the sense of 'climate deniers', although there are obvious differences between scepticism and denial (see Shermer, 2010; Kemp, et al., 2010). However, 'climate sceptic' and 'climate scepticism' were commonly used during the 'climategate' debate as meaning 'climate denier'."

[12] Oreskes, Naomi (2007). "The Scientific Consensus on Climate Change: How Do We Know We're Not Wrong?". In DiMento, Joseph F. C.; Doughman, Pamela M. *Climate Change: What It Means for Us, Our Children, and Our Grandchildren*. The MIT Press. pp. 65–66. ISBN 978-0-262-54193-0.

[13] "CLIMATE CHANGE 2014: Synthesis Report. Summary for Policymakers" (PDF). IPCC. Retrieved 7 March 2015. The evidence for human influence on the climate system has grown since the Fourth Assessment Report (AR4). It is extremely likely that more than half of the observed increase in global average surface temperature from 1951 to 2010 was caused by the anthropogenic increase in greenhouse gas concentrations and other anthropogenic forcings together

[14] Dunlap 2013: "Even though climate science has now firmly established that global warming is occurring, that human activities contribute to this warming... a significant portion of the American public remains ambivalent or unconcerned, and many policymakers (especially in the United States) deny the necessity of taking steps to reduce carbon emissions...From the outset, there has been an organized "disinformation" campaign... to generate skepticism and denial concerning AGW."

[15] Jacques, Dunlap & Freeman 2008, p. 351: "Conservative think tanks...and their backers launched a full-scale counter-movement... We suggest that this counter-movement has been central to the reversal of US support for environmental protection, both do- mestically and internationally. Its major tactic has been disputing the seriousness of environmental problems and undermining environmental science by promoting what we term 'environmental scepticism.'"

[16] Painter & Ashe 2012: "Despite a high degree of consensus amongst publishing climate researchers that global warming is occurring, and that it is anthropogenic, this discourse, promoted largely by non-scientists, has had a significant impact on public perceptions of the issue, fostering the impression that elite opinion is divided as to the nature and extent of the threat."

[17] Hoofnagle, Mark (April 30, 2007). "Hello Science blogs (Welcome to Denialism blog)".

[18] Diethelm & McKee 2009

[19] Klein, Naomi (November 9, 2011). "Capitalism vs. the Climate". *The Nation*. Retrieved 2 January 2012.

[20] Dunlap 2013: "The campaign has been waged by a loose coalition of industrial (especially fossil fuels) interests and conservative foundations and think tanks... These actors are greatly aided by conservative media and politicians, and more recently by a bevy of skeptical bloggers."

[21] David Michaels (2008) *Doubt is Their Product: How Industry's Assault on Science Threatens Your Health.*

[22] Hoggan, James; Littlemore, Richard (2009). *Climate Cover-Up: The Crusade to Deny Global Warming.* Vancouver: Greystone Books. ISBN 978-1-55365-485-8. Retrieved 2010-03-19. See, e.g., p31 *ff*, describing industry-based advocacy strategies in the context of climate change denial, and p73 *ff*, describing involvement of free-market think tanks in climate-change denial.

[23] Goldenberg, Suzanne (14 February 2013). "Secret funding helped build vast network of climate denial thinktanks". *The Guardian* (London). Retrieved 1 March 2013.

[24] Pilkington, Ed (14 November 2013). "Facebook and Microsoft help fund rightwing lobby network, report finds". *The Guardian*. Retrieved 17 November 2013.

[25] Painter & Ashe 2012: "'Climate scepticism' and 'climate denial' are readily used concepts, referring to a discourse that has become important in public debate since climate change was first put firmly on the policy agenda in 1988. This discourse challenges the views of mainstream climate scientists and environmental policy advocates, contending that parts, or all, of the scientific treatment and political interpretation of climate change are unreliable."

[26] National Center for Science Education 2012: "There is debate...about how to refer to the positions that reject, and to the people who doubt or deny, the scientific community's consensus on...climate change. Many such people prefer to call themselves skeptics and describe their position as climate change skepticism. Their opponents, however, often prefer to call such people climate change deniers and to describe their position as climate change denial... "Denial" is the term preferred even by many deniers."

[27] Rennie 2009: "Within the community of scientists and others concerned about anthropogenic climate change, those whom Inhofe calls skeptics are more commonly termed contrarians, naysayers and denialists."

[28] Brown 1996, pp. 9, 11 "Indeed, the 'skeptic' scientists[14] were perceived to be all the more credible precisely *because* their views were contrary to the consensus of peer-reviewed science.
14. All scientists are skeptics because the scientific process demands continuing questioning. In this report, however, the scientists we refer to as 'skeptics' are those who have taken a highly visible public role in criticizing the scientific consensus on ozone depletion and climate change through publications and statements addressed more to the media and the public than to the scientific community."

[29] Gelbspan 1998, pp. 69–70, 246 At the 16 Nov 1995 United States House Science Subcommittee on Energy hearing, Pat Michaels testified of "a small minority" opposing the IPCC assessment, and said "that the so-called skeptics were right".

[30] Antilla 2005, p. footnote 5

[31] Gelbspan 1995

[32] Painter & Ashe 2012: "The term 'climate scepticism' emerged in around 1995, the year journalist Ross Gelbspan authored perhaps the first book focusing directly on what would retrospectively be understood as climate scepticism."

[33] Gelbspan 1998 p. 3 "But some individuals do not want the public to know about the immediacy and extent of the climate threat. They have been waging a persistent campaign of denial and suppression that has been lamentably effective."
pp. 33–34 "The campaign to keep the climate change off the public agenda involves more than the undisclosed funding of these 'greenhouse skeptics.' In their efforts to challenge the consensus scientific view....."
p. 35 "If the climate skeptics have succeeded in confusing the general public, their influence on decision makers has been, if anything, even more effective
p. 173 "pervasive denial of global warming"

[34] CBC News: the fifth estate 2007: "*The Denial Machine* investigates the roots of the campaign to negate the science and the threat of global warming. It tracks the activities of a group of scientists, some of whom previously consulted for Big Tobacco, and who are now receiving donations from major coal and oil companies. ... The documentary shows how fossil fuel corporations have kept the global warming debate alive long after most scientists believed that global warming was real and had potentially catastrophic consequences. ... *The Denial Machine* also explores how the arguments supported by oil companies were adopted by policy makers in both Canada and the U.S. and helped form government policy."

[35] Orlóci 2008, pp. 86, 97: "The ideological justification for this came from the sceptics (e.g., Lomborg 2001a,b) and from the industrial 'denial machine'. ... CBC Television Fifth Estate, November 15, 2006, The Climate Denial Machine, Canada.

[36] Begley 2007: "If you think those who have long challenged the mainstream scientific findings about global warming recognize that the game is over, think again. ... outside Hollywood, Manhattan and other habitats of the chattering classes, the denial machine is running at full throttle—and continuing to shape both government policy and public opinion. Since the late 1980s, this well-coordinated, well-funded campaign by contrarian scientists, free-market think tanks and industry has created a paralyzing fog of doubt around climate change. Through advertisements, op-eds, lobbying and media attention, greenhouse doubters (they hate being called deniers) argued first that the world is not warming; measurements indicating otherwise are flawed, they said. Then they claimed that any warming is natural, not caused by human activities. Now they contend that the looming warming will be minuscule and harmless. 'They patterned what they did after the tobacco industry,' says former senator Tim Wirth"

[37] Norgaard, Kari (2011). *Living in Denial Climate Change, Emotions, and Everyday Life*. Cambridge, Mass: MIT Press. pp. 1–4. ISBN 978-0-262-01544-8.

[38] Washington 2013, p. 2: "Many climate change deniers call themselves climate 'skeptics'...However, refusing to accept the overwhelming 'preponderance of evidence' is not skepticism, it is *denial* and should be called by its true name... The use of the term 'climate skeptic' is a distortion of reality...Skepticism is healthy in both science and society; denial is not."

[39] O'Neill, Saffron J.; sjoneill@unimelb.edu.au; Boykoff, Max (28 Sep 2010). "Climate denier, skeptic, or contrarian?". *Proceedings of the National Academy of Sciences* **107** (39): E151–E151. Bibcode:2010PNAS..107E.151O. doi:10.1073/pnas.1010507107. ISSN 0027-8424. PMID 20807754. Retrieved 2 Jun 2015. Using the language of denialism brings a moralistic tone into the climate change debate that we would do well to avoid. Further, labeling views as denialist has the potential to inappropriately link such views with Holocaust denial... However, skepticism forms an integral part of the scientific method, and, thus, the term is frequently misapplied in such phrases as "climate change skeptic."

[40] Mann, Michael E. (2013). *The Hockey Stick and the Climate Wars: Dispatches from the Front Lines*. Columbia University Press. ISBN 0231526385. Skepticism plays an essential role in the progress of science... Yet...in the context of the climate change denial movement... the term *skeptic* has often been co-opted to describe those who simply deny, rather than appraise critically.

[41] Jenkins 2015, p. 229: "many who deny the consensus on climate change are not really skeptics but rather contrarians who practice "a kind of one-sided skepticism that entails simply rejecting evidence that challenges one's preconceptions" (Mann 2012:26)"

[42] NCSE 2012: "Recognizing that no terminological choice is entirely unproblematic, NCSE — in common with a number of scholarly and journalistic observers of the social controversies surrounding climate change — opts to use the terms "climate changer deniers" and "climate change denial""

[43] Weart 2015 footnote 136a: "I do not mean to use the term "denier" pejoratively—it has been accepted by some of the group as a self-description—but simply to designate those who deny any likelihood of future danger from anthropogenic global warming."

[44] Anderegg, William R. L.; anderegg@stanford.edu; Prall, James W.; Harold, Jacob (19 Jul 2010). "Reply to O'Neill and Boykoff: Objective classification of climate experts". *Proceedings of the National Academy of Sciences* **107** (39): E152–E152. Bibcode:2010PNAS..107E.152A. doi:10.1073/pnas.1010824107. ISSN 0027-8424. PMID 20807739. Retrieved 2 Jun 2015.

[45] Gillis, Justin (12 February 2015). "Verbal Warming: Labels in the Climate Debate". *The New York Times*. Retrieved 30 June 2015.

[46] Timmer 2014: "some of the people who deserve that label are offended by it, thinking it somehow lumps them in with holocaust deniers. But that in its own way is a form of denial; the word came into use before the holocaust, and... denialism has been used as a label for people who refuse to accept the evidence for all sorts of things: HIV causing AIDS, vaccines being safe, etc."

[47] Boslough 2014

[48] Face the Facts petition

[49] "NY Times Public Editor: We're "Moving In A Good Direction" On Properly Describing Climate Deniers". *Media Matters for America*. 22 June 2015. Retrieved 2 July 2015.

[50] Conway & Oreskes 2010, p. 170: "The doubts and confusion of the American people are particularly peculiar when put into historical perspective"

[51] Powell 2012, pp. 36–39

[52] Weart 2015a: "From the late 1940s into the 1960s, many of the papers cited in these essays carried a thought-provoking footnote: "This work was supported by the 'Office of Naval Research.' "

[53] Weart 2007

[54] Weart 2015a: Charney Report quote p. viii in the Foreword by Climate Research Board chair Verner E. Suomi.

[55] Weart 2015a: Global Warming Becomes a Political Issue (1980-1983); "In 1981, Ronald Reagan took the presidency with an administration that openly scorned their concerns. He brought with him a backlash that had been building against the environmental movement. Many conservatives denied nearly every environmental worry, global warming included. They lumped all such concerns together as the rants of business-hating liberals, a Trojan Horse for government regulation." For details, see Money for Keeling: Monitoring CO2

[56] Weart 2015: Breaking into Politics (1980-1988), "Sherwood Idso, who published arguments that greenhouse gas emissions would not warm the Earth or bring any other harm to climate. Better still, by fertilizing crops, the increase of CO2 would bring tremendous benefits."

[57] Weart 2015 The Summer of 1988: "A new breed of interdisciplinary studies was showing that even a few degrees of warming might have harsh consequences, both for fragile natural ecosystems and for certain agricultural systems and other human endeavours.... The timing was right, and the media leaped on the story. Hansen's statements, especially that severe warming was likely within the next 50 years, got on the front pages of newspapers and were featured in television news and radio talk shows..... The story grew as the summer of 1988 wore on. Reporters descended unexpectedly upon an international conference of scientists held in Toronto at the end of June. Their stories prominently reported how the world's leading climate scientists declared that atmospheric changes were already causing harm, and might cause much more; the scientists called for vigorous government action to restrict greenhouse gases.

[58] Weart 2015: "Environmentalist organizations continued... lobbying and advertising efforts to argue for restrictions on emissions. The environmentalists were opposed, and greatly outspent, by industries that produced or relied on fossil fuels. Industry groups not only mounted a sustained and professional public relations effort, but also channeled considerable sums of money to individual scientists and small conservative organizations and publications that denied any need to act against global warming."

[59] Begley 2007: "Through advertisements, op-eds, lobbying and media attention, greenhouse doubters (they hate being called deniers) argued first that the world is not warming... Then they claimed that any warming is natural... Now they contend that the looming warming will be minuscule and harmless. 'They patterned what they did after the tobacco industry,' says former senator Tim Wirth... 'Both figured, sow enough doubt, call the science uncertain and in dispute. That's had a huge impact on both the public and Congress.'"

[60] Weart 2015: "The technical criticism most widely noted in the press came in several brief "reports" — not scientific papers in the usual sense — published between 1989 and 1992 by the conservative George C. Marshall Institute. The anonymously authored pamphlets... [claimed] that proposed government regulation would be "extraordinarily costly to the U.S. economy," they insisted it would be unwise to act on the basis of the existing global warming theories... In 1989 some of the biggest corporations in the petroleum, automotive, and other industries created a Global Climate Coalition, whose mission was to disparage every call for action against global warming."

[61] Conway & Oreskes 2010: "Millions of pages of documents released during tobacco litigation...show the crucial role that scientists played in sowing doubt about the links between smoking and health risks. These documents...also show that the same strategy was applied not only to global warming, but to a laundry list of environmental and health concerns, including asbestos, secondhand smoke, acid rain, and the ozone hole."

[62] Weart 2015: "Scientists noticed something that the public largely overlooked: the most outspoken scientific critiques of global warming predictions did not appear in the standard peer-reviewed scientific publications. The critiques tended to appear in venues funded by industrial groups, or in conservative media like the Wall Street Journal."

[63] Weart 2011, p. 46: "At some point they were no longer skeptics — people who would try to see every side of a case — but deniers, that is, people whose only interest was in casting doubt upon what other scientists agreed was true."

[64] Weart 2011, pp. 47: "As the deniers found ever less scientific ground to stand on, they turned to political arguments. Some of these policy arguments were straightforward, raising serious questions about the efficacy and expense of proposed carbon taxes and emission-regulation schemes. But leading deniers also resorted to ad hominem tactics... On each side, some people were coming to believe that they faced a dishonest conspiracy, driven by ideological bias and naked self-interest"

[65] Jacques, Dunlap & Freeman 2008, pp. 349–385: "Environmental scepticism encompasses several themes, but denial of the authenticity of environmental problems, particularly problems such as biodiversity loss or climate change that threaten ecological sustainability, is its defining feature"

[66] (Hamilton 2011, pp. 104–106): "the tactics, personnel and organisations mobilised to serve the interests of the tobacco lobby in the 1980s were seamlessly transferred to serve the interests of the fossil-fuel lobby in the 1990s. Frederick Seitz... the task of the climate sceptics in the think tanks and PR companies hired by fossil fuel companies was to engage in 'consciousness lowering activities', to 'de-problematise' global warming by describing it as a form of politically driven panicmongering." For the tobacco company memo, see "Original "Doubt is our product..." memo". University of California, San Francisco. 21 August 1969. Retrieved 19 March 2010.

[67] Conway & Oreskes 2010

[68] Conway & Oreskes 2010, p. 105: "As recently as 2007, the George Marshall Institute continued to insist that the damages associated with acid rain were always "largely hypothetical," and that "further scientific investigation revealed that most of them were not in fact occurring." The Institute cited no studies to support this extraordinary claim."

[69] Weart 2015: "Public support for environmental concerns in general seems to have waned after 1988."

[70] Weart 2015: "A study of American media found that in 1987 most items that mentioned the greenhouse effect had been feature stories about the science, whereas in 1988 the majority of the stories addressed the politics of the controversy. It was not that the number of science stories declined, but rather that as media coverage doubled and redoubled, the additional stories moved into social and political areas...Before 1988, the journalists had drawn chiefly on scientists for their information, but afterward they relied chiefly on sources who were identified with political positions or special interest groups."

[71] Wald, Matthew L. (1991-07-08). "Pro-Coal Ad Campaign Disputes Warming Idea". New York Times. Retrieved 1 March 2013.

[72] Begley 2007: "Individual companies and industry associations—representing petroleum, steel, autos and utilities, for instance— formed lobbying groups...[the Information Council on the Environment's] game plan called for enlisting greenhouse doubters to "reposition global warming as theory rather than fact," and to sow doubt about climate research just as cigarette makers had about smoking research.... The coal industry's Western Fuels Association paid Michaels to produce a newsletter called World Climate Report, which has regularly trashed mainstream climate science."

[73] Cox, Robert (2009). Environmental Communication and the Public Sphere. Sage. pp. 311–312. to recruit a cadre of scientists who share the industry's views of climate science and to train them in public relations so they can help convince journalists, politicians and the public that the risk of global warming is too uncertain to justify controls on greenhouse gases

[74] Cushman, John, "Industrial Group Plans to Battle Climate Treaty", The New York Times, April 25, 1998. Retrieved March 10, 2010.

[75] Gelbspan 1998, pp. 3, 35, 46, 197.

[76] Michael A. Milburn; Sheree D. Conrad (January 1998). The Politics of Denial. MIT Press. pp. 216–. ISBN 978-0-262-63184-6. Here again, as in the case of ozone depletion, economic and psychological forces are operating to produce a level of denial that threatens future generations.

[77] Painter & Ashe 2012: "Academics took note of the discourse when they began to analyse media representations of climate change knowledge and its effect on public perceptions and policy-making, but in the 1990s, they did not yet focus on it as a coherent and defined phenomenon. This changed in the 2000s, when McCright and Dunlap played an important role in deepening the concept of climate scepticism."

[78] Painter & Ashe 2012: "McCright and Dunlap played an important role in deepening the concept of climate scepticism. Examining what they termed a 'conservative countermovement' to undermine climate change policy...McCright and Dunlap went beyond the study of media representations of climate change knowledge to give a coherent picture of the movement behind climate scepticism in the US."

[79] Gelbspan, Ross (22 Jul 2004). "An excerpt from Boiling Point by Ross Gelbspan". *Grist*. Retrieved 1 Jun 2015.

[80] Wayne A. White (18 October 2012). *Biosequestration and Ecological Diversity: Mitigating and Adapting to Climate Change and Environmental Degradation*. CRC Press. p. 206. ISBN 978-1-4398-5363-4. Climate change denial and discrediting climate science have become pivotal to the antiregulatory cause of neoliberals.

[81] Antilla 2005: "At the centre of this climate backlash is a group of dissident scientists. The number of these climate sceptics is greater in the US than in any other country. Although the peer-reviewed scientific literature agrees with the IPCC, within the media—wherefrom the majority of adults in the US are informed about science—claims that are dismissive of anthropogenic climate change are prominently featured."

[82] Jenkins 2015, p. 243: "the community of climate change contrarians also includes a host of amateurs, from talk radio hosts to newspaper columnists to bloggers. In particular, the tremendous growth of the Internet has given sustenance to the contrarian movement"

[83] Global Warming 'Hiatus' Challenged by NOAA Research, New York Times, JUN 4, 2015. Quote: Russell S. Vose, chief of the climate science division at NOAA's Asheville center, pointed out in an interview that while the corrections do eliminate the recent warming slowdown, the overall effect of the agency's adjustments has long been to raise the reported global temperatures in the late 19th and early 20th centuries by a substantial margin. That makes the temperature increase of the past century appear less severe than it does in the raw data. If you just wanted to release to the American public our uncorrected data set, it would say that the world has warmed up about 2.071 degrees Fahrenheit since 1880," Dr. Vose said. "Our corrected data set says things have warmed up about 1.65 degrees Fahrenheit. Our corrections lower the rate of warming on a global scale."

[84] Rennie 2009: "Claim 1: Anthropogenic CO_2 can't be changing climate, because CO_2 is only a trace gas in the atmosphere and the amount produced by humans is dwarfed by the amount from volcanoes and other natural sources. Water vapor is by far the most important greenhouse gas, so changes in CO_2 are irrelevant."

[85] Rennie 2009: " According to the U.S. Geological Survey, anthropogenic CO_2 amounts to about 30 billion tons annually—more than 130 times as much as volcanoes produce."

[86] Rennie 2009: "from Arrhenius on, climatologists have incorporated water vapor into their models. In fact, water vapor is why rising CO_2 has such a big effect on climate... Nevertheless, within this dynamic, the CO_2 remains the main driver... of the greenhouse effect."

[87] Rennie 2009: "Claim 3: Global warming stopped a decade ago; Earth has been cooling since then."

[88] Rennie 2009: "Claim 4: The sun or cosmic rays are much more likely to be the real causes of global warming. After all, Mars is warming up, too."

[89] Rennie 2009: "But in defiance of the naysayers who want to chalk the recent warming up to natural cycles, there is insufficient evidence that enough extra solar energy is reaching our planet to account for the observed rise in global temperatures."

[90] Rennie 2009: "Claim 5: Climatologists conspire to hide the truth about global warming by locking away their data. Their so-called "consensus" on global warming is scientifically irrelevant because science isn't settled by popularity.... Claim 6: Climatologists have a vested interest in raising the alarm because it brings them money and prestige."

[91] White, Rob (2012). *Climate Change from a Criminological Perspective*. Springer Science & Business Media. p. 49. ISBN 1461436400. many Americans, including many American politicians and decision-makers, are increasingly viewing climate change as a "left-wing plot"–part of the "one-world socialist agenda" or a "conspiracy to impose world government and a sweeping redistribution of wealth." Just as Republican Senator James Inhofe of Oklahoma proclaimed on the Senate floor that "[g]lobal warming is the greatest hoax ever perpetrated on the American people", many Americans believe that climate change is "a cynical hoax perpetrated by climate scientists... greedy for grants."

[92] Rennie 2009: "If there were a massive conspiracy to defraud the world on climate (and to what end?), surely the thousands of e-mails and other files stolen from the University of East Anglia's Climatic Research Unit and distributed by hackers on November 20 would bear proof of it. So far, however, none has emerged. Most of the few statements that critics claim as evidence of malfeasance seem to have more innocent explanations that make sense in the context of scientists conversing privately and informally."

[93] Eight major investigations on the leaked emails include: House of Commons Science and Technology Committee (UK); Independent Climate Change Review (UK); International Science Assessment Panel (UK); Pennsylvania State University first panel and second panel (US); United States Environmental Protection Agency (US); Department of Commerce (US); National Science Foundation (US)

[94] Anderegg, William R L; Prall, James W.; Harold, Jacob; Schneider, Stephen H. (2010). "Expert credibility in climate change" (PDF). *Proc. Natl. Acad. Sci. U.S.A.* **107** (27): 12107–9. Bibcode:2010PNAS..10712107A. doi:10.1073/pnas.1003187107. PMC 2901439. PMID 20566872. (i) 97–98% of the climate researchers most actively publishing in the field support the tenets of ACC (Anthropogenic Climate Change) outlined by the Intergovernmental Panel on Climate Change, and (ii) the relative climate expertise and scientific prominence of the researchers unconvinced of ACC are substantially below that of the convinced researchers.

[95] Rennie 2009: "Climatologists are frequently frustrated by accusations that they are hiding their data or the details of their models because, as Gavin Schmidt points out, much of the relevant information is in public databases or otherwise accessible—a fact that contrarians conveniently ignore when insisting that scientists stonewall their requests."

[96] Lewandowsky, Stephan; Oberauer, Klaus (2013). "NASA Faked the Moon Landing—Therefore, (Climate) Science Is a Hoax". *Psychological Science* (Sage Publications) **24** (5): 622–633. doi:10.1177/0956797612457686.

[97] Rennie 2009: "Claim 7: Technological fixes, such as inventing energy sources that don't produce CO2 or geoengineering the climate, would be more affordable, prudent ways to address climate change than reducing our carbon footprint."

[98] Rahmstorf, S., 2004, The climate sceptics: Weather Catastrophes and Climate Change—Is There Still Hope For Us? (Munich: PG Verlag) pp 76–83

[99] Painter & Ashe 2012: "We focused on the marked differences in what climate sceptics are sceptical about... (1) trend sceptics (who deny the global warming trend), (2) attribution sceptics (who accept the trend, but either question the anthropogenic contribution saying it is overstated, negligent or non-existent compared to other factors like natural variation, or say it is not known with sufficient certainty what the main causes are) and (3) impact sceptics (who accept human causation, but claim impacts may be benign or beneficial, or that the models are not robust enough) and/or question the need for strong regulatory policies or interventions. "

[100] Dunlap & Jacques 2013, p. 702: "These books reject evidence that global warming is occurring, that human actions are the predominant cause of global warming, and/or that global warming will have negative impacts on human and natural systems. These arguments have been labelled trend, attribution, and impact denial (Rahmstorf, 2004). ... We located 108 books espousing one or more of these versions of climate change denial published through 2010"

[101] Michael E. Mann (13 August 2013). *The Hockey Stick and the Climate Wars: Dispatches from the Front Lines.* Columbia University Press. p. 23. ISBN 978-0-231-52638-8.

[102] Liu, D. W. C. (2012). "Science Denial and the Science Classroom". *CBE- Life Sciences Education* (American Society for Cell Biology) **11** (2): 129–134. doi:10.1187/cbe.12-03-0029. Retrieved 30 June 2015.

[103] Hoofnagle, Mark (11 March 2009). "Climate change deniers: failsafe tips on how to spot them". *the Guardian.* Retrieved 30 June 2015.

[104] "NCSE Tackles Climate Change Denial". *National Center for Science Education.* January 13, 2012. Retrieved July 2015. Science education is under attack... by climate change deniers, who ignore a mountain of evidence gathered over the last fifty years that the planet is warming and that humans are largely responsible. These deniers attempt to sabotage science education with fringe ideas, pseudoscience, and outright lies.

[105] Lahsen, Myanna (Winter 2005). "Technocracy, Democracy, and the U.S. Climate Politics: The Need for Demarcations". *Science, Technology, & Human Values* **30**: 137–169. doi:10.1177/0162243904270710. Numerous high-ranked officers in the Clinton-Gore administration sought to dismiss all critics of the climate paradigm as "pseudoscientists"

[106] Brown, Michael. Adversaries, zombies and NIPCC climate pseudoscience, "Phys.org", Sep 26, 2013

[107] Brown 1996, p. 28: "As the scientific fringe has become institutionalized, professionalized, and lionized... One finds that a fundamental difference between the traditional scientific establishment and the emerging "skeptic" establishment relates to their ultimate scientific goals. The former has traditionally emphasized the generation of new knowledge as a measure of productivity. That is, the collection of original data, construction of new mathematical techniques, and generation and validation of testable hypotheses have been the hallmarks of the traditional scientific community... On the other hand, the emerging culture profiled in these hearings emphasizes the generation of new perspectives. Productivity is measured on the ability to alter public opinion - through opinion pieces aimed not at their fellow scientists but at policymakers, the media, and the general public - and funding flows accordingly."

[108] Morrison, David. The Parameters of Pseudoscience, *Skeptical Inquirer*, Volume 37.2, March/April 2013. Book review of *The Pseudoscience Wars: Immanuel Velikovsky and the Birth of the Modern Fringe*, by Michael D. Gordin.

[109] Monbiot, George (2006-09-19). "The denial industry". London: Guardian Unlimited.

[110] Ellen Goodman (2007-02-09). "No change in political climate". The Boston Globe. Retrieved 2008-08-30.

[111] George Monbiot (2009-02-27). "Climate change: The semantics of denial". The Guardian. Retrieved 2015-05-27.

[112] Dunlap 2013: "From the outset, there has been an organized "disinformation" campaign... to "manufacture uncertainty" over AGW, especially by attacking climate science and scientists. This appears an effective strategy given that confidence in climate science and trust in climate scientists are key factors influencing the public's views of AGW."

[113] Boykoff, M.; Boykoff, J. (July 2004). "Balance as bias: global warming and the US prestige press" (PDF). *Global Environmental Change Part A* **14** (2): 125–136. doi:10.1016/j.gloenvcha.2003.10.001.

[114] Antilla 2005: "One problematic trend of the US media has been the suggestion that substantive disagreement exists within the international scientific community as to the reality of anthropogenic climate change; however, this concept is false...Although the science of climate change does not appear to be a prime news topic for most of the 255 newspapers included in this study...articles that framed climate change in terms of debate, controversy, or uncertainty were plentiful."

[115] Painter & Ashe 2012: "Media analysis of climate change reporting was always of interest to academics but from the mid-2000s, it became one of the key areas of research interest, highlighting a tendency to give undue weight to voices questioning the science of climate change."

[116] Antilla 2005: "Not only were there many examples of journalistic balance that led to bias, but some of the news outlets repeatedly used climate sceptics—with known fossil fuel industry ties—as primary definers"

[117] Dispensa, Jaclyn Marisa; Brulle, Robert J. "International Journal of Sociology and Social Policy". *International Journal of Sociology and Social Policy* **23** (10). doi:10.1108/01443330310790327. ISSN 0144-333X.

[118] Painter & Ashe 2012: "news coverage of scepticism is mostly limited to the USA and the UK...the type of sceptics who question whether global temperatures are warming are almost exclusively found in the US and UK newspapers. Sceptics who challenge the need for robust action to combat climate change also have a much stronger presence in the media of the same two countries."

[119] David, Adam (20 Sep 2006). "Royal Society tells Exxon: stop funding climate change denial". London: The Guardian. Retrieved 12 January 2009.

[120] Sandell, Clayton (3 January 2007). "Report: Big Money Confusing Public on Global Warming". ABC News. Retrieved 12 January 2009.

[121] Painter & Ashe 2012: "in the USA and the UK... sceptical voices generally appear in much higher numbers... in France, the UK and the USA... right-leaning newspapers are much more likely to include uncontested sceptical voices."

[122] Saad, Lydia (21 March 2007). "Did Hollywood's Glare Heat Up Public Concern About Global Warming?". Gallup. Retrieved 12 January 2010.

[123] Gifford R. (2011). "The dragons of inaction: psychological barriers that limit climate change mitigation and adaptation". *Am Psychol.* **66** (4): 290–302. doi:10.1037/a0023566. PMID 21553954.

[124] Jacques, Dunlap & Freeman 2008, p. 352: "While these CTTs sometimes joined corporate America in directly lobbying against environmental policies, their primary tactic in combating environmentalism has been to challenge the need for protective environmental policy by questioning the seriousness of environmental problems and the validity of environmental science."

[125] Dryzek, John S.; Norgaard, Richard B.; Schlosberg, David (2011). *The Oxford Handbook of Climate Change and Society*. Oxford University Press. p. 154. ISBN 9780199683420.

[126] Borowy, Iris (2014). *Defining Sustainable Development for Our Common Future: A History of the World Commission on Environment and Development*. Routledge. p. 44. Corporations and conservative think tanks such as the Heritage Foundation, Marshall Institute], the Cato Institute and the American Enterprise Institute waged campaigns to obscure scientific evidence about acid rain, ozone depletion and climate change and, thereby, to prevent or rollback environmental, health and safety regulations.

[127] Manjit, Kumar (2010-10-18). "Merchants of Doubt, By Naomi Oreskes & Erik M Conway". London: *The Independent*. Retrieved 17 February 2013.

[128] Hertsgaard, Mark (May 2006). "While Washington Slept". Vanity Fair. Retrieved 2007-08-02.

[129] Painter & Ashe 2012: "The work by McCright and Dunlap has highlighted the effectiveness of organized climate sceptic groups in influencing US policy making in the 1990s and early 2000s, including their central role in the rejection of the Kyoto Protocol by the US Congress"

[130] Brulle, Robert J. (December 21, 2013). "Institutionalizing delay: foundation funding and the creation of U.S. climate change counter-movement organizations". *Climatic Change* **122** (4): 681–694. doi:10.1007/s10584-013-1018-7.

[131] Goldenberg, Suzanne (December 20, 2013). "Conservative groups spend up to $1bn a year to fight action on climate change". *The Guardian. Retrieved 29 January 2015.*

[132] Fischer, Douglas (December 23, 2013). ""Dark Money" Funds Climate Change Denial Effort". *Scientific American.* Retrieved January 29, 2015.

[133] Goldenberg, Suzanne (February 14, 2013). "Secret funding helped build vast network of climate denial thinktanks". *The Guardian.* Retrieved February 7, 2015.

[134] "Robert Brulle: Inside the Climate Change "Countermovement"". Frontline (PBS). October 23, 2012. Retrieved February 21, 2015.

[135] Antilla 2005: "A number of large corporations that profit substantially from fossil fuel consumption, such as ExxonMobil, provide financial support to their political allies in an effort to undermine public trust in climate science."

[136] Sample, Ian (2007-02-02). "Scientists offered cash to dispute climate study". London: The Guardian. Retrieved 2007-08-16. The AEI has received more than $1.6m from ExxonMobil and more than 20 of its staff have worked as consultants to the Bush administration. Lee Raymond, a former head of ExxonMobil, is the vice-chairman of AEI's board of trustees.

[137] Begley 2007

[138] Adams, David (2006-09-20). "Royal Society tells Exxon: stop funding climate change denial". London: The Guardian. Retrieved 2007-08-02.

[139] Ward, Bob (2006-09-04). "Letter to Nick Thomas, Director, Corporate affairs, Esso UK Ltd. (ExxonMobil)" (PDF). London: Royal Society. Retrieved 2007-08-06.

[140] "Interfaith Stewardship Alliance Newsletter" (PDF). *Moyers on America.* 2006. Retrieved 2014-12-10.

[141] "Gore takes aim at corporately funded climate research". CBC News from Associated Press. 2007-08-07. Retrieved 2007-08-16.

[142] Jennings, Katie, Grandoni, Dino, & Rust, Susanne. (23 October 2015) "How Exxon went from leader to skeptic on climate change research". Los Angeles Times. Retrieved 26 October 2015. LA Times website

[143] Revkin, Andrew C. Industry Ignored Its Scientists on Climate, *New York Times.* April 23, 2009.

[144] Bradsher, Keith (1999-12-07). "Ford Announces Its Withdrawal From Global Climate Coalition". New York Times. Retrieved 2013-07-21. the Ford Motor Company said today that it would pull out of the Global Climate Coalition, a group of big manufacturers and oil and mining companies that lobbies against restrictions on emissions of gases linked to global warming.

[145] "GCC Suffers Technical Knockout, Industry defections decimate Global Climate Coalition".

[146] Gillis, Justin; Schartz, John (21 February 2015). "Deeper Ties to Corporate Cash for Doubtful Climate Researcher". *The New York Times.* Retrieved 7 March 2015. newly released documents show the extent to which Dr. Soon's work has been tied to funding he received from corporate interests. He has accepted more than $1.2 million in money from the fossil-fuel industry over the last decade while failing to disclose that conflict of interest in most of his scientific papers. At least 11 papers he has published since 2008 omitted such a disclosure, and in at least eight of those cases, he appears to have violated ethical guidelines of the journals that published his work. The documents show that Dr. Soon, in correspondence with his corporate funders, described many of his scientific papers as "deliverables" that he completed in exchange for their money.

[147] Goldenberg, Suzanne (21 February 2015). "Work of prominent climate change denier was funded by energy industry". *The Guardian.* Retrieved 7 March 2015. Over the last 14 years Willie Soon, a researcher at the Harvard-Smithsonian Centre for Astrophysics, received a total of $1.25m from Exxon Mobil, Southern Company, the American Petroleum Institute (API) and a foundation run by the ultra-conservative Koch brothers... the biggest single funder was Southern Company, one of the country's biggest electricity providers that relies heavily on coal.

[148] Schwartz, John (25 February 2015). "Lawmakers Seek Information on Funding for Climate Change Critics". *The New York Times*. Retrieved 7 March 2015. Charles R. Alcock, director of the Harvard-Smithsonian Center for Astrophysics, said last week that a contract provision with funders of Dr. Soon's work that appeared to prohibit disclosure of funding sources "was a mistake." "We will not permit similar wording in future grant agreements"

[149] "Frontline: Hot Politics: Interviews: Frank Luntz". PBS. 13 November 2006. Retrieved 2010-03-19.

[150] Revkin, Andrew C. (2005-06-08). "Bush Aide Edited Climate Reports". New York Times. Retrieved 2007-08-03.

[151] Andrew Revkin (10 June 2005). "Editor of Climate Report Resigns". *The New York Times*. Retrieved 2008-04-23.

[152] Andrew Revkin (15 June 2005). "Ex-Bush Aide Who Edited Climate Reports to Join ExxonMobil". *The New York Times*. Retrieved 2008-04-23.

[153] "Obama's Catastrophic Climate-Change Denial" by Bill McKibben, NY Times op-ed, May 12, 2015.

[154] Justin Gillis; Leslie Kaufman (February 15, 2012). "Leak Offers Glimpse of Campaign Against Climate Science". *The New York Times*. Retrieved February 16, 2012. plans to promote a curriculum that would cast doubt on the scientific finding that fossil fuel emissions endanger the long-term welfare of the planet.

[155] Stephanie Pappas; LiveScience (February 15, 2012). "Leaked: Conservative Group Plans Anti-Climate Education Program". Scientific American. Retrieved 2012-02-15.

[156] Suzanne Goldenberg (February 15, 2012). "Heartland Institute claims fraud after leak of climate change documents". The Guardian. Retrieved 2014-10-23.

[157] Lever-Tracy 2010, p. 255: "In sum, we see that manufacturing uncertainty over climate change is the fundamental strategy of the denial machine [...] As we reflect on the evolution of climate science and policy-making over the past few decades, we believe the denial machine has achieved considerable success – especially in the US but internationally as well. Public concern over global warming and support for climate policy-making in the US is low relative to other nations (see Chapter 10, this volume), contributing to inaction by the US government.

[158] Corcoran, Terence (6 January 2010). "The cool down in climate polls". *Financial Post*. Angus Reid surveyed people...before and after Copenhagen. The drop off in public support for the idea that global warming is a fact mostly caused by human activity looks most pronounced in Canada. In November, 63% of Canadians supported global warming as a man-made phenomenon. By Dec. 23, that support had fallen 52%... A similar trend has been noted in the United States, where confidence in global warming theory has dropped to 46%... down from 51% in July last year. In Britain, only 43% believe man-made global warming is a fact, down from... 55% in July. In all three countries, there are signs of growing skepticism.

[159] White, Rob (2012). *Climate Change from a Criminological Perspective*. Springer Science & Business Media. ISBN 1461436400. belief that climate change is "real" and confidence in climate science has surprisingly decreased... Angus Reid polls conducted in December 2009 found declining support for climate change...in Britain, Canada, and the United States.

[160] Rasmussen Reports (2009, December 03). Americans Skeptical of Science Behind Global Warming.

[161] Rasmussen Reports. (2009, February 06). 54% Say Media Hype Global Warming Dangers.

[162] Antilla 2005: "the popular press uses a number of methods to frame climate science as uncertain, including 'through the practice of interjecting and emphasizing controversy or disagreement among scientists'... In order to provide balance while reporting on climate change, some journalists include rebuttals by *experts* who, often through think-tanks, are affiliated with the fossil fuel industry. Regrettably, this creates the impression that scientific opinion is evenly divided or completely unsettled"

[163] Begley 2007: "polls found that 64 percent of Americans thought there was "a lot" of scientific disagreement on climate change; only one third thought planetary warming was "mainly caused by things people do." In contrast, majorities in Europe and Japan recognize a broad consensus among climate experts"

[164] Begley 2007: "A new NEWSWEEK Poll finds that the influence of the denial machine remains strong. Although the figure is less than in earlier polls, 39 percent of those asked say there is "a lot of disagreement among climate scientists" on the basic question of whether the planet is warming; 42 percent say there is a lot of disagreement that human activities are a major cause of global warming. Only 46 percent say the greenhouse effect is being felt today."

[165] Holt, Rush (13 July 2007). "Trying to Get Us to Change Course" (film review.)". *Science* **317** (5835): 198–9. doi:10.1126/science.1142810.

[166] Cook, John; et al. (15 May 2013). "Quantifying the consensus on anthropogenic global warming in the scientific literature". *Environmental Research Letters* **8** (2). Bibcode:2013ERL.....8b4024C. doi:10.1088/1748-9326/8/2/024024. there is a significant gap between public perception and reality, with 57% of the US public either disagreeing or unaware that scientists overwhelmingly agree that the earth is warming due to human activity (Pew 2012). Contributing to this 'consensus gap' are campaigns designed to confuse the public about the level of agreement among climate scientists....The narrative presented by some dissenters is that the scientific consensus is '...on the point of collapse' while '...the number of scientific "heretics" is growing with each passing year' A systematic, comprehensive review of the literature provides quantitative evidence countering this assertion. The number of papers rejecting AGW is a minuscule proportion of the published research, with the percentage slightly decreasing over time. Among papers expressing a position on AGW, an overwhelming percentage (97.2% based on self-ratings, 97.1% based on abstract ratings) endorses the scientific consensus on AGW.

1.1.10   Bibliography

- Antilla, Liisa (2005). "Climate of scepticism: US newspaper coverage of the science of climate change". *Global Environmental Change* **15**: 338–352. doi:10.1016/j.gloenvcha.2005.08.003.

- Begley, Sharon (13 August 2007). "The Truth About Denial". *Newsweek*. Archived from the original on 21 October 2007. (MSNBC single page version, archived 20 August 2007)

- Boslough, Mark (5 December 2014). "Deniers are not Skeptics". *Committee for Skeptical Inquiry*. Retrieved 7 July 2015.

- Brown, R. G. E., Jr. (23 October 1996). "Environmental science under siege: Fringe science and the 104th Congress, U. S. House of Representatives." (PDF). *Report, Democratic Caucus of the Committee on Science* (Washington, D. C.: U. S. House of Representatives).

- CBC News: the fifth estate (2007). "The Denial Machine". Archived from the original on 12 March 2007. Retrieved 29 July 2015.

- Conway, Erik; Oreskes, Naomi (2010). *Merchants of Doubt: How a Handful of Scientists Obscured the Truth on Issues from Tobacco Smoke to Global Warming*. USA: Bloomsbury. ISBN 1-59691-610-9.

- Diethelm, Pascal; McKee, Martin (January 2009). "Denialism: what is it and how should scientists respond?" (PDF). *European Journal of Public Health* **19** (1): 2–4. doi:10.1093/eurpub/ckn139. PMID 19158101.

- Dunlap, Riley E; McCright, Aaron M. (2011). *Climate Change Denial: Sources, actors, and strategies*. Taylor & Francis. ISBN 0-415-54478-5.

- Dunlap, R. E. (2013). "Climate Change Skepticism and Denial: An Introduction" (PDF). *American Behavioral Scientist* (SAGE) **57** (6). doi:10.1177/0002764213477097. Retrieved 27 May 2015.

- Dunlap, Riley E.; McCright, Aaron M. (2011). "Organised Climate Change Denial". In Dryzek, John S.; Norgaard, Richard B.; Schlosberg, David. *The Oxford Handbook of Climate Change and Society*. Oxford University Press. p. 153. ISBN 0199566607.

- Dunlap, R. E.; Jacques, P. J. (2013). "Climate Change Denial Books and Conservative Think Tanks: Exploring the Connection". *American Behavioral Scientist* (SAGE) **57** (6). doi:10.1177/0002764213477096. Retrieved 31 May 2015.

- Gelbspan, Ross (December 1995). "The heat is on: The warming of the world's climate sparks a blaze of denial". Harper's Magazine. Retrieved 2015-06-02.

- Gelbspan, Ross (1 January 1997). *The Heat is on: The High Stakes Battle Over Earth's Threatened Climate*. Addison-Wesley Publishing Company. ISBN 978-0-201-13295-3.

- Gelbspan, Ross (1998). *The heat is on : the climate crisis, the cover-up, the prescription*. Reading, Mass: Perseus Books. ISBN 0-7382-0025-5.

- Hamilton, Clive (7 April 2011). *Requiem for a Species: Why We Resist the Truth about Climate Change*. Routledge. ISBN 978-1-84977-498-7.

- Jacques, P.J.; Dunlap, Riley E.; Freeman, M. (2008). "The organisation of denial: Conservative think tanks and environmental scepticism". *Environmental Politics* (Routledge) **17** (3). doi:10.1080/09644010802055576.

- Jenkins, Stephen H. (2015). *Tools for Critical Thinking in Biology*. Oxford University Press. pp. 229, 243.

- Kemp, Jeremy; Milne, Richard; Reay, Dave S. (2010). "Sceptics and deniers of climate change not to be confused". *Nature* (Nature Publishing Group) **464** (7289): 673–673. Bibcode:2010Natur.464..673K. doi:10.1038/464673a. Retrieved 7 September 2015. pdf

- Lever-Tracy, Constance (2010). *Routledge Handbook of Climate Change and Society*. Taylor & Francis. ISBN 9780203876213.

- Mooney, Chris (2005). *The Republican war on science*. New York: Basic Books. ISBN 0-465-04675-4.

- National Center for Science Education (5 January 2012). "Why Is It Called Denial?". *National Center for Science Education*. Retrieved 2 Jun 2015.

- National Center for Science Education (4 June 2010). "Climate change is good science". *National Center for Science Education*. Retrieved 21 June 2015.

- Nerlich, Brigitte (2010). "'Climategate': Paradoxical Metaphors and Political Paralysis". *Environmental Values* (White Horse Press) **19** (4): 419–442. doi:10.3197/096327110x531543. Retrieved 4 July 2015.

- Orlóci, L. (2008). "Vegetation displacement issues and transition statistics in climate warming cycle" (PDF). *Community Ecology* (Akademiai Kiado) **9** (1): 83–98. doi:10.1556/comec.9.2008.1.10. Retrieved 29 July 2015.

- Painter, James; Ashe, Teresa (2012). "Cross-national comparison of the presence of climate scepticism in the print media in six countries, 2007". *Environ. Res. Lett.* (IOP) **7** (4): 044005. Bibcode:2012ERL.....7d4005P. doi:10.1088/1748-9326/7/4/044005. Retrieved 27 May 2015.

- Powell, James Lawrence (1 December 2012). *The Inquisition of Climate Science*. Columbia University Press. ISBN 978-0-231-15719-3.

- Rennie, John (30 November 2009). "Seven Answers to Climate Contrarian Nonsense". *Scientific American*.

- Timmer, John (16 December 2014). "Skeptics, deniers, and contrarians: The climate science label game". *Ars Technica*.

- Washington, Haydn (2013). *Climate Change Denial: Heads in the Sand*. Routledge. ISBN 1136530045.

- Weart, Spencer R. (July 2007). "Roger Revelle's Discovery". *The Discovery of Global Warming*. American Institute of Physics. Retrieved 18 July 2015.

- Weart, Spencer R. (February 2015). "The Public and Climate, cont.". *The Discovery of Global Warming*. American Institute of Physics. Retrieved 2 Jun 2015.

- Weart, Spencer R. (June 2015a). "Government: The View from Washington, DC". *The Discovery of Global Warming*. American Institute of Physics. Retrieved 18 July 2015.

- Weart, Spencer (2011). "Global warming: How skepticism became denial" (PDF). *Bulletin of the Atomic Scientists* (SAGE) **67** (1). doi:10.1177/0096340210392966.

1.1.11   Further reading

- "Frontline: Climate of Doubt". PBS. 23 October 2012. Retrieved 2012-10-25.

- Greenpeace USA (2013). "Dealing in Doubt: The Climate Denial Machine Vs Climate Science" (PDF). Retrieved 2014-10-28.

- Bowen, Mark (2008). Censoring Science: Dr. James Hansen and the Truth of Global Warming. Plume. ISBN 0-452-28962-9

- McCright, Aaron M.; Dunlap, Riley E. (2003). "Defeating Kyoto: The Conservative Movement's Impact on U.S. Climate Change Policy" (PDF). *Social Problems* **50** (3): 348–373. doi:10.1525/sp.2003.50.3.348.

- Shearer, Christine (2011). "Kivalina: A Climate Change Story" Haymarket Books. ISBN 978-1-60846-128-8

1.2   Global warming controversy

This article is about the public debate over scientific conclusions on climate change. For views which undermine public confidence in scientific opinion, see Climate change denial.

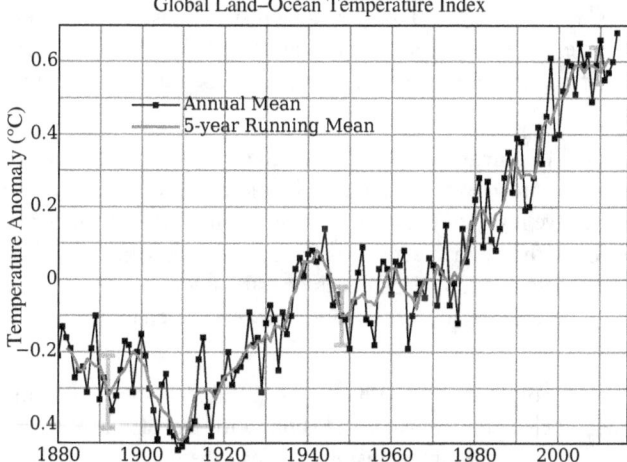

Global mean land-ocean temperature changes from 1880, relative to the 1951–1980 mean. The black line is the annual mean and the red line is the 5-year running mean. The green bars show uncertainty estimates. Source: NASA GISS.

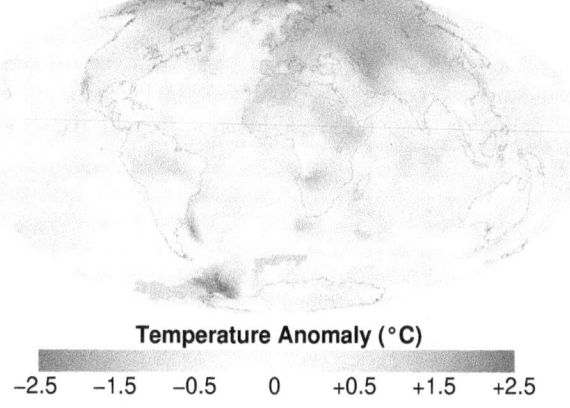

The map shows the 10-year average (2000–2009) global mean temperature anomaly relative to the 1951–1980 mean. The most extreme warming was in the Arctic. Source: NASA Earth Observatory[1]

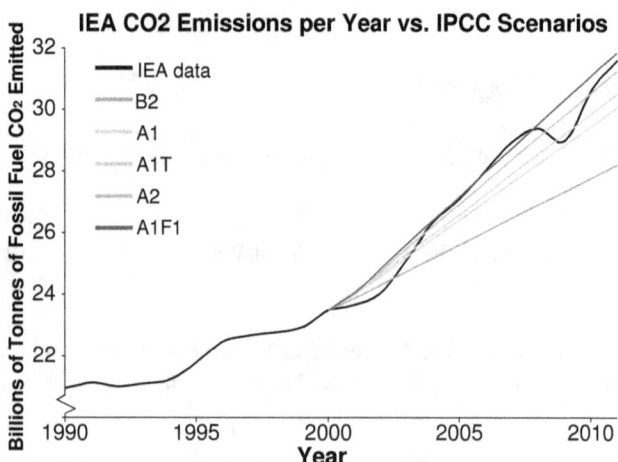

Fossil fuel related CO_2 emissions compared to five of the IPCC's "SRES" emissions scenarios. The dips are related to global recessions. Image source: Skeptical Science.

The **global warming controversy** concerns the public debate over whether global warming is occurring, how much has occurred in modern times, what has caused it, what its effects will be, whether any action should be taken to curb it, and if so what that action should be. In the scientific literature, there is a strong consensus that global surface temperatures have increased in recent decades and that the trend is caused by human-induced emissions of greenhouse gases.[2][3][4][5][6][7] No scientific body of national or international standing disagrees with this view,[8] though a few organizations with members in extractive industries hold non-committal positions.[9] Disputes over the key scientific facts of global warming are now more prevalent in the popular media than in the scientific literature, where such issues are treated as resolved, and more prevalent in the United States than globally.[10][11]

Political and popular debate concerning the existence and cause of climate change includes the reasons for the increase seen in the instrumental temperature record, whether the warming trend exceeds normal climatic variations, and whether human activities have contributed significantly to it. Scientists have resolved many of these questions decisively in favour of the view that the current warming trend exists and is ongoing, that human activity is the primary cause, and that it is without precedent in at least 2000 years.[12] Disputes that also reflect scientific debate include estimates of how responsive the climate system might be to any given level of greenhouse gases (climate sensitivity), and what the consequences of global warming will be.

Global warming remains an issue of widespread political debate, often split along party political lines, especially in the United States.[13] Many of the largely settled scientific issues, such as the human responsibility for global warming, remain the subject of politically or economically motivated attempts to downplay, dismiss or deny them – an ideological phenomenon categorised by academics and scientists as climate change denial. The sources of funding for those involved with climate science – both supporting and opposing mainstream scientific positions – have been questioned by both sides. There are debates about the best policy responses to the science, their cost-effectiveness and their urgency. Climate scientists, especially in the United States, have reported official and oil-industry pressure to censor or suppress their work and hide scientific data, with directives not to discuss the subject in public communications. Legal cases regarding global warming, its effects, and measures to reduce it have reached American courts. The fossil fuels lobby and free market think tanks have often been identified as overtly or covertly supporting efforts to undermine or discredit the scientific consensus on global warming.

1.2.1 History

Public opinion

Main article: Public opinion on climate change

In the United States, the mass media devoted little coverage to global warming until the drought of 1988, and James E.

Hansen's testimony to the Senate, which explicitly attributed "the abnormally hot weather plaguing our nation" to global warming.[14]

The British press also changed its coverage at the end of 1988, following a speech by Margaret Thatcher to the Royal Society advocating action against human-induced climate change.[15] According to Anabela Carvalho, an academic analyst, Thatcher's "appropriation" of the risks of climate change to promote nuclear power, in the context of the dismantling of the coal industry following the 1984–1985 miners' strike was one reason for the change in public discourse. At the same time environmental organizations and the political opposition were demanding "solutions that contrasted with the government's".[16] In May 2013 Prince Charles took a strong stance criticising both climate change deniers and corporate lobbyists by likening the Earth to a dying patient. "A scientific hypothesis is tested to absolute destruction, but medicine can't wait. If a doctor sees a child with a fever, he can't wait for [endless] tests. He has to act on what is there."[17]

Many European countries took action to reduce greenhouse gas emissions before 1990. West Germany started to take action after the Green Party took seats in Parliament in the 1980s. All countries of the European Union ratified the 1997 Kyoto Protocol. Substantial activity by NGOs took place as well.[18] The United States Energy Information Administration reports that, in the United States, "The 2012 downturn means that emissions are at their lowest level since 1994 and over 12 percent below the recent 2007 peak".[19]

In Europe, the notion of human influence on climate gained wide acceptance more rapidly than in the United States and other countries.[20][21] A 2009 survey found that Europeans rated climate change as the second most serious problem facing the world, between "poverty, the lack of food and drinking water" and "a major global economic downturn". 87% of Europeans considered climate change to be a very serious or serious problem, while ten per cent did not consider it a serious problem.[22]

In 2007 the BBC announced the cancellation of a planned television special Planet Relief, which would have highlighted the global warming issue and included a mass electrical switch-off.[23] The editor of BBC's Newsnight current affairs show said: "It is absolutely not the BBC's job to save the planet. I think there are a lot of people who think that, but it must be stopped."[24] Author Mark Lynas said "The only reason why this became an issue is that there is a small but vociferous group of extreme right-wing climate 'sceptics' lobbying against taking action, so the BBC is behaving like a coward and refusing to take a more consistent stance."[25]

The authors of the 2010 book *Merchants of Doubt* provide documentation for the assertion that professional deniers have tried to sow seeds of doubt in public opinion in order to halt any meaningful social or political progress to reduce the impact of human carbon emissions. The fact that only half of the American population believe that global warming is caused by human activity could be seen as a victory for these deniers.[11] One of the authors' main arguments is that most prominent scientists who have been voicing opposition to the near-universal consensus are being funded by industries, such as automotive and oil, that stand to lose money by government actions to regulate greenhouse gases.

A compendium of poll results on public perceptions about global warming is below.[26][27][28]

In 2007 a report on public perceptions in the UK by Ipsos MORI[31] reported that

- There is widespread recognition that the climate, irrespective of the cause, is changing—88% believe this to be true.

- However, the public is out of step with the scientific community, with 41% believing that climate change is being caused by both human activity and natural processes. 46% believe human activity is the main cause.

- Only a small minority reject anthropogenic climate change, while almost half (44%) are very concerned. However, there remains a large proportion who are not fully persuaded and hold doubts about the extent of the threat.

- There is still a strong appetite among the public for more information, and 63% say they need this to come to a firm view on the issue and what it means for them.

- The public continue to externalize climate change to other people, places and times. It is increasingly perceived as a major global issue with far-reaching consequences for future generations—45% say it is the most serious threat facing the World today and 53% believe it will impact significantly on future generations. However, the issue features less prominently nationally and locally, indeed only 9% believe climate change will have a significant impact upon them personally.

The Canadian science broadcaster and environmental activist, David Suzuki, reports that focus groups organized by the David Suzuki Foundation in 2006 showed that the public has a poor understanding of the science behind global warming.[32] This is despite publicity through different means, including the films *An Inconvenient Truth* and *The 11th Hour*.

An example of the poor understanding is public confusion between global warming and ozone depletion or other environmental problems.[33][34]

A 15-nation poll conducted in 2006 by Pew Global found that there "is a substantial gap in concern over global warming—roughly two-thirds of Japanese (66%) and Indians (65%) say they personally worry a great deal about global warming. Roughly half of the populations of Spain (51%) and France (46%) also express great concern over global warming, based on those who have heard about the issue. But there is no evidence of alarm over global warming in either the United States or China—the two largest producers of greenhouse gases. Just 19% of Americans and 20% of the Chinese who have heard of the issue say they worry a lot about global warming—the lowest percentages in the 15 countries surveyed. Moreover, nearly half of Americans (47%) and somewhat fewer Chinese (37%) express little or no concern about the problem".[35]

A 47-nation poll by Pew Global Attitudes conducted in 2007 found that "Substantial majorities 25 of 37 countries say global warming is a 'very serious' problem".[36]

There are differences between the opinion of scientists and that of the general public. A 2009 poll in the US by Pew Research Center found that "[w]hile 84% of scientists say the earth is getting warmer because of human activity such as burning fossil fuels, just 49% of the public agrees".[30] A 2010 poll in the UK for the BBC showed "Climate scepticism on the rise".[37] Robert Watson found this "very disappointing" and said that "We need the public to understand that climate change is serious so they will change their habits and help us move towards a low carbon economy".

A 2012 Canadian poll found that 32% of Canadians said they believe climate change is happening because of human activity, while 54% said they believe it's because of human activity and partially due to natural climate variation. 9% believe climate change is occurring due to natural climate variation, and only 2% said they don't believe climate change is occurring at all.[38]

Related controversies

Many of the critics of the consensus view on global warming have disagreed, in whole or part, with the scientific consensus regarding other issues, particularly those relating to environmental risks, such as ozone depletion, DDT, and passive smoking.[39][40] Chris Mooney, author of *The Republican War on Science*, has argued that the appearance of overlapping groups of skeptical scientists, commentators and think tanks in seemingly unrelated controversies results from an organized attempt to replace scientific analysis with political ideology. Mooney says that the promotion of doubt regarding issues that are politically, but not scientifically, controversial became increasingly prevalent under the Bush Administration, which, he says, regularly distorted and/or suppressed scientific research to further its own political aims. This is also the subject of a 2004 book by environmental lawyer Robert F. Kennedy, Jr. entitled *Crimes Against Nature: How George W. Bush and Corporate Pals are Plundering the Country and Hijacking Our Democracy* (ISBN 978-0060746872). Another book on this topic is *The Assault on Reason* by former U.S. Vice-President Al Gore. Earlier instances of this trend are also covered in the book *The Heat Is On* by Ross Gelbspan.

Some critics of the scientific consensus on global warming have argued that these issues should not be linked and that reference to them constitutes an unjustified ad hominem attack.[41] Political scientist Roger Pielke, Jr., responding to Mooney, has argued that science is inevitably intertwined with politics.[42]

1.2.2 Mainstream scientific position, and challenges to it

Main article: Scientific opinion on climate change
The finding that the climate has warmed in recent decades and that human activities are producing global climate change has been endorsed by every national science academy that has issued a statement on climate change, including the science academies of all of the major industrialized countries.[46]

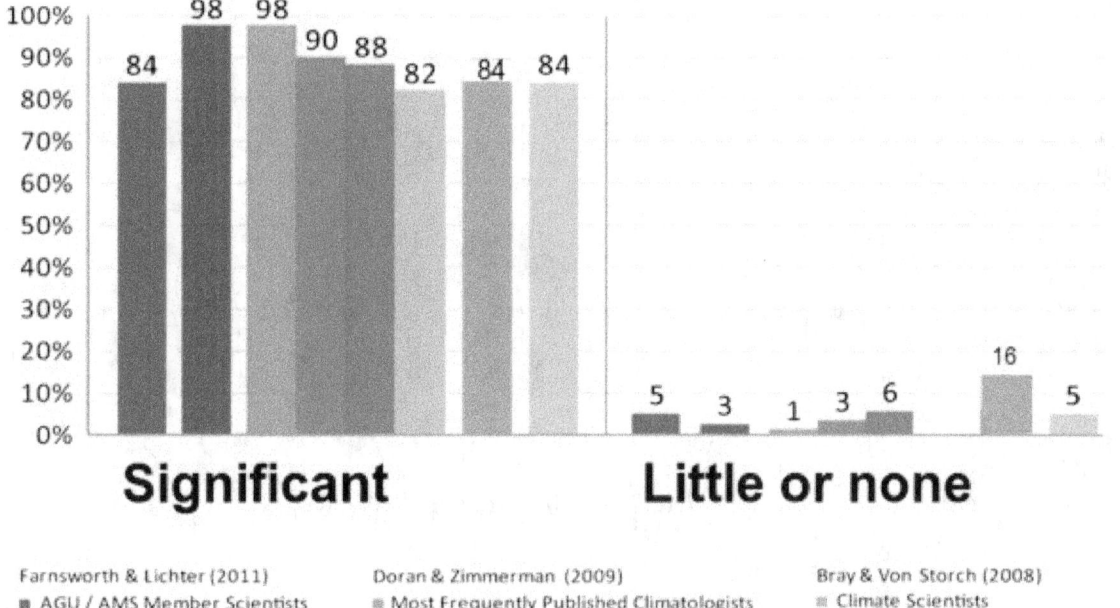

Summary of opinions from climate and earth scientists regarding climate change. Click to see a more detailed summary of the sources.

Attribution of recent climate change discusses how global warming is attributed to anthropogenic greenhouse gases (GHGs).

Scientific consensus

Scientific consensus is normally achieved through communication at conferences, publication in the scientific literature, replication (reproducible results by others), and peer review. In the case of global warming, many governmental reports, the media in many countries, and environmental groups, have stated that there is virtually unanimous scientific agreement that human-caused global warming is real and poses a serious concern.[47][48][49] According to the United States National Research Council,

> [T]here is a strong, credible body of evidence, based on multiple lines of research, documenting that climate is changing and that these changes are in large part caused by human activities. While much remains to be learned, the core phenomenon, scientific questions, and hypotheses have been examined thoroughly and have stood firm in the face of serious scientific debate and careful evaluation of alternative explanations. * * * Some scientific conclusions or theories have been so thoroughly examined and tested, and supported by so many independent observations and results, that their likelihood of subsequently being found to be wrong is vanishingly small. Such conclusions and theories are then regarded as settled facts. This is the case for the conclusions that the Earth system is warming and that much of this warming is very likely due to human activities.[50]

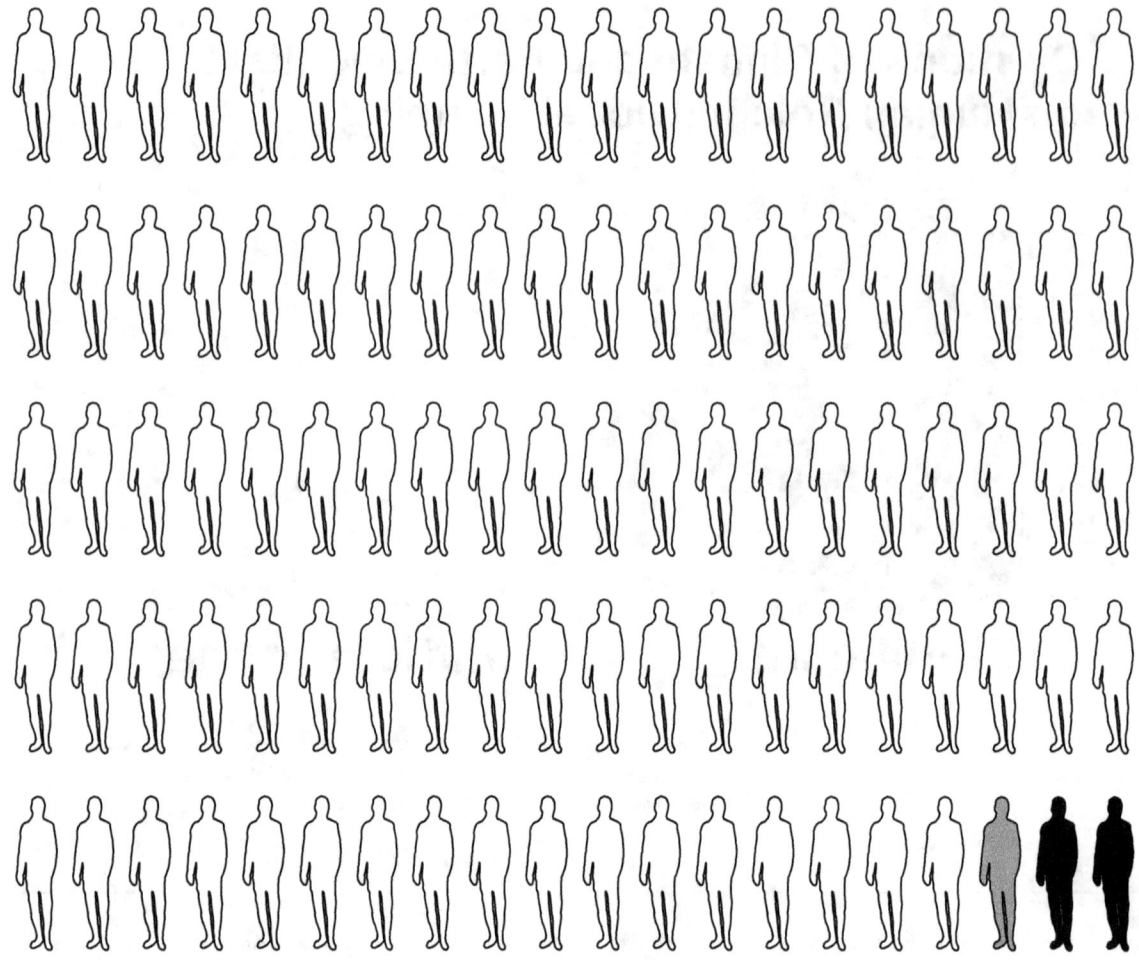

Just over 97% of climate researchers say humans are causing most global warming.[43][44][45]

Among opponents of the mainstream scientific assessment, some say that while there is agreement that humans do have an effect on climate, there is no universal agreement about the quantitative magnitude of anthropogenic global warming (AGW) relative to natural forcings and its harm to benefit ratio.[51] Other opponents assert that some kind of ill-defined "consensus argument" is being used, and then dismiss this by arguing that science is based on facts rather than consensus.[52] Some highlight the dangers of focusing on only one viewpoint in the context of what they say is unsettled science, or point out that science is based on facts and not on opinion polls or consensus.[53][54]

Dennis T. Avery, a food policy analyst at the Hudson Institute wrote an article entitled "500 Scientists Whose Research Contradicts Man-Made Global Warming Scares"[55] published in 2007 by The Heartland Institute. After the publishing of this article, numerous scientists who had been included in the list demanded their names be removed after the list was immediately called into question for misunderstanding and distorting the conclusions of many of the named studies and/or citing outdated, flawed studies that had long been abandoned and deemed inaccurate.[56][57][58] The Heartland Institute refused requests by scientists to have their names removed, stating that the scientists "have no right—legally or ethically—to demand that their names be removed from a bibliography composed by researchers with whom they disagree"[59] despite the aforementioned falsification and refutation of much of the list.[60]

A 2010 paper in the Proceedings of the National Academy of Sciences analysed "1,372 climate researchers and their publication and citation data to show that (i) 97–98% of the climate researchers most actively publishing in the field support the tenets of ACC outlined by the Intergovernmental Panel on Climate Change, and (ii) the relative climate expertise and scientific prominence of the researchers unconvinced of ACC are substantially below that of the convinced researchers".[61][62] Judith Curry has said "This is a completely unconvincing analysis", whereas Naomi Oreskes said that the paper shows that "the vast majority of working [climate] research scientists are in agreement [on climate change]...

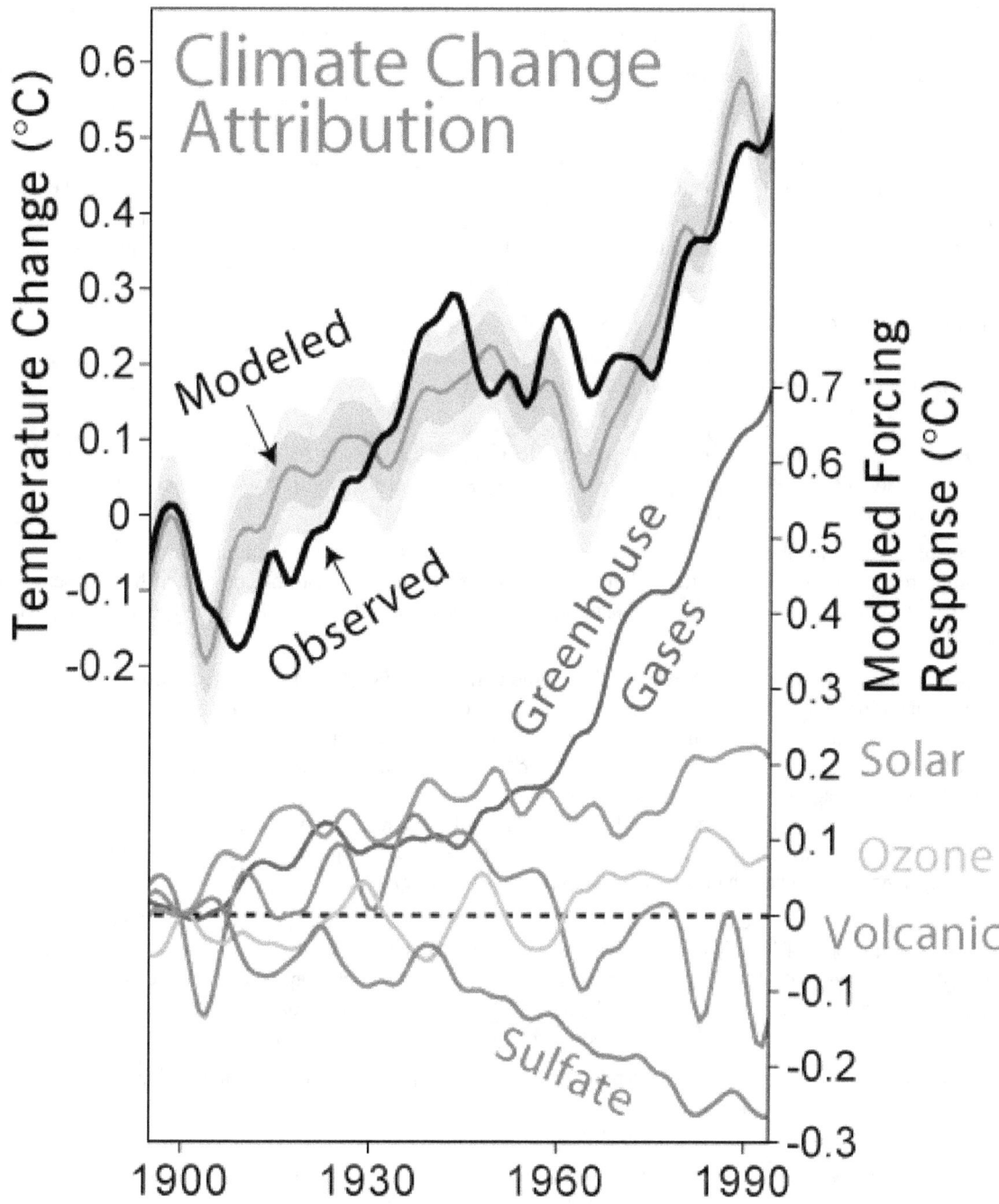

Reproduction of the temperature record using historical forcings

Those who don't agree, are, unfortunately—and this is hard to say without sounding elitist—mostly either not actually climate researchers or not very productive researchers".[62][63] Jim Prall, one of the coauthors of the study, acknowledged "it would be helpful to have lukewarm [as] a third category".[62]

A 2013 study published in the peer-reviewed journal *Environmental Research Letters* analyzed 11,944 abstracts from papers published in the peer-reviewed scientific literature between 1991 and 2011, identified by searching the ISI Web of Science citation index engine for the text strings "global climate change" or "global warming". The authors found that 3974 of the abstracts expressed a position on anthropogenic global warming, and that 97.1% of those endorsed the consensus that humans are causing global warming. The authors found that of the 11,944 abstracts, 3896 endorsed that consensus, 7930 took no position on it, 78 rejected the consensus, and 40 expressed uncertainty about it.[45]

In 2014 a letter from 52 leading skeptics was published by the Committee for Skeptical Inquiry supporting the scientific consensus and asking the media to stop referring to deniers as "skeptics." The letter clarified the skeptical opinion on climate and denial: "As scientific skeptics, we are well aware of political efforts to undermine climate science by those who deny reality but do not engage in scientific research or consider evidence that their deeply held opinions are wrong. The most appropriate word to describe the behavior of those individuals is 'denial.' Not all individuals who call themselves climate change skeptics are deniers. But virtually all deniers have falsely branded themselves as skeptics. By perpetrating this misnomer, journalists have granted undeserved credibility to those who reject science and scientific inquiry."[64]

Authority of the IPCC

Main article: Intergovernmental Panel on Climate Change

The "standard" view of climate change has come to be defined by the reports of the IPCC, which is supported by many other science academies and scientific organizations. In 2001, sixteen of the world's national science academies made a joint statement on climate change, and gave their support for the IPCC[46]

Opponents have generally attacked either the IPCC's processes, people[65] or the Synthesis and Executive summaries; the scientific reports attract less attention. Some of the controversy and criticism has originated from experts invited by the IPCC to submit reports or serve on its panels. For example, Richard Lindzen has publicly dissented from IPCC positions.[66]

Christopher Landsea, a hurricane researcher, said of "the part of the IPCC to which my expertise is relevant" that "I personally cannot in good faith continue to contribute to a process that I view as both being motivated by pre-conceived agendas and being scientifically unsound",[67] because of comments made at a press conference by Kevin Trenberth of which Landsea disapproved. Trenberth said that "Landsea's comments were not correct";[68] the IPCC replied that "individual scientists can do what they wish in their own rights, as long as they are not saying anything on behalf of the IPCC" and offered to include Landsea in the review phase of the AR4.[69] Roger Pielke, Jr. commented that "Both Landsea and Trenberth can and should feel vindicated... the IPCC accurately reported the state of scientific understandings of tropical cyclones and climate change in its recent summary for policy makers".[68]

In 2005, the House of Lords Economics Committee wrote that "We have some concerns about the objectivity of the IPCC process, with some of its emissions scenarios and summary documentation apparently influenced by political considerations". It doubted the high emission scenarios and said that the IPCC had "played-down" what the committee called "some positive aspects of global warming".[70] The main statements of the House of Lords Economics Committee were rejected in the response made by the United Kingdom government[71] and by the Stern Review.

Speaking to the difficulty of establishing scientific consensus on the precise extent of human action on climate change, John Christy, a contributing author, wrote:

> Contributing authors essentially are asked to contribute a little text at the beginning and to review the first two drafts. We have no control over editing decisions. Even less influence is granted the 2,000 or so reviewers. Thus, to say that 800 contributing authors or 2,000 reviewers reached consensus on anything describes a situation that is not reality.[72]

On 10 December 2008, a report was released by the U.S. Senate Committee on Environment and Public Works Minority members, under the leadership of the Senate's most vocal global warming skeptic Jim Inhofe. The timing of the report coincided with the UN global warming conference in Poznań, Poland. It says it summarizes scientific dissent from the IPCC.[73] Many of its statements about the numbers of individuals listed in the report, whether they are actually scientists, and whether they support the positions attributed to them, have been disputed.[74][75][76]

While some critics have argued that the IPCC overstates likely global warming, others have made the opposite criticism. David Biello, writing in the Scientific American, argues that, because of the need to secure consensus among governmental representatives, the IPCC reports give conservative estimates of the likely extent and effects of global warming.[77] *Science* editor Brooks Hanson states in a 2010 editorial: "The IPCC reports have underestimated the pace of climate change while overestimating societies' abilities to curb greenhouse gas emissions".[78] Climate scientist James E. Hansen argues that the IPCC's conservativeness seriously underestimates the risk of sea-level rise on the order of meters—enough to inundate many low-lying areas, such as the southern third of Florida.[79] Roger A. Pielke Sr. has also stated that "Humans are significantly altering the global climate, but in a variety of diverse ways beyond the radiative effect of carbon dioxide. The IPCC assessments have been too conservative in recognizing the importance of these human climate forcings as they alter regional and global climate".[80]

Henderson-Sellers has collected comments from IPCC authors in a 2007 workshop revealing a number of concerns. She concluded, "Climate change research entered a new and different regime with the publication of the IPCC Fourth Assessment Report. There is no longer any question about "whether" human activities are changing the climate; instead research must tackle the urgent questions of: "how fast?"; "with what impacts?'; and "what responses are needed?""[81]

Greenhouse gases

Attribution of recent climate change discusses the evidence for recent global warming. Correlation of CO_2 and temperature is not part of this evidence. Nonetheless, one argument against global warming says that rising levels of carbon dioxide (CO_2) and other greenhouse gases (GHGs) do not correlate with global warming.[82]

- Studies of the Vostok ice core show that at the "beginning of the deglaciations, the CO_2 increase either was in phase or lagged by less than ~1000 years with respect to the Antarctic temperature, whereas it clearly lagged behind the temperature at the onset of the glaciations".[83] Recent warming is followed by carbon dioxide levels with only a 5 months delay.[84] The time lag has been used to argue that the current rise in CO_2 is a *result* of warming and not a cause. While it is generally agreed that variations before the industrial age are mostly timed by astronomical forcing,[85] a main part of current warming is found to be timed by anthropogenic releases of CO_2, having a much closer time relation not observed in the past (thus returning the argument to the importance of human CO_2 emissions). Analysis of carbon isotopes in atmospheric CO_2 shows that the recent observed CO_2 increase cannot have come from the oceans, volcanoes, or the biosphere, and thus is not a response to rising temperatures as would be required if the same processes creating past lags were active now.[86]

- Carbon dioxide accounts for about 390 parts per million by volume (ppm) of the Earth's atmosphere, increasing from 284 ppm in the 1830s to 387 ppm in 2009.[87][88] Carbon dioxide contributes between 9 and 26% of the natural greenhouse effect.[89]

- In the Ordovician period of the Paleozoic era (about 450 million years ago), the Earth had an atmospheric CO_2 concentration estimated at 4400ppm (or 0.44% of the atmosphere), while also having evidence of some glaciation. Modeling work has shown that it is possible for local areas at elevations greater than 300–500 meters to contain year-round snow cover even with high atmospheric CO_2 concentrations.[90] A 2006 study suggests that the elevated CO_2 levels and the glaciation are not synchronous, but rather that weathering associated with the uplift and erosion of the Appalachian Mountains greatly reduced atmospheric greenhouse gas concentrations and permitted the observed glaciation.[91]

As noted above, climate models are only able to simulate the temperature record of the past century when GHG forcing is included, being consistent with the findings of the IPCC which has stated that: "Greenhouse gas forcing, largely the result of human activities, has very likely caused most of the observed global warming over the last 50 years"[92]

The "standard" set of scenarios for future atmospheric greenhouse gases are the IPCC SRES scenarios. The purpose of the range of scenarios is not to predict what exact course the future of emissions will take, but what it may take under a range of possible population, economic and societal trends.[93] Climate models can be run using any of the scenarios as inputs to illustrate the different outcomes for climate change. No one scenario is officially preferred, but in practice the "A1b" scenario roughly corresponding to 1%/year growth in atmospheric CO_2 is often used for modelling studies.

There is debate about the various scenarios for fossil fuel consumption. Global warming skeptic Fred Singer stated that "some good experts believe" that atmospheric CO_2 concentration will not double since economies are becoming less reliant on carbon.[94]

However, the Stern report,[95] like many other reports, notes the past correlation between CO_2 emissions and economic growth and then extrapolates using a "business as usual" scenario to predict GDP growth and hence CO_2 levels, concluding that:

> Increasing scarcity of fossil fuels alone will not stop emissions growth in time. The stocks of hydrocarbons that are profitable to extract are more than enough to take the world to levels of CO_2 well beyond 750 ppm with very dangerous consequences for climate change impacts.

According to a 2006 paper from Lawrence Livermore National Laboratory, "the earth would warm by 8 degrees Celsius (14.4 degrees Fahrenheit) if humans use the entire planet's available fossil fuels by the year 2300".[96]

Solar variation

Main article: Solar variation

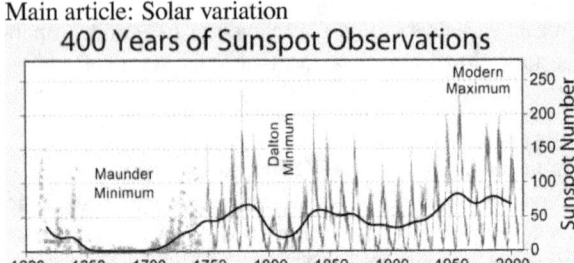

400 year history of sunspot numbers.

Last 30 years of solar variability.

Scientists opposing the mainstream scientific assessment of global warming express varied opinions concerning the cause of global warming. Some say only that it has not yet been ascertained whether humans are the primary cause of global warming; others attribute global warming to natural variation; ocean currents; increased solar activity or cosmic rays. The consensus position is that solar radiation may have increased by 0.12 W/m² since 1750, compared to 1.6 W/m² for the net anthropogenic forcing.[97] The TAR said, "The combined change in radiative forcing of the two major natural factors (solar variation and volcanic aerosols) is estimated to be negative for the past two, and possibly the past four, decades".[98] The AR4 makes no direct assertions on the recent role of solar forcing, but the previous statement is consistent with the AR4's figure 4.

A few studies say that the present level of solar activity is historically high as determined by sunspot activity and other factors. Solar activity could affect climate either by variation in the Sun's output or, more speculatively, by an indirect effect on the amount of cloud formation. Solanki and co-workers suggest that solar activity for the last 60 to 70 years may be at its highest level in 8,000 years; Muscheler *et al.* disagree, suggesting that other comparably high levels of activity have occurred several times in the last few thousand years.[99] Muscheler *et al.* concluded that "solar activity reconstructions tell us that only a minor fraction of the recent global warming can be explained by the variable Sun".[100]

Solanki *et al.* concluded "that solar variability is unlikely to have been the dominant cause of the strong warming during the past three decades", and that "at the most 30% of the strong warming since then can be of solar origin".[101]

Another point of controversy is the correlation of temperature with solar variation.[102]

Mike Lockwood and Claus Fröhlich reject the statement that the warming observed in the global mean surface temperature record since about 1850 is the result of solar variations.[103] Lockwood and Fröhlich conclude that "the observed rapid rise in global mean temperatures seen after 1985 cannot be ascribed to solar variability, whichever of the mechanisms is invoked and no matter how much the solar variation is amplified."

Aerosols forcing

The hiatus in warming from the 1940s to 1960s is generally attributed to cooling effect of sulphate aerosols.[104][105] More recently, this forcing has (relatively) declined, which may have enhanced warming, though the effect is regionally varying. See global dimming. Another example of this is in Ruckstuhl's paper who found a 60% reduction in aerosol concentrations over Europe causing solar brightening:[106]

> [...] the direct aerosol effect had an approximately five times larger impact on climate forcing than the indirect aerosol and other cloud effects. The overall aerosol and cloud induced surface climate forcing is ~ 1 $W\ m^{-2}\ decade^{-1}$ and has most probably strongly contributed to the recent rapid warming in Europe.

Analysis of temperature records

Instrumental record of surface temperature   Main articles: Instrumental temperature record and Urban heat island
There have been attempts to raise public controversy over the accuracy of the instrumental temperature record on the basis

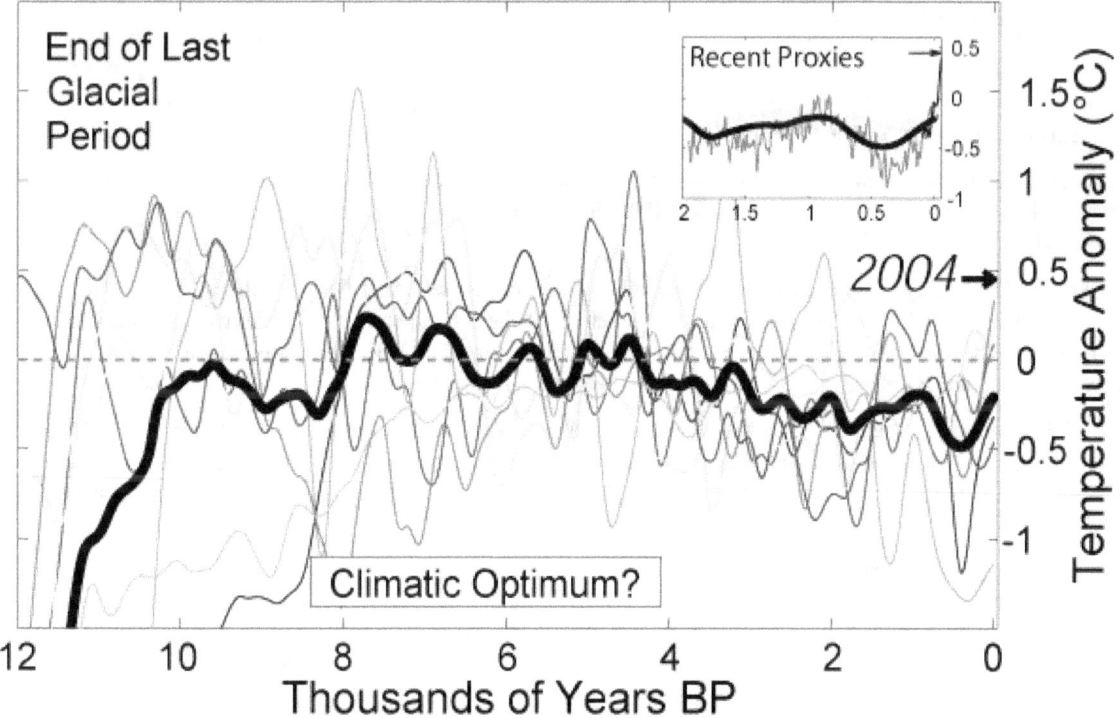

Temperature variations during the present geological age

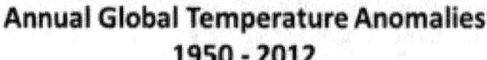

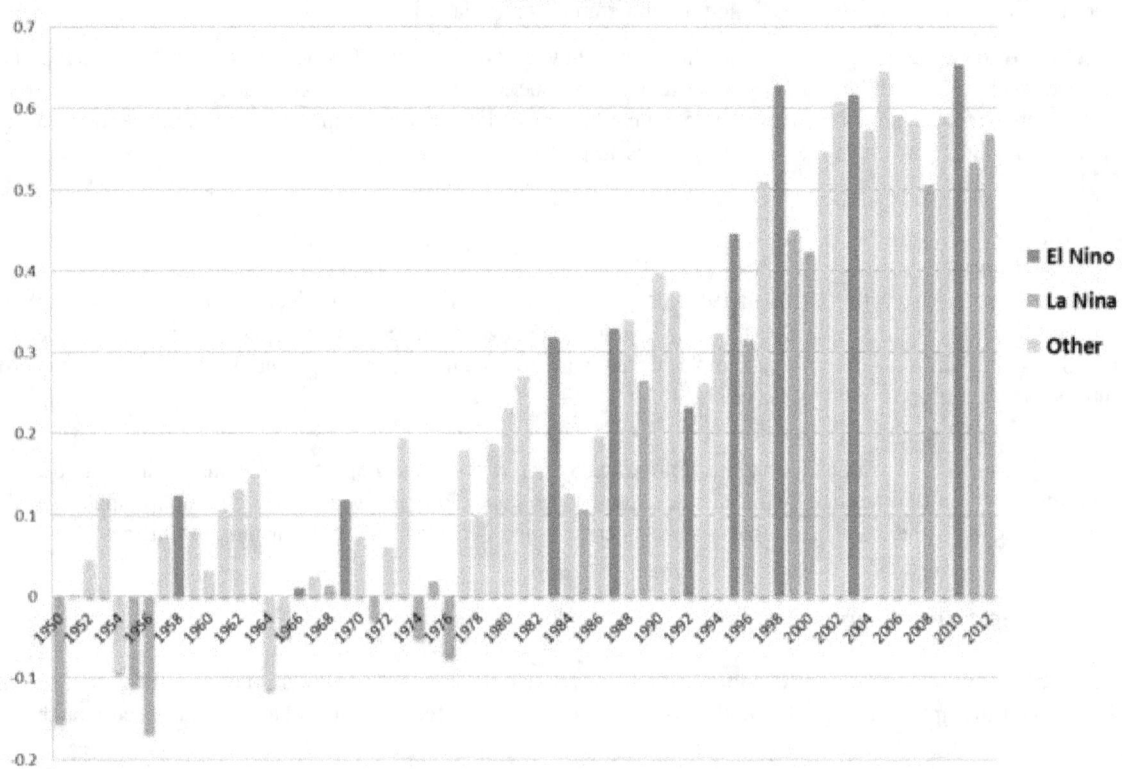

NOAA graph of Global Annual Temperature Anomalies 1950–2012

of the urban heat island effect, the quality of the surface station network, and assertions that there have been unwarranted adjustments to the temperature record.

Weather stations that are used to compute global temperature records are not evenly distributed over the planet. There were a small number of weather stations in the 1850s, and the number didn't reach the current 3000+ until the 1951 to 1990 period [107]

The 2001 IPCC Third Assessment Report (TAR) acknowledged that the urban heat island is an important *local* effect, but cited analyses of historical data indicating that the effect of the urban heat island on the *global* temperature trend is no more than 0.05 °C (0.09 °F) degrees through 1990.[108] Peterson (2003) found no difference between the warming observed in urban and rural areas.[109]

Parker (2006) found that there was no difference in warming between calm and windy nights. Since the urban heat island effect is strongest for calm nights and is weak or absent on windy nights, this was taken as evidence that global temperature trends are not significantly contaminated by urban effects.[110] Pielke and Matsui published a paper disagreeing with Parker's conclusions.[111]

In 2005, Roger A. Pielke and Stephen McIntyre criticized the US instrumental temperature record and adjustments to it, and Pielke and others criticized the poor quality siting of a number of weather stations in the United States.[112][113] In 2007, Anthony Watts began a volunteer effort to photographically document the siting quality of these stations.[114] The *Journal of Geophysical Research – Atmospheres* subsequently published a study by Menne et al. which examined the record of stations picked out by Watts' *Surfacestations.org* and found that, if anything, the poorly sited stations showed a slight cool bias rather than the warm bias which Watts had anticipated.[115][116]

The Berkeley Earth Surface Temperature group carried out an independent assessment of land temperature records, which examined issues raised by skeptics, such as the urban heat island effect, poor station quality, and the risk of data selection bias. The preliminary results, made public in October 2011, found that these factors had not biased the results obtained by NOAA, the Hadley Centre together with the Climatic Research Unit (HadCRUT) and NASA's GISS in earlier studies. The group also confirmed that over the past 50 years the land surface warmed by 0.911 °C, and their results closely matched those obtained from these earlier studies. The four papers they had produced had been submitted for peer review.[117][118][119][120]

Instrumental record of tropospheric temperature   General circulation models and basic physical considerations predict that in the tropics the temperature of the troposphere should increase more rapidly than the temperature of the surface. A 2006 report to the U.S. Climate Change Science Program noted that models and observations agreed on this amplification for monthly and interannual time scales but not for decadal time scales in most observed data sets. Improved measurement and analysis techniques have reconciled this discrepancy: corrected buoy and satellite surface temperatures are slightly cooler and corrected satellite and radiosonde measurements of the tropical troposphere are slightly warmer.[121] Satellite temperature measurements show that tropospheric temperatures are increasing with "rates similar to those of the surface temperature", leading the IPCC to conclude that this discrepancy is reconciled.[122]

Geologic temperature records   Before humans learned to record the temperature of earth's climate system, various biological and geological processes left clues to past climate conditions. The analysis of these clues is the focus of the science of paleoclimatology. The field still has a variety of uncertainties.

Antarctica cooling   Main article: Antarctica cooling controversy
There has been a public dispute regarding the apparent contradiction in the observed behavior of Antarctica, as opposed to the global rise in temperatures measured elsewhere in the world. This became part of the public debate in the global warming controversy, particularly between advocacy groups of both sides in the public arena, as well as the popular media.[123][124][125][126][127][128][129][130]

In contrast to the popular press, there is no evidence of a corresponding controversy in the scientific community. Observations unambiguously show the Antarctic Peninsula to be warming. The trends elsewhere show both warming and cooling but are smaller and dependent on season and the timespan over which the trend is computed.[131] A study released in 2009, combined historical weather station data with satellite measurements to deduce past temperatures over large regions of the continent, and these temperatures indicate an overall warming trend. One of the paper's authors stated *"We now see warming is taking place on all seven of the earth's continents in accord with what models predict as a response to greenhouse gases."*[132] According to 2011 paper by Ding, et al., *"The Pacific sector of Antarctica, including both the Antarctic Peninsula and continental West Antarctica, has experienced substantial warming in the past 30 years."*[133][134]

This controversy began with the misinterpretation of the results of a 2002 paper by Doran et al.,[135][136] which found that *"Although previous reports suggest slight recent continental warming, our spatial analysis of Antarctic meteorological data demonstrates a net cooling on the Antarctic continent between 1966 and 2000, particularly during summer and autumn."*[135] Later the controversy was popularized by Michael Crichton's 2004 fiction novel State of Fear,[137] who advocated skepticism in global warming.[138][139] This novel has a docudrama plot based upon the idea that there is a deliberately alarmist conspiracy behind global warming activism. One of the characters argues that *"data show that one relatively small area called the Antarctic Peninsula is melting and calving huge icebergs... but the continent as a whole is getting colder, and the ice is getting thicker."* As a basis for this plot twist, Crichton cited the peer reviewed scientific article by Doran, et al.[135] Peter Doran, the lead author of the paper cited by Crichton, stated that *"... our results have been misused as 'evidence' against global warming by Crichton in his novel 'State of Fear'... Our study did find that 58 percent of Antarctica cooled from 1966 to 2000. But during that period, the rest of the continent was warming. And climate models created since our paper was published have suggested a link between the lack of significant warming in Antarctica and the ozone hole over that continent."*[140]

Climate sensitivity

As defined by the IPCC, climate sensitivity is the "equilibrium temperature rise that would occur for a doubling of CO_2 concentration above pre-industrial levels."[141] In its 2007 Fourth Assessment Report, IPCC said that climate sensitivity is "likely to be in the range 2 to 4.5 °C with a best estimate of about 3 °C".[142]

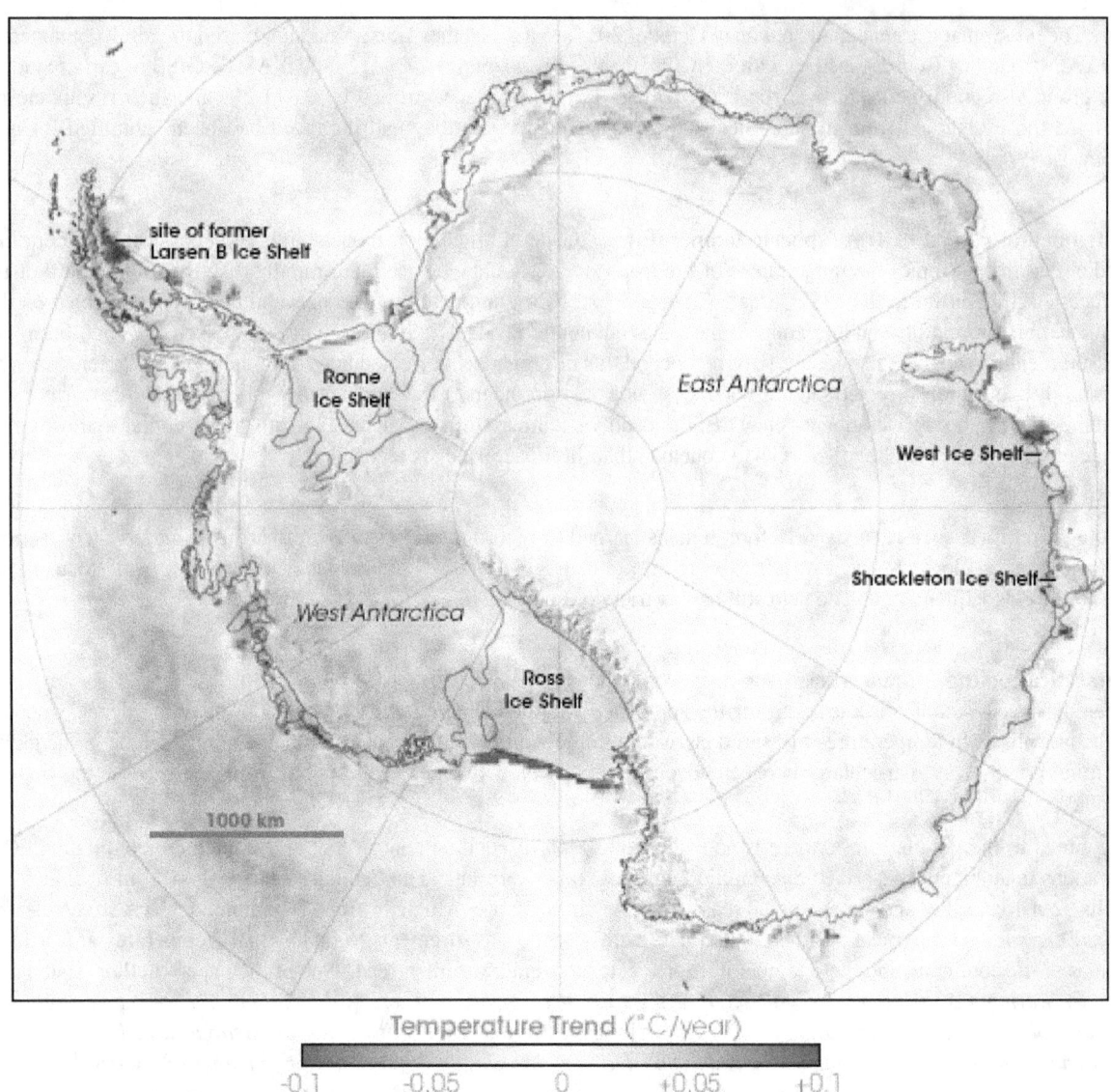

Antarctic Skin (the roughly top millimeter of land, sea, snow, or ice) Temperature Trends between 1981 and 2007, based on thermal infrared observations made by a series of NOAA satellite sensors; note that they do not necessarily reflect air temperature trends.

Using a combination of surface temperature history and ocean heat content, Stephen E. Schwartz has proposed an estimate of climate sensitivity of 1.9 ± 1.0 K for doubled CO_2.,[143] revised upwards from 1.1 ± 0.5 K.[144] Grant Foster, James Annan, Gavin Schmidt, and Michael E. Mann[145][146] argue that there are errors in both versions of Schwartz's analysis. Petr Chylek and co-authors have also proposed low climate sensitivity to doubled CO_2, estimated to be 1.6 K ± 0.4 K.[147]

In January 2013 widespread publicity was given to work led by Terje Berntsen of the University of Oslo, Julia Hargreaves of the Research Institute for Global Change in Yokohama, and Nic Lewis, an independent climate scientist, which reportedly found lower climate sensitivities than IPCC estimates and the suggestion that there is a 90% probability that doubling CO_2 emissions will increase temperatures by lower values than those estimated by the climate models used by the IPCC was featured in news outlets including *The Economist*.[148][149] This premature announcement came from a preliminary news release about a study which had not yet been peer reviewed.[150] The Center for International Climate and Environmental Research, Oslo (CICERO) issued a statement that they were involved with the relevant research project, and the news story was based on a report submitted to the research council which included both published and unpublished material. The highly publicised figures came from work still undergoing peer review, and CICERO would wait until they had been published in a journal before disseminating the results.[151]

Infrared iris hypothesis

In 2001, Richard Lindzen proposed a system of compensating meteorological processes involving clouds that tend to stabilize climate change; he tagged this the "Iris hypothesis, or "Infrared Iris.""[152] This work has been discussed in a number of papers[153]

Roy Spencer *et al.* suggested that "a net reduction in radiative input into the ocean-atmosphere system" in tropical intraseasonal oscillations "may potentially support" the idea of an "Iris" effect, although they point out that their work is concerned with much shorter time scales.[154]

Other analyses have found that the iris effect is a *positive* feedback rather than the negative feedback proposed by Lindzen.[155]

Internal radiative forcing

Roy Spencer hypothesized in 2008 that there is an "internal radiative forcing" affecting climate variability,[156][157]

> [...] mixing up of cause and effect when observing natural climate variability can lead to the mistaken conclusion that the climate system is more sensitive to greenhouse gas emissions than it really is. [...] it provides a quantitative mechanism for the (minority) view that global warming is mostly a manifestation of natural internal climate variability.

> [...] low frequency, internal radiative forcing amounting to little more than 1 W/m^2, assumed to be proportional to a weighted average of the southern oscillation and Pacific decadal oscillation indices since 1900, produces ocean temperature behavior similar to that observed: warming from 1900 to 1940, then slight cooling through the 1970s, then resumed warming up to the present, as well as 70% of the observed centennial temperature trend.

Temperature projections

James Hansen's 1984 climate model projections versus observed temperatures are updated each year by Dr Mikako Sato of Columbia University.[159] The RealClimate website provides an annual update comparing both Hansen's 1988 model projections and the IPCC Fourth Assessment Report (AR4) climate model projections with observed temperatures recorded by GISS and HadCRUT. The measured temperatures show continuing global warming.[158]

Conventional projections of future temperature rises depend on estimates of future anthropogenic GHG emissions (see SRES), those positive and negative climate change feedbacks that have so far been incorporated into the models, and the climate sensitivity. Models referenced by the Intergovernmental Panel on Climate Change (IPCC) predict that global temperatures are likely to increase by 1.1 to 6.4 °C (2.0 to 11.5 °F) between 1990 and 2100. Others have proposed that temperature increases may be higher than IPCC estimates. One theory is that the climate may reach a "tipping point" where positive feedback effects lead to runaway global warming; such feedbacks include decreased reflection of solar radiation as sea ice melts, exposing darker seawater, and the potential release of large volumes of methane from thawing permafrost.[160] In 1959 Dr. Bert Bolin, in a speech to the National Academy of Sciences, predicted that by the year 2000 there would be a 25% increase in carbon dioxide in the atmosphere compared to the levels in 1859. The actual increase by 2000 was about 29%.[161]

David Orrell or Henk Tennekes[162] say that climate change cannot be accurately predicted. Orrell says that the range of future increase in temperature suggested by the IPCC rather represents a social consensus in the climate community, but adds that "we are having a dangerous effect on the climate".[163]

A 2007 study by David Douglass and coworkers concluded that the 22 most commonly used global climate models used by the IPCC were unable to accurately predict accelerated warming in the troposphere although they did match actual surface warming, concluding that "projections of future climate based on these models should be viewed with much caution". This result went against a similar study of 19 models which found that discrepancies between model predictions and actual temperature were likely due to measurement errors.[164]

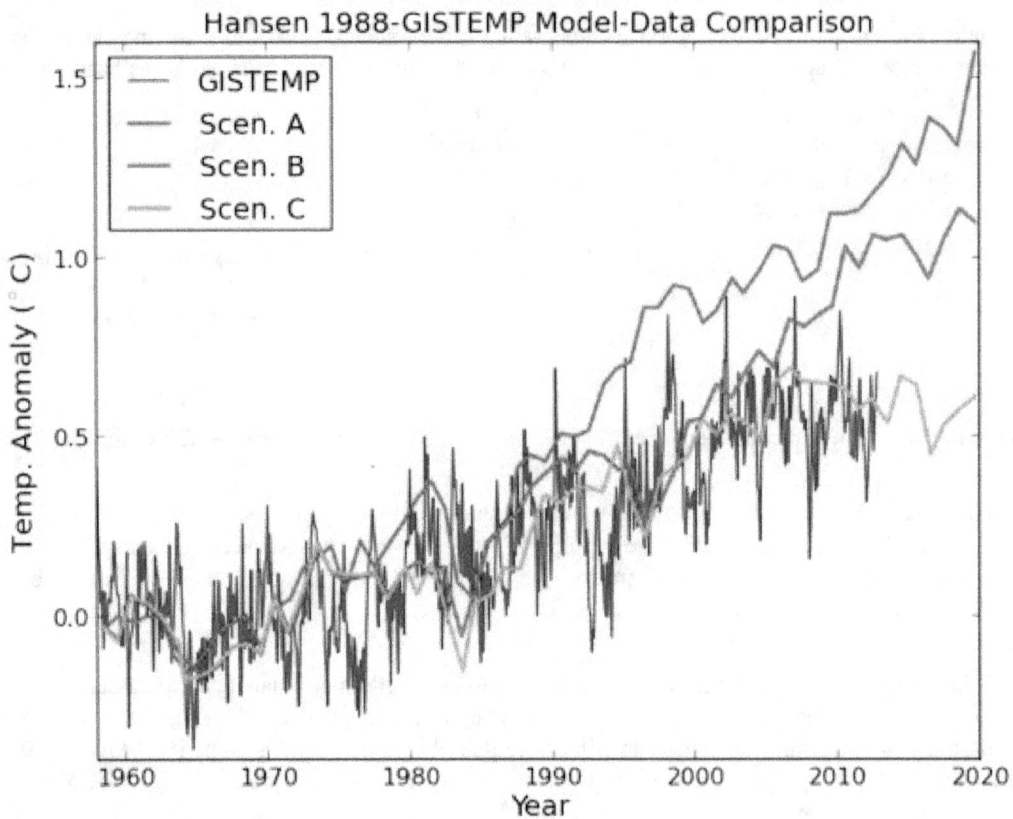

James Hansen's 1988 climate model projections compared with the GISS measured temperature record.[158]

In a NASA report published in January 2013, Hansen and Sato noted that "the 5-year mean global temperature has been flat for a decade, which we interpret as a combination of natural variability and a slowdown in the growth rate of the net climate forcing."[148][165] According to several papers published in 2012, previous projections by the main climate simulation models have failed to predict this lack of additional warming that took place between 2000 and 2010. Ed Hawkins, of the University of Reading, the "surface temperatures since 2005 are already at the low end of the range of projections derived from 20 climate models. If they remain flat, they will fall outside the models' range within a few years".[148][166] Using the long-term temperature trends for the earth scientists and statisticians conclude that it continues to warm through time.[167]

Forecasts confidence

The IPCC states it has increased confidence in forecasts coming from General Circulation Models or GCMs. Chapter 8 of AR4 reads:

> *There is considerable confidence that climate models provide credible quantitative estimates of future climate change, particularly at continental scales and above. This confidence comes from the foundation of the models in accepted physical principles and from their ability to reproduce observed features of current climate and past climate changes. Confidence in model estimates is higher for some climate variables (e.g., temperature) than for others (e.g., precipitation). Over several decades of development, models have consistently provided a robust and unambiguous picture of significant climate warming in response to increasing greenhouse gases.*[168]

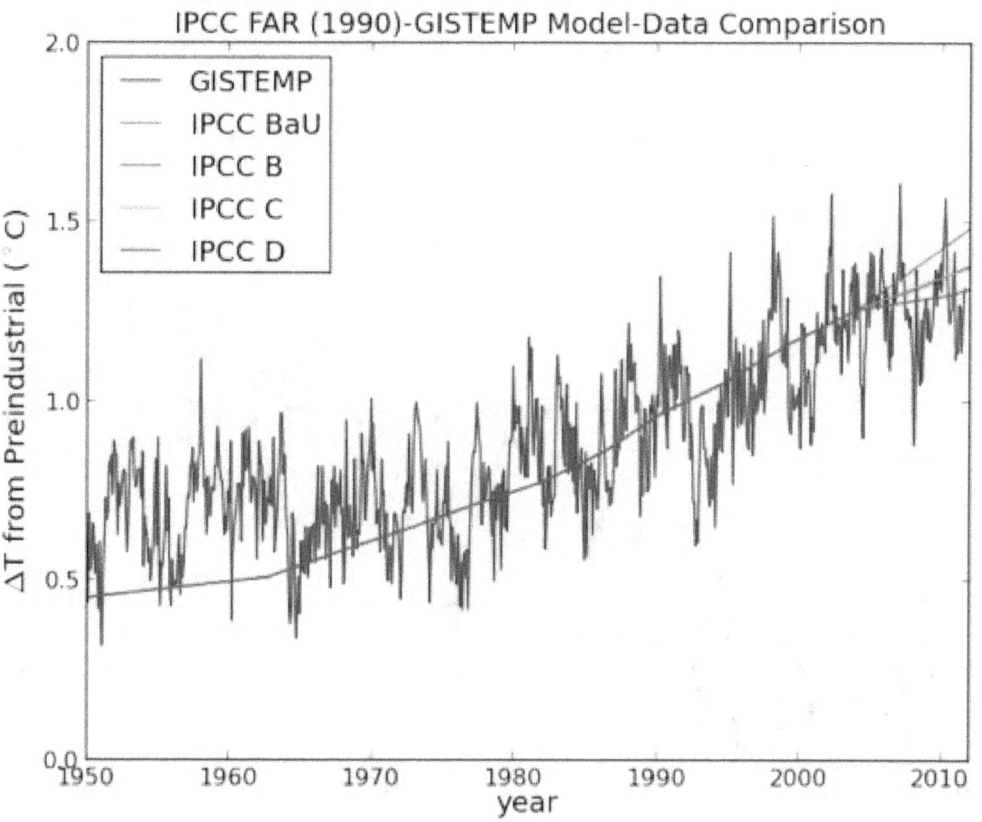

IPCC AR4 projections compared to the GISS temperature record.[158]

Certain scientists, skeptics and otherwise, believe this confidence in the models' ability to predict future climate is not earned.[169][170][171][172]

Arctic sea ice decline

Main article: Arctic sea ice decline

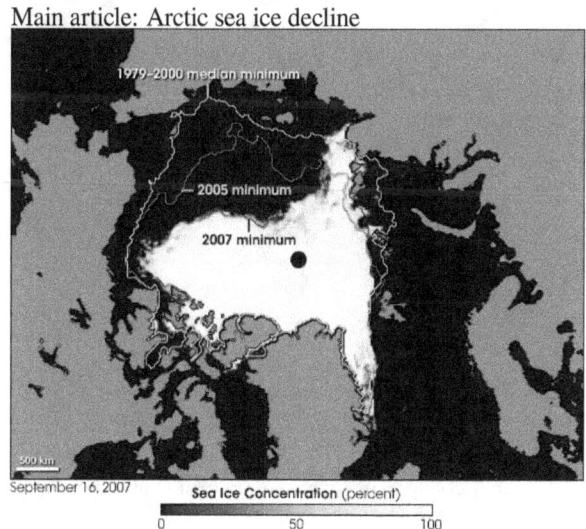

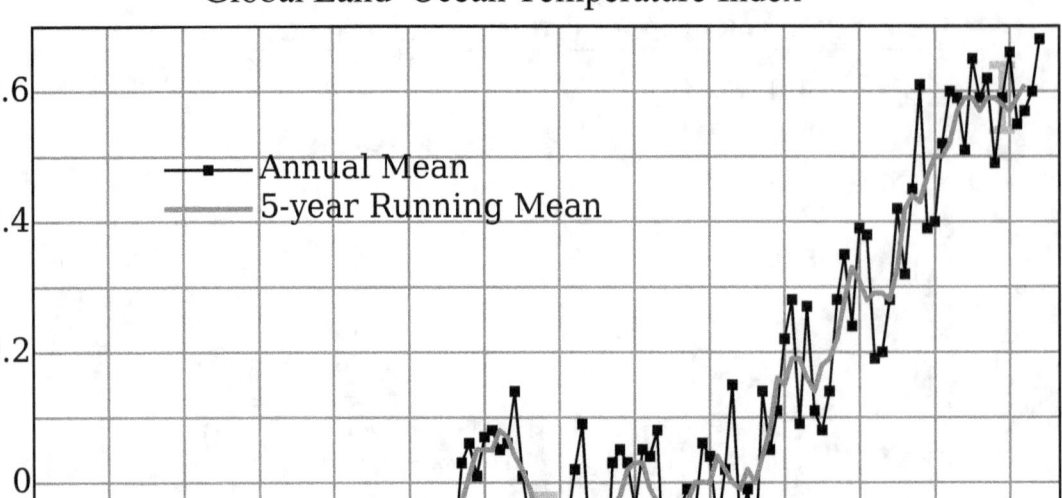

Global Land–Ocean Temperature Index

Global mean land-ocean temperature changes from 1880, relative to the 1951–1980 mean. Source: NASA GISS.

Arctic Sea ice as of 2007 compared to 2005 and also compared to 1979–2000 average

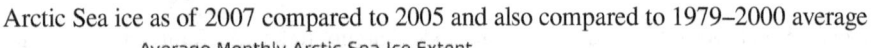

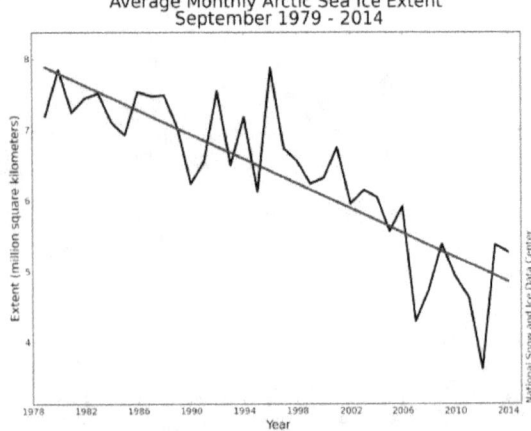

Northern Hemisphere ice trends

Following the (then) record low of the arctic sea ice extend in 2007,[173] Mark Serreze, the director of US National Snow and Ice Data Center, stated "If you asked me a couple of years ago when the Arctic could lose all of its ice then I would have said 2100, or 2070 maybe. But now I think that 2030 is a reasonable estimate".[174] In 2012, during another record low, Peter Wadhams of Cambridge University predicted a possible final collapse of Arctic sea ice in summer months around 2016.[175]

Antarctic and Arctic sea ice extent are available on a daily basis from the National Snow & Ice Data Center.[176]

Data archiving and sharing

Main article: Hockey stick controversy

Scientific journals and funding agencies generally require authors of peer-reviewed research to provide information on archives of data and share sufficient data and methods necessary for a scientific expert on the topic to reproduce the work.

In political controversy over the 1998 and 1999 historic temperature reconstructions widely publicised as the "hockey stick graphs", Mann, Bradley and Hughes as authors of the studies were sent letters on 23 June 2005 from Rep. Joe Barton, chairman of the House Committee on Energy and Commerce and Ed Whitfield, Chairman of the Subcommittee on Oversight and Investigations, demanding full records on the research.[177] [178] [179] The letters told the scientist to provide not just data and methods, but also personal information about their finances and careers, information about grants provided to the institutions they had worked for, and the exact computer codes used to generate their results.[180]

Sherwood Boehlert, chairman of the House Science Committee, told his fellow Republican Joe Barton it was a "misguided and illegitimate investigation" seemingly intended to "intimidate scientists rather than to learn from them, and to substitute congressional political review for scientific review." The U.S. National Academy of Sciences (NAS) president Ralph J. Cicerone wrote to Barton proposing that the NAS should appoint an independent panel to investigate. Barton dismissed this offer.[181][182]

On 15 July, Mann wrote giving his detailed response to Barton and Whitfield. He emphasised that the full data and necessary methods information was already publicly available in full accordance with National Science Foundation (NSF) requirements, so that other scientists had been able to reproduce their work. NSF policy was that computer codes are considered the intellectual property of researchers and are not subject to disclosure, but notwithstanding these property rights, the program used to generate the original MBH98 temperature reconstructions had been made available at the Mann et al. public FTP site. [183]

Many scientists protested about Barton's demands.[181][184] Alan I. Leshner wrote to him on behalf of the American Association for the Advancement of Science stating that the letters gave "the impression of a search for some basis on which to discredit these particular scientists and findings, rather than a search for understanding," He stated that Mann, Bradley and Hughes had given out their full data and descriptions of methods.[185][186] A *Washington Post* editorial on 23 July which described the investigation as harassment quoted Bradley as saying it was "intrusive, far-reaching and intimidating", and Alan I. Leshner of the AAAS describing it as unprecedented in the 22 years he had been a government scientist; he thought it could "have a chilling effect on the willingness of people to work in areas that are politically relevant."[180] Congressman Boehlert said the investigation was as "at best foolhardy" with the tone of the letters showing the committee's inexperience in relation to science.[185]

Barton was given support by global warming sceptic Myron Ebell of the Competitive Enterprise Institute, who said "We've always wanted to get the science on trial ... we would like to figure out a way to get this into a court of law", and "this could work".[185] In his *Junk Science* column on *Fox News*, Steven Milloy said Barton's inquiry was reasonable.[187] In September 2005 David Legates alleged in a newspaper op-ed that the issue showed climate scientists not abiding by data access requirements and suggested that legislators might ultimately take action to enforce them.[188]

1.2.3   Political questions

See also: Politics of global warming and Economics of global warming

 In the U.S. global warming is often a partisan political issue. Republicans tend to oppose action against a threat that they regard as unproven, while Democrats tend to support actions that they believe will reduce global warming and its effects through the control of greenhouse gas emissions.[189] A bipartisan measure was introduced in the US House of Representatives as recently as 2007.[190]

Climatologist Kevin E. Trenberth stated:

> The SPM was approved line by line by governments[...] The argument here is that the scientists deter-

mine what can be said, but the governments determine how it can best be said. Negotiations occur over wording to ensure accuracy, balance, clarity of message, and relevance to understanding and policy. The IPCC process is dependent on the good will of the participants in producing a balanced assessment. However, in Shanghai, it appeared that there were attempts to blunt, and perhaps obfuscate, the messages in the report, most notably by Saudi Arabia. This led to very protracted debates over wording on even bland and what should be uncontroversial text... The most contentious paragraph in the IPCC (2001) SPM was the concluding one on attribution. After much debate, the following was carefully crafted: "In the light of new evidence, and taking into account the remaining uncertainties, most of the observed warming over the last 50 years is likely to have been due to the increase in greenhouse-gas concentrations".[191]

As more evidence has become available over the existence of global warming debate has moved to further controversial issues, including:

1. The social and environmental impacts

2. The appropriate response to climate change

3. Whether decisions require less uncertainty

The single largest issue is the importance of a few degrees rise in temperature:

> Most people say, "A few degrees? So what? If I change my thermostat a few degrees, I'll live fine." ... [The] point is that one or two degrees is about the experience that we have had in the last 10,000 years, the era of human civilization. There haven't been—globally averaged, we're talking—fluctuations of more than a degree or so. So we're actually getting into uncharted territory from the point of view of the relatively benign climate of the last 10,000 years, if we warm up more than a degree or two. (Stephen H. Schneider[192])

The other point that leads to major controversy—because it could have significant economic impacts—is whether action (usually, restrictions on the use of fossil fuels to reduce carbon-dioxide emissions) should be taken now, or in the near future; and whether those restrictions would have any meaningful effect on global temperature.

Because of the economic ramifications of such restrictions, there are those, including the Cato Institute, a libertarian think tank, who argue that the negative economic effects of emission controls outweigh the environmental benefits.[193] They state that even if global warming is caused solely by the burning of fossil fuels, restricting their use would have more damaging effects on the world economy than the increases in global temperature.[194]

> The linkage between coal, electricity, and economic growth in the United States is as clear as it can be. And it is required for the way we live, the way we work, for our economic success, and for our future. Coal-fired electricity generation. It is necessary.(Fred Palmer, President of Western Fuels Association[194])

Conversely, others argue that early action to reduce emissions would help avoid much greater economic costs later, and would reduce the risk of catastrophic, irreversible change.[195] In his December 2006 book, *Hell and High Water*, Joseph J. Romm

> discusses the urgency to act and the sad fact that America is refusing to do so...

On a local or regional level, some specific effects of global warming might be considered beneficial.[196]

Ultimately, however, a strictly economic argument for or against action on climate change is limited at best, failing to take into consideration other potential impacts of any change.

Council on Foreign Relations senior fellow Walter Russell Mead argues that the 2009 Copenhagen Summit failed because environmentalists have changed from "Bambi to Godzilla". According to Mead, environmentalist used to represent the skeptical few who made valid arguments against big government programs which tried to impose simple but massive solutions on complex situations. Environmentalists' more recent advocacy for big economic and social intervention against global warming, according to Mead, has made them, "the voice of the establishment, of the tenured, of the technocrats" and thus has lost them the support of a public which is increasingly skeptical of global warming.[197]

Various campaigns such as 350.org and many Greenpeace projects have been started in an effort to push the world's leaders towards changing laws and policies that would effectively reduce the world's carbon emissions and use of non-renewable energy resources.

Kyoto Protocol

Main article: Kyoto Protocol

The Kyoto protocol is the most prominent international agreement on climate change, and is also highly controversial. Some argue that it goes too far[198] or not nearly far enough[199] in restricting emissions of greenhouse gases. Another area of controversy is the fact that China and India, the world's two most populous countries, both ratified the protocol but are not required to reduce or even limit the growth of carbon emissions under the present agreement even though when listed by greenhouse gas emissions per capita, they have rankings of 121st largest per capita emitter at 3.9 Tonnes of CO_2e and 162nd largest per capita emitter at 1.8 Tonnes of CO_2e respectively, compared with for example the US at position of the 14th largest per capita CO_2e emitter at 22.9 Tonnes of CO_2e. Nevertheless, China is the world's second largest producer of greenhouse gas emissions, and India 4th (see: countries by greenhouse emissions). Various predictions see China overtaking the US in total greenhouse emissions between late 2007 and 2010,[200][201][202] and according to many other estimates, this already occurred in 2006.[203][204]

Additionally, high costs of decreasing emissions may cause significant production to move to countries that are not covered under the treaty, such as India and China, says Fred Singer.[205] As these countries are less energy efficient, this scenario is said to cause additional carbon emissions.

In May 2010 the Hartwell Paper was published by the London School of Economics in collaboration with the University of Oxford.[206] This paper was written by 14 academics from various disciplines in the sciences and humanities, and also some policies thinkers, and they argued that the Kyoto Protocol crashed in late 2009 and "*has failed to produce any discernable real world reductions in emissions of greenhouse gases in fifteen years.*"[206] They argued that this failure opened an opportunity to set climate policy free from Kyoto and the paper advocates a controversial and piecemeal approach to decarbonization of the global economy.[207][208] The Hartwell paper proposes that "*the organising principle of our effort should be the raising up of human dignity via three overarching objectives: ensuring energy access for all; ensuring that we develop in a manner that does not undermine the essential functioning of the Earth system; ensuring that our societies are adequately equipped to withstand the risks and dangers that come from all the vagaries of climate, whatever their cause may be*".[206][207][208]

The only major developed nation which has signed but not ratified the Kyoto protocol is the US (see signatories). The countries with no official position on Kyoto are mainly African countries with underdeveloped scientific infrastructure or are oil producers .

Funding

See also: Global warming denial

The Global Climate Coalition was an industry coalition that funded several scientists who expressed skepticism about global warming. In the year 2000, several members left the coalition when they became the target of a national divestiture campaign run by John Passacantando and Phil Radford at Ozone Action. According to the New York Times, when Ford Motor Company was the first company to leave the coalition, it was "the latest sign of divisions within heavy industry over how to respond to global warming."[209][210] After that, between December, 1999 and early March, 2000, the GCC was deserted by Daimler-Chrysler, Texaco, energy firm the Southern Company and General Motors.[211] The Global Climate Coalition closed in 2002, or in their own words, 'deactivated'.[212]

Documents obtained by Greenpeace under the US Freedom of Information Act show that the Charles G. Koch Foundation gave climate change writer Willie Soon two grants totaling $175,000 in 2005/6 and again in 2010. Multiple grants to

Soon from the American Petroleum Institute between 2001 and 2007 totalled $274,000, and from ExxonMobil totalled $335,000 between 2005 and 2010. Other coal and oil industry sources which funded him include the Mobil Foundation, the Texaco Foundation and the Electric Power Research Institute. Soon, acknowledging that he received this money, stated unequivocally that he has "never been motivated by financial reward in any of my scientific research."[213] In February 2015, Greenpeace disclosed papers documenting that Soon failed to disclose to academic journals funding including more than $1.2 million from fossil fuel industry related interests including ExxonMobil, the American Petroleum Institute, the Charles G. Koch Charitable Foundation and the Southern Company.[214][215][216] To investigate how widespread such hidden funding was, senators Barbara Boxer, Edward Markey and Sheldon Whitehouse wrote to a number of companies. Koch general counsel refused the request and said it would infringe the company's first amendment rights.[217]

The Greenpeace research project ExxonSecrets, and George Monbiot writing in *The Guardian*, as well as various academics,[218][219] have linked several skeptical scientists—Fred Singer, Fred Seitz and Patrick Michaels—to organizations funded by ExxonMobil and Philip Morris for the purpose of promoting global warming skepticism. These organizations include the Cato Institute and the Heritage Foundation.[220] Similarly, groups employing global warming skeptics, such as the George C. Marshall Institute, have been criticized for their ties to fossil fuel companies.[221]

On 2 February 2007, *The Guardian* stated[222][223] that Kenneth Green, a Visiting Scholar with AEI, had sent letters[224] to scientists in the UK and the U.S., offering US$10,000 plus travel expenses and other incidental payments in return for essays with the purpose of "highlight[ing] the strengths and weaknesses of the IPCC process", specifically regarding the IPCC Fourth Assessment Report.

A furor was raised when it was revealed that the Intermountain Rural Electric Association (an energy cooperative that draws a significant portion of its electricity from coal-burning power plants) donated $100,000 to Patrick Michaels and his group, New Hope Environmental Services, and solicited additional private donations from its members.[225][226][227]

The Union of Concerned Scientists have produced a report titled 'Smoke, Mirrors & Hot Air',[228] that criticizes Exxon-Mobil for "underwriting the most sophisticated and most successful disinformation campaign since the tobacco industry" and for "funnelling about $16 million between 1998 and 2005 to a network of ideological and advocacy organizations that manufacture uncertainty on the issue". In 2006 Exxon said that it was no longer going to fund these groups[229] though that statement has been challenged by Greenpeace.[230]

The Center for the Study of Carbon Dioxide and Global Change, a skeptic group, when confronted about the funding of a video they put together ($250,000 for "The Greening of Planet Earth" from an oil company) stated, "We applaud Western Fuels for their willingness to publicize a side of the story that we believe to be far more correct than what at one time was 'generally accepted'. But does this mean that they fund The Center? Maybe it means that we fund them!"[231]

Donald Kennedy, editor-in-chief of *Science*, has said that skeptics such as Michaels are lobbyists more than researchers, and that "I don't think it's unethical any more than most lobbying is unethical", he said. He said donations to skeptics amounts to "trying to get a political message across".[232]

Global warming skeptic Reid Bryson said in June 2007 that "There is a lot of money to be made in this... If you want to be an eminent scientist you have to have a lot of grad students and a lot of grants. You can't get grants unless you say, 'Oh global warming, yes, yes, carbon dioxide'."[233] Similar positions have been advanced by University of Alabama, Huntsville climate scientist Roy Spencer, Spencer's University of Alabama, Huntsville colleague and IPCC contributor John Christy, University of London biogeographer Philip Stott,[234] Accuracy in Media,[235] and Ian Plimer in his 2009 book *Heaven and Earth — Global Warming: The Missing Science*.

Richard Lindzen, the Alfred P. Sloan Professor of Meteorology at MIT, said that "[in] the winter of 1989 Reginald Newell, a professor of meteorology [at MIT], lost National Science Foundation funding for data analyses that were failing to show net warming over the past century". Lindzen also suggested that four other scientists "apparently" lost their funding or positions after questioning the scientific underpinnings of global warming.[236] Lindzen himself has been the recipient of money from energy interests such as OPEC and the Western Fuels Association, including "$2,500 a day for his consulting services",[237] as well as funding from US federal sources including the National Science Foundation, the Department of Energy, and NASA.[238]

46   *CHAPTER 1. CLIMATE CHANGE DENIAL*

Debate over most effective response to warming

See also: List of scientists opposing the mainstream scientific assessment of global warming

In recent years some skeptics have changed their positions regarding global warming. Ronald Bailey, author of *Global Warming and Other Eco-Myths* (published by the Competitive Enterprise Institute in 2002), stated in 2005, "Anyone still holding onto the idea that there is no global warming ought to hang it up".[239] By 2007, he wrote "Details like sea level rise will continue to be debated by researchers, but if the debate over whether or not humanity is contributing to global warming wasn't over before, it is now.... as the new IPCC Summary makes clear, climate change Pollyannaism is no longer looking very tenable".[240]

"There are alternatives to its [the climate-change crusade's] insistence that the only appropriate policy response is steep and immediate emissions reductions.... a greenhouse-gas-emissions cap ultimately would constrain energy production. A sensible climate policy would emphasize building resilience into our capacity to adapt to climate changes.... we should consider strategies of adaptation to a changing climate. A rise in the sea level need not be the end of the world, as the Dutch have taught us". says Steven F. Hayward of American Enterprise Institute, a conservative think-tank.[241] Hayward also advocates the use of "orbiting mirrors to rebalance the amounts of solar radiation different parts of the earth receive"—the space sunshade example of so-called geoengineering for solar radiation management.

In 2001 Richard Lindzen, asked whether it was necessary to try to reduce CO_2 emissions, said that responses needed to be prioritised. "You can't just say, 'No matter what the cost, and no matter how little the benefit, we'll do this'. If we truly believe in warming, then we've already decided we're going to adjust...The reason we adjust to things far better than Bangladesh is that we're richer. Wouldn't you think it makes sense to make sure we're as robust and wealthy as possible? And that the poor of the world are also as robust and wealthy as possible?"[242]

Others argue that if developing nations reach the wealth level of the United States this could greatly increase CO_2 emissions and consumption of fossil fuels. Large developing nations such as India and China are predicted to be major emitters of greenhouse gases in the next few decades as their economies grow.[243][244]

The conservative National Center for Policy Analysis whose "Environmental Task Force" contains a number of climate change skeptics including Sherwood Idso and S. Fred Singer[245] says, "The growing consensus on climate change policies is that adaptation will protect present and future generations from climate-sensitive risks far more than efforts to restrict CO_2 emissions".[246]

The adaptation-only plan is also endorsed by oil companies like ExxonMobil, "ExxonMobil's plan appears to be to stay the course and try to adjust when changes occur. The company's plan is one that involves adaptation, as opposed to leadership",[247] says this Ceres report.[248]

Gregg Easterbrook characterized himself as having "a long record of opposing alarmism". In 2006, he stated, "based on the data I'm now switching sides regarding global warming, from skeptic to convert".[249]

The Bush administration also voiced support for an adaptation-only policy in the US in 2002. "In a stark shift for the Bush administration, the United States has sent a climate report [*U.S. Climate Action Report 2002*] to the United Nations detailing specific and far-reaching effects it says global warming will inflict on the American environment. In the report, the administration also for the first time places most of the blame for recent global warming on human actions—mainly the burning of fossil fuels that send heat-trapping greenhouse gases into the atmosphere". The report however "does not propose any major shift in the administration's policy on greenhouse gases. Instead it recommends adapting to inevitable changes instead of making rapid and drastic reductions in greenhouse gases to limit warming".[250] This position apparently precipitated a similar shift in emphasis at the COP 8 climate talks in New Delhi several months later,[251] "The shift satisfies the Bush administration, which has fought to avoid mandatory cuts in emissions for fear it would harm the economy. 'We're welcoming a focus on more of a balance on adaptation versus mitigation', said a senior American negotiator in New Delhi. 'You don't have enough money to do everything'".[252][253] The White House emphasis on adaptation was not well received however:

> Despite conceding that our consumption of fossil fuels is causing serious damage and despite implying that current policy is inadequate, the Report fails to take the next step and recommend serious alternatives. Rather, it suggests that we simply need to accommodate to the coming changes. For example, reminiscent of former Interior Secretary Hodel's proposal that the government address the hole in the ozone layer by

encouraging Americans to make better use of sunglasses, suntan lotion and broad-brimmed hats, the Report suggests that we can deal with heat-related health impacts by increased use of air-conditioning ... Far from proposing solutions to the climate change problem, the Administration has been adopting energy policies that would actually increase greenhouse gas emissions. Notably, even as the Report identifies increased air conditioner use as one of the 'solutions' to climate change impacts, the Department of Energy has decided to roll back energy efficiency standards for air conditioners.
— Letter from 11 State Attorneys General to George W. Bush., [254]

Some find this shift and attitude disingenuous and indicative of an inherent bias against prevention (i.e. reducing emissions/consumption) and for the prolonging of profits to the oil industry at the expense of the environment. "Now that the dismissal of climate change is no longer fashionable, the professional deniers are trying another means of stopping us from taking action. It would be cheaper, they say, to wait for the impacts of climate change and then adapt to them" says writer and environmental activist George Monbiot[255] in an article addressing the supposed economic hazards of addressing climate change. Others argue that adaptation alone will not be sufficient.[256] See also Copenhagen Consensus.

Though not emphasized to the same degree as mitigation, adaptation to a climate certain to change has been included as a necessary component in the discussion as early as 1992,[257] and has been all along.[258] However it was not to the *exclusion*, advocated by the skeptics, of *preventative* mitigation efforts, and therein, say carbon cutting proponents, lies the difference.[259]

Another highly debated potential climate change mitigation strategy is Cap and Trade due to its direct relationship with the economy.

Political pressure on scientists

Many climate scientists state that they are put under enormous pressure to distort or hide any scientific results which suggest that human activity is to blame for global warming. A survey of climate scientists which was reported to the US House Oversight and Government Reform Committee in 2007 noted that "Nearly half of all respondents perceived or personally experienced pressure to eliminate the words 'climate change', 'global warming' or other similar terms from a variety of communications". These scientists were pressured to tailor their reports on global warming to fit the Bush administration's climate change scepticism. In some cases, this occurred at the request of former oil-industry lobbyist Phil Cooney, who worked for the American Petroleum Institute before becoming chief of staff at the White House Council on Environmental Quality (he resigned in 2005 before being hired by ExxonMobil).[260] In June 2008, a report by NASA's Office of the Inspector General concluded that NASA staff appointed by the White House had censored and suppressed scientific data on global warming in order to protect the Bush administration from controversy close to the 2004 presidential election.[261]

U.S. officials, such as Philip Cooney, have repeatedly edited scientific reports from US government scientists,[262] many of whom, such as Thomas Knutson, have been ordered to refrain from discussing climate change and related topics.[263][264][265] Attempts to suppress scientific information on global warming and other issues have been described by journalist Chris Mooney in his book *The Republican War on Science.*

Climate scientist James E. Hansen, director of NASA's Goddard Institute for Space Studies, wrote in a widely cited *New York Times* article[266] in 2006 that his superiors at the agency were trying to "censor" information "going out to the public". NASA denied this, saying that it was merely requiring that scientists make a distinction between personal, and official government, views in interviews conducted as part of work done at the agency. Several scientists working at the National Oceanic and Atmospheric Administration have made similar complaints;[267] once again, government officials said they were enforcing long-standing policies requiring government scientists to clearly identify personal opinions as such when participating in public interviews and forums.

The BBC's long-running current affairs series *Panorama* recently investigated the issue, and was told that "scientific reports about global warming have been systematically changed and suppressed".[268]

Scientists who agree with the consensus view have sometimes expressed concerns over what they view as sensationalism of global warming by interest groups and the press. For example, Mike Hulme, director of the Tyndall Centre for Climate Research, wrote how increasing use of pejorative terms like "catastrophic", "chaotic" and "irreversible", had altered the public discourse around climate change: "This discourse is now characterised by phrases such as 'climate change is worse than we thought', that we are approaching 'irreversible tipping in the Earth's climate', and that we are 'at the point of no return'. I have found myself increasingly chastised by climate change campaigners when my public statements and lectures on climate change have not satisfied their thirst for environmental drama and exaggerated rhetoric".[269]

According to an Associated Press release on 30 January 2007,

> Climate scientists at seven government agencies say they have been subjected to political pressure aimed at downplaying the threat of global warming.

> The groups presented a survey that shows two in five of the 279 climate scientists who responded to a questionnaire complained that some of their scientific papers had been edited in a way that changed their meaning. Nearly half of the 279 said in response to another question that at some point they had been told to delete reference to "global warming" or "climate change" from a report".[270]

Critics writing in the *Wall Street Journal* editorial page state that the survey[271] was itself unscientific.[272]

In addition to the pressure from politicians, many prominent scientists working on climate change issues have reported increasingly severe harassment from members of the public. The harassment has taken several forms. The US FBI told *ABC News* that it was looking into a spike in threatening emails sent to climate scientists, while a white supremacist website posted pictures of several climate scientists with the word "Jew" next to each image. One climate scientist interviewed by *ABC News* had a dead animal dumped on his doorstep and now frequently has to travel with bodyguards.[273]

In April 2010, Virginia Attorney General Ken Cuccinelli claimed that leading climate scientist Michael E. Mann had possibly violated state fraud laws, and without providing any evidence of wrongdoing, filed the Attorney General of Virginia's climate science investigation as a civil demand that the University of Virginia provide a wide range of records broadly related to five research grants Mann had obtained as an assistant professor at the university from 1999 to 2005. This litigation was widely criticized in the academic community as politically motivated and likely to have a chilling effect on future research.[274][275] The university filed a court petition and the judge dismissed Cuccinelli's demand on the grounds that no justification had been shown for the investigation.[276] Cuccinelli issued a revised subpoena, and appealed the case to the Virginia Supreme Court which ruled in March 2012 that Cuccinelli did not have the authority to make these demands. The outcome was hailed as a victory for academic freedom.[277][278]

Exxon Mobil is also notorious for skewing scientific evidence through their private funding of scientific organizations. In 2002, Exxon Mobil contributed $10,000 to The Independent Institute and then $10,000 more in 2003. In 2003, The Independent Institute release a study that reported the evidence for imminent global warming found during the Clinton administration was based on now-dated satellite findings and wrote off the evidence and findings as a product of "bad science."[279]

This is not the only consortium of skeptics that Exxon Mobil has supported financially. The George C. Marshall Institute received $630,000 in funding for climate change research from ExxonMobil between 1998 and 2005. Exxon Mobil also gave $472,000 in funding to The Board of Academic and Scientific Advisors for the Committee for a Constructive Tomorrow from 1998 to 2005. Dr. Frederick Seitz, well known as "the god father of global warming skepticism," served as both Chairman Emeritus of The George C. Marshall Institute and a board member of the Committee for a Constructive Tomorrow from 1998 to 2005.[280]

Litigation

Several lawsuits have been filed over global warming. For example, Massachusetts v. Environmental Protection Agency before the Supreme Court of the United States allowed the EPA to regulate greenhouse gases under the Clean Air Act. A similar approach was taken by California Attorney General Bill Lockyer who filed a lawsuit California v. General Motors Corp. to force car manufacturers to reduce vehicles' emissions of carbon dioxide. This lawsuit was found to lack legal merit and was tossed out.[281][282] A third case, Comer v. Murphy Oil USA, Inc., a class action lawsuit filed by Gerald Maples, a trial attorney in Mississippi, in an effort to force fossil fuel and chemical companies to pay for damages caused by global warming. Described as a nuisance lawsuit, it was dismissed by District Court.[283] However, the District Court's decision was overturned by the United States Court of Appeals for the Fifth Circuit, which instructed the District Court to reinstate several of the plaintiffs' climate change-related claims on 22 October 2009.[284] The Sierra Club sued the U.S. government over failure to raise automobile fuel efficiency standards, and thereby decrease carbon dioxide emissions.[285][286]

1.2.4 See also

- Attitude polarization

- Climatic Research Unit documents

- Climatic Research Unit email controversy

- Criticism of the IPCC Fourth Assessment Report

- Global warming conspiracy theory

- How Global Warming Works

- Science wars

- Global cooling

- Skeptical Science (website, information resource)

1.2.5 Notes

[1] 2009 Ends Warmest Decade on Record. NASA Earth Observatory Image of the Day, 22 January 2010.

[2] Oreskes, Naomi (December 2004). "BEYOND THE IVORY TOWER: The Scientific Consensus on Climate Change". *Science* **306** (5702): 1686. doi:10.1126/science.1103618. PMID 15576594. Such statements suggest that there might be substantive disagreement in the scientific community about the reality of anthropogenic climate change. This is not the case. [...] Politicians, economists, journalists, and others may have the impression of confusion, disagreement, or discord among climate scientists, but that impression is incorrect.

[3] America's Climate Choices: Panel on Advancing the Science of Climate Change; National Research Council (2010). *Advancing the Science of Climate Change*. Washington, D.C.: The National Academies Press. ISBN 0-309-14588-0. Retrieved 19 February 2014. (p1) ... there is a strong, credible body of evidence, based on multiple lines of research, documenting that climate is changing and that these changes are in large part caused by human activities. While much remains to be learned, the core phenomenon, scientific questions, and hypotheses have been examined thoroughly and have stood firm in the face of serious scientific debate and careful evaluation of alternative explanations. * * * (p21-22) Some scientific conclusions or theories have been so thoroughly examined and tested, and supported by so many independent observations and results, that their likelihood of subsequently being found to be wrong is vanishingly small. Such conclusions and theories are then regarded as settled facts. This is the case for the conclusions that the Earth system is warming and that much of this warming is very likely due to human activities.

[4] "Understanding and Responding to Climate Change" (PDF). United States National Academy of Sciences. 2008. Archived from the original (PDF) on 23 April 2013. Retrieved 30 May 2010. Most scientists agree that the warming in recent decades has been caused primarily by human activities that have increased the amount of greenhouse gases in the atmosphere.

[5] Lovejoy, Shaun; Chipello, Chris (11 April 2014). "Is global warming just a giant natural fluctuation?". McGill University. Retrieved 17 April 2014.

[6] Lovejoy, S. (April 2014). "Scaling fluctuation analysis and statistical hypothesis testing of anthropogenic warming". *Climate Dynamics* **42**: 2339–2351. Bibcode:2014ClDy...42.2339L. doi:10.1007/s00382-014-2128-2. Retrieved 17 April 2014.

[7] *Climate Change 2013: The Physical Science Basis*. IPCC Fifth Assessment Report, Working Group I, Summary for Policymakers. "The best estimate of the human-induced contribution to warming is similar to the observed warming over this period."

[8] Julie Brigham-Grette; et al. (September 2006). "Petroleum Geologists' Award to Novelist Crichton Is Inappropriate". *Eos* (PDF) **87** (36): 364. Bibcode:2006EOSTr..87..364B. doi:10.1029/2006EO360008. The AAPG stands alone among scientific societies in its denial of human-induced effects on global warming.

[9] DiMento, Joseph F. C.; Doughman, Pamela M. (2007). *Climate Change: What It Means for Us, Our Children, and Our Grandchildren*. The MIT Press. p. 68. ISBN 978-0-262-54193-0.

[10] Boykoff, M.; Boykoff, J. (July 2004). "Balance as bias: global warming and the US prestige press" (PDF). *Global Environmental Change Part A* **14** (2): 125–136. doi:10.1016/j.gloenvcha.2003.10.001.

[11] Oreskes, Naomi; Conway, Erik. *Merchants of Doubt: How a Handful of Scientists Obscured the Truth on Issues from Tobacco Smoke to Global Warming* (first ed.). Bloomsbury Press. ISBN 978-1-59691-610-4.

[12] Committee on Surface Temperature Reconstructions for the Last 2,000 Years and National Research Council (2006). *Surface Temperature Reconstructions for the Last 2,000 Years*. Washington, D.C.: The National Academies. ISBN 978-0-309-10225-4. Retrieved 4 May 2013.

[13] *Public Support for Climate and Energy Policies in March 2012* (PDF). Yale Project on Climate Change Communication:. 2012.

[14] McCright, A.M.; Dunlap R.E. (2000). "Challenging global warming as a social problem: An analysis of the conservative movement's counter-claims" (PDF). *Social problems* **47** (4): 499–522. doi:10.2307/3097132. JSTOR 3097132. See p. 500.

[15] Speech to the Royal Society (27 September 1988), Public Statement, Speech Archive, *Margaret Thatcher Foundation*. Retrieved 9 April 2007.

[16] Carvalho, Anabela (2007). "Ideological cultures and media discourses on scientific knowledge". *Public Understanding of Science* **16** (2): 223–43. doi:10.1177/0963662506066775 (inactive 2015-01-09).

[17] Harvey, Fiona (9 May 2013). "Charles: 'Climate change sceptics are turning Earth into dying patient'". London: The Guardian. Retrieved 10 May 2013.

[18] Mintzer, Irving M. (1992). *Confronting climate change*. Cambridge University Press. pp. 265–272. ISBN 978-0-521-42091-4.

[19] "USEIA U.S. Energy-Related Carbon Dioxide Emissions, 2012". 21 October 2013.

[20] Crampton, Thomas (4 January 2007). "More in Europe worry about climate than in U.S., poll shows". International Herald Tribune. Retrieved 14 April 2007.

[21] "Little Consensus on Global Warming – Partisanship Drives Opinion – Summary of Findings". Pew Research Center for the People and the Press. 12 July 2006. Retrieved 14 April 2007.

[22] TNS Opinion and Social (December 2009). "Europeans' Attitudes Towards Climate Change" (Full free text). European Commission. Retrieved 24 December 2009.

[23] Black, Richard (5 September 2007). "BBC switches off climate special". BBC. Retrieved 15 December 2011.

[24] BBC drops climate change special. *The Guardian*. 5 September 2007. Retrieved 15 December 2011.

[25] McCarthy, Michael, Global Warming: Too Hot to Handle for the BBC, the *Independent*, 6 September 2007

[26] Weart, Spencer (2006). "The Public and Climate Change". In Weart, Spencer. *The Discovery of Global Warming*. American Institute of Physics. ISBN 978-0-674-01157-1. Retrieved 14 April 2007.

[27] Langer, Gary (26 March 2006). "Poll: Public Concern on Warming Gains Intensity". ABC News. Retrieved 12 April 2007.

[28] GlobeScan and the Program on International Policy Attitudes at University of Maryland (25 September 2007). "Man causing climate change – poll". BBC World Service. Retrieved 25 September 2007.

[29] Program on International Policy Attitudes (5 April 2006). "30-Country Poll Finds Worldwide Consensus that Climate Change is a Serious Problem". Program on International Policy Attitudes. Retrieved 20 April 2007.

[30] Pew Research Center: "Public Praises Science; Scientists Fault Public, Media" 9 July 2009.

[31] Tipping Point or Turning Point? Social Marketing & Climate Change (3Mb pdf), by Ipsos Mori, July 2007.

[32] David Suzuki (18 August 2006). "Public doesn't understand global warming". David Suzuki Foundation. Retrieved 18 August 2007.

[33] Richard J. Bord, Ann Fisher & Robert E. O'Connor (1997). "Is Accurate Understanding of Global Warming Necessary to Promote Willingness to Sacrifice?". Archived from the original on 27 September 2007. Retrieved 29 February 2008.

[34] Richard J. Bord, Robert E. O'Connor, Ann Fischer; O'Connor; Fisher (1 July 2000). "In what sense does the public need to understand global climate change?". *Public Understanding of Science* **9** (3): 205–218. doi:10.1088/0963-6625/9/3/301.

[35] No Global Warming Alarm in the U.S., China – 15-Nation Pew Global Attitudes Survey. Released 13 June 2006.

[36] Rising Environmental Concern in 47-Nation Survey. Pew Global Attitudes. Released 27 June 2007.

[37] "Climate scepticism 'on the rise', BBC poll shows". BBC News. 7 February 2010.

[38] "Only 2 per cent of Canadians don't believe in climate change: poll". *The Globe and Mail* (Toronto). 15 August 2012.

[39] Peter Jacques (2009). *Environmental skepticism: ecology, power and public life.* Global environmental governance series. Ashgate Publishing, Ltd. ISBN 978-0-7546-7102-2.

[40] George E. Brown (March 1997). "Environmental Science Under Siege in the U.S. Congress". *Environment: Science and Policy for Sustainable Development* **39** (2): 12–31. doi:10.1080/00139159709604359.

[41] "NEW ON THE SEPP WEB". Archived from the original on 15 February 2007. Retrieved 23 May 2007.

[42] Pielke Jr., Roger A. (10 January 2005). "Accepting Politics In Science". Washington Post. p. A17. Retrieved 24 April 2007.

[43] Anderegg, William R L; James W. Prall, Jacob Harold, and Stephen H. Schneider; Harold, J.; Schneider, S. H. (2010). "Expert credibility in climate change" (PDF). *Proc. Natl. Acad. Sci. U.S.A.* **107** (27): 12107–9. Bibcode:2010PNAS..10712107A. doi:10.1073/pnas.1003187107. PMC 2901439. PMID 20566872. Retrieved 22 August 2011.

[44] Doran, P.T., Zimmerman, M.K. (2009). "Examining the Scientific Consensus on Climate Change" (PDF). *Eos, Transactions American Geophysical Union* **30** (3).

[45] John Cook, Dana Nuccitelli, Sarah A Green, Mark Richardson, Bärbel Winkler, Rob Painting, Robert Way, Peter Jacobs. Andrew Skuce; Nuccitelli; Green; Richardson; Winkler; Painting; Way; Jacobs; Skuce (15 May 2013). "Quantifying the consensus on anthropogenic global warming in the scientific literature" (PDF). *Environ. Res. Lett.* **8** (2): 024024. Bibcode:2013ERL.....8b4024C. doi:10.1088/1748-9326/8/2/024024.

[46] Joint statement of sixteen National Academies of Science (18 May 2001). "The Science of Climate Change". The Royal Society. Retrieved 20 May 2009. "The work of the Intergovernmental Panel on Climate Change (IPCC) represents the consensus of the international scientific community on climate change science. We recognise IPCC as the world's most reliable source of information on climate change and its causes, and we endorse its method of achieving this consensus. Despite increasing consensus on the science underpinning predictions of global climate change, doubts have been expressed recently about the need to mitigate the risks posed by global climate change. We do not consider such doubts justified".

[47] Union of Concerned Scientists. "World Scientists Call For Action". Archived from the original on 12 October 2007. Projections indicate that demand for food in Asia will exceed the supply by 2010.

[48] Union Of Concerned Scientists (2 October 1997). "World's Nobel Laureates And Preeminent Scientists Call On Government Leaders To Halt Global Warming". ScienceDaily.com. Retrieved 9 February 2010.

[49] "List of Selected Prominent Signatories with awards and affiliations". Dieoff.org. Retrieved 29 August 2010.

[50] America's Climate Choices: Panel on Advancing the Science of Climate Change; National Research Council (2010). *Advancing the Science of Climate Change.* Washington, D.C.: The National Academies Press. pp. 1 & 21–22. ISBN 0-309-14588-0. Retrieved 19 February 2014.

[51] Dr. Roy W., Spencer (2010). *The Great Global Warming Blunder.* Encounter Books. ISBN 1-59403-373-0.

[52] Carter, Professor Robert M. (2010). *Climate: The Counter Consensus.* pp. 191–210.

[53] "New York Global Warming Conference Considers 'Manhattan Declaration' – by Heartland Institute staff – The Heartland Institute". Heartland.org. 4 March 2008. Retrieved 29 August 2010.

[54] Crichton, Michael (17 January 2003). "Lecture at CalTech: "Aliens Cause Global Warming"". Archived from the original on 10 January 2006. Retrieved 14 April 2007.

[55] "500 Scientists Whose Research Contradicts Man-Made Global Warming Scares – by Dennis T. Avery – The Heartland Institute". Heartland.org. 14 September 2007. Archived from the original on 14 July 2010. Retrieved 29 August 2010.

[56] Heartland's bogus list of 500 scientists – Deltoid

[57] Monbiot, George (7 December 2009). "The Real Climate Scandal". Archived from the original on 12 December 2009. Published in The Guardian 8 December 2009

[58] Monbiot, George (9 December 2009). "The climate denial industry seeks to dupe the public. It's working". *The Hindu*. Retrieved 3 September 2010.

[59] "Controversy Arises Over Lists of Scientists Whose Research Contradicts Man-Made Global Warming Scares – by Joseph Bast – News Releases". Heartland.org. Retrieved 29 August 2010.

[60] Do 500 scientists refute the consensus?

[61] Anderegg W.R.L., Prall J.W., Harold J., Schneider S.H. (21 June 2010). "Expert credibility in climate change". *Proc. Natl. Acad. Sci. U.S.A.* **107** (27): 12107–9. Bibcode:2010PNAS..10712107A. doi:10.1073/pnas.1003187107. PMC 2901439. PMID 20566872.

[62] Kintisch E. (21 June 2010). "Scientists 'Convinced' of Climate Consensus More Prominent Than Opponents, Says Paper". *Science Insider*. AAAS.

[63] Collins, Nick (22 June 2010). "Climate change sceptic scientists 'less prominent and authoritative'". *The Telegraph* (London: Telegraph Media Group Limited). p. 1. Retrieved 22 June 2010.

[64] "Deniers are not Skeptics". Skeptical Inquirer. 2014.

[65] Oreskes, Naomi (20 December 2007). "The American Denial of Global Warming—The Truth About Denial". *Perspectives on Ocean Science—UCSD-TV*. YouTube. Retrieved 29 August 2010. In 1995, the IPCC concluded that the human effect on climate is now discernible. The lead author of the key chapter on detection and attribution...was a scientist of the Lawrence Livermore National Laboratory named Benjamin J. Santer.
When the IPCC report came out, Seitz, Nierenberg, and now a 4th physicist—a man by the name of S. Fred Singer—launched a highly personal attack on Santer. In an open letter to the IPCC, which they sent to numerous members of the US Congress, Singer, Seitz, and Nierenberg accused Santer of making "unauthorized" changes to the IPCC report [...]
They followed this with an op-ed piece in the *Wall Street Journal* entitled "A Major Deception on Global Warming". This piece was written by Seitz, in which he states that the effect of the changes was "to deceive policy makers and the public".
Now Santer replied, in a letter to the editor of the *Wall Street Journal*, and in the response he explained that he had made changes, but those changes were in response to the peer review process. In other words, totally normal scientific practice...This account was corroborated by the Chairman of the IPCC and by all of the other authors of the chapters. In fact, over 40 scientists were co-authors of this chapter. This letter was signed by Santer and 40 others and published in the *Wall Street Journal* in June 1996. And Santer was also formally defended by the American Meteorological Society.
But neither Seitz nor Singer ever retracted the charges, which was then repeated—many times, over and over again—by industry groups and think-tanks. And in fact, if you google "Ben Santer", these same charges are still in the Internet today. In fact, one site said that it was *proven* in 1996 that Santer had fraudulently altered the IPCC report.

[66] "A Case Against Precipitous Climate Action". 15 January 2011. Retrieved 16 February 2011.

[67] "An Open Letter to the Community from Chris Landsea". Archived from the original on 18 February 2007. Retrieved 28 April 2007.

[68] "Prometheus: Final Chapter, Hurricanes and IPCC, Book IV Archives". Sciencepolicy.colorado.edu. 14 February 2007. Retrieved 29 August 2010.

[69] "Hurricanes and Global Warming for IPCC" (PDF). Washington. Reuters. 21 October 2004. Retrieved 30 December 2008.

[70] "Final Climate Change Report" (PDF). Retrieved 29 December 2008.

[71] The Committee Office, House of Lords (28 November 2005). "House of Lords — Economic Affairs — Third Report". Publications.parliament.uk. Retrieved 29 August 2010.

[72] "Written testimony of John R. Christy Ph.D. before House Committee on Energy and Commerce on March 7, 2007" (PDF). Archived from the original (PDF) on 2007-11-28. Retrieved 29 December 2008.

[73] "UN Blowback: More Than 650 International Scientists Dissent Over Man-Made Global Warming Claims".

[74] "How many on Inhofe's list are IPCC authors?". Archived from the original on 27 January 2012.

[75] "More on Inhofe's alleged list of 650 scientists". Archived from the original on 22 January 2012.

[76] "Inhofe's 650 "dissenters" (make That 649... 648...)".

[77] Biello D (April 2007). "Conservative Climate". *Sci. Am.* **296** (4): 16, 19. doi:10.1038/scientificamerican0407-16. PMID 17479619.

[78] Hanson, Brooks (7 May 2010). "Stepping Back; Moving Forward". American Association for the Advancement of Science. Retrieved 23 May 2010.

[79] Hansen JE (April–June 2007). "Scientific reticence and sea level rise". *Environ. Res. Lett.* **2** (2): 024002. arXiv:physics/0703220. Bibcode:2007ERL.....2b4002H. doi:10.1088/1748-9326/2/2/024002.

[80] "Climate Science: Main Conclusions". Archived from the original on 11 December 2008. Retrieved 11 December 2008.

[81] Henderson-Sellers, Ann. "The IPCC report: what the lead authors really think — environmentalresearchweb". Retrieved 24 December 2009.

[82] Idso, C. D.; K. E. Idso. "Carbon Dioxide and Global Warming — Where We Stand on the Issue". CO2science. Archived from the original on 10 April 2007. Retrieved 13 April 2007.

[83] Barkov, N.I. (February 2003). "Historical carbon dioxide record from the Vostok ice core". Carbon Dioxide Information Analysis Center. Archived from the original on 6 March 2007. Retrieved 13 March 2007.

[84] Kuo, C.; Lindberg, C.; Thomson, D.J. (22 February 1990). "Coherence established between atmospheric carbon dioxide and global temperature". *Nature* **343** (6260): 709–714. Bibcode:1990Natur.343..709K. doi:10.1038/343709a0.

[85] Weart, Spencer (2006). "Past Cycles: Ice Age Speculations". In Weart, Spencer. *The Discovery of Global Warming*. American Institute of Physics. ISBN 978-0-674-01157-1. Retrieved 14 April 2007.

[86] "More Notes on Global Warming". Physics Today. May 2005. Archived from the original on 11 August 2007. Retrieved 10 September 2007.

[87] "Historical CO_2 record derived from a spline fit (20 year cutoff) of the Law Dome DE08 and DE08-2 ice cores". Retrieved 12 June 2007.

[88] Tans, Pieter. "Trends in Carbon Dioxide". NOAA/ESRL. Retrieved 11 December 2009.

[89] "Water vapour: feedback or forcing?".

[90] Crowley, Thomas J.; Baum, Steven K. (1995). "Reconciling Late Ordovician (440 Ma) glaciation with very high (14X) CO_2 levels". *Journal of Geophysical Research* **100** (D1): 1093–1102. Bibcode:1995JGR...100.1093C. doi:10.1029/94JD02521.

[91] Gorder, Pam Frost (25 October 2006). "Appalachian Mountains, carbon dioxide caused long-ago global cooling". Ohio State University Research news. Retrieved 13 April 2007.

[92] Hegerl; et al. "Chapter 9 Understanding and Attributing Climate Change" Missing or empty |title= (help) (pdf), in (IPCC AR4 WG1 2007).

[93] IPCC (2001). "2.3 Global Futures Scenarios". *Climate Change 2001:Synthesis Report*.

[94] "Dr Fred Singer".

[95] Stern, Nicholas Herbert (2007). *The Economics of Climate Change — The Stern Review*. Cambridge, UK: Cambridge University Press. ISBN 978-0-521-70080-1. Retrieved 19 February 2014.

[96] "Modeling of long-term fossil fuel consumption shows 14.5-degree hike in Earth's temperature". 1 November issue of the American Meteorological Society's *Journal of Climate*. Archived from the original on 8 October 2006.

[97] IPCC. "Summary for Policymakers". Human and Natural Drivers of Climate Change Missing or empty |title= (help), in IPCC AR4 SYR 2007.

[98] IPCC. "Summary for Policymakers". Natural factors have made small contributions to radiative forcing over the past century Missing or empty |title= (help), in IPCC TAR WG1 2001.

[99] Muscheler, Raimund; Joos, Fortunat; Müller, Simon A.; Snowball, Ian (2005). "How unusual is today's solar activity? Arising from: S. K. Solanki, I. G. Usoskin, B. Kromer, M. Schüssler and J. Beer, Nature, 2004, 431, 1084–1087" (PDF). *Nature* **436** (7050): E3–E4. Bibcode:2005Natur.436E...3M. doi:10.1038/nature04045. PMID 16049429. Archived from the original (PDF) on 8 January 2006.

[100] Leidig, Michael; Nikkhah, Roya (17 July 2004). "The truth about global warming – it's the Sun that's to blame". London: The Daily Telegraph. Retrieved 12 April 2007.

[101] Solanki, Sami K.; Usoskin, Ilya G.; Kromer, Bernd; Schüssler, Manfred; Beer, Jürg (2004). "Unusual activity of the Sun during recent decades compared to the previous 11,000 years" (PDF). *Nature* **431** (7012): 1084–7. Bibcode:2004Natur.431.1084S. doi:10.1038/nature02995. PMID 15510145.

[102] "Space Weather/Solar Activity and Climate". DMI Solar-Terrestrial Physics Division. 19 October 1998. Retrieved 13 April 2007.

[103] Lockwood, Mike; Lockwood, Claus (2007). "Recent oppositely directed trends in solar climate forcings and the global mean surface air temperature" (PDF). *Proceedings of the Royal Society A* **463** (2086): 2447–2460. Bibcode:2007RSPSA.463.2447L. doi:10.1098/rspa.2007.1880. Archived from the original (PDF) on 26 September 2007. Retrieved 21 July 2007. There are many interesting palaeoclimate studies that suggest that solar variability had an influence on pre-industrial climate. There are also some detection–attribution studies using global climate models that suggest there was a detectable influence of solar variability in the first half of the twentieth century and that the solar radiative forcing variations were amplified by some mechanism that is, as yet, unknown. However, these findings are not relevant to any debates about modern climate change. Our results show that the observed rapid rise in global mean temperatures seen after 1985 cannot be ascribed to solar variability, whichever of the mechanisms is invoked and no matter how much the solar variation is amplified.

[104] Mitchell; et al. "Chapter 12. Detection of Climate Change and Attribution of Causes". Sec. 12.4.3.3 Space-time studies Missing or empty |title= (help), in IPCC TAR WG1 2001.

[105] Mitchell, J. F. B.; Johns, T. C. (1997). "On Modification of Global Warming by Sulfate Aerosols". *Journal of Climate* **10** (2): 245–267. Bibcode:1997JCli...10..245M. doi:10.1175/1520-0442(1997)010<0245:OMOGWB>2.0.CO;2. ISSN 1520-0442. Retrieved 14 April 2007.

[106] Ruckstuhl, C.; et al. (2008). "Aerosol and cloud effects on solar brightening and the recent rapid warming". *Geophysical Research Letters* **35** (12): L12708. Bibcode:2008GeoRL..3512708R. doi:10.1029/2008GL034228.

[107] http://www.cru.uea.ac.uk/cru/data/temperature/#faq

[108] Folland; et al. "Chapter 2: Observed Climate Variability and Change". Sec. 2.2 How Much is the World Warming? Missing or empty |title= (help), in IPCC TAR WG1 2001.

[109] Peterson, Thomas C. (2003). "Assessment of urban versus rural in situ surface temperatures in the contiguous United States: no difference found. Journal of Climate". *Journal of Climate* **16** (18): 2941–59. Bibcode:2003JCli...16.2941P. doi:10.1175/1520-0442(2003)016<2941:AOUVRI>2.0.CO;2. ISSN 1520-0442.

[110] David, Parker (2006). "A demonstration that large-scale warming is not urban". *Journal of Climate* **19** (12): 2882–95. Bibcode:2006JCli...19.2882P. doi:10.1175/JCLI3730.1.

[111] Pielke Sr., R.A., and T. Matsui (2005). "Should light wind and windy nights have the same temperature trends at individual levels even if the boundary layer averaged heat content change is the same?" (PDF). *Geophys. Res. Letts.* **32** (21): L21813. Bibcode:2005GeoRL..3221813P. doi:10.1029/2005GL024407. Archived from the original (PDF) on 10 September 2008.

[112] Davey, Christopher A.; Pielke Sr., Roger A. (2005). "Microclimate Exposures of Surface-Based Weather Stations: Implications For The Assessment of Long-Term Temperature Trends" (PDF). *Bull. Am. Meteor. Soc.* **86** (4): 497–504. Bibcode:2005BAMS...86..497D. doi:10.1175/BAMS-86-4-497. Archived from the original (PDF) on 10 September 2008.

[113] Mahmood, Rezaul; Stuart A. Foster, David Logan (2006). "The GeoProfile metadata, exposure of instruments, and measurement bias in climatic record revisited". *International Journal of Climatology* **26** (8): 1091–1124. Bibcode:2006IJCli..26.1091M. doi:10.1002/joc.1298.

[114] "Fiddler On The Roof". Investor's Business Daily. 22 June 2007. Archived from the original on 15 August 2007.

[115] Menne, Matthew J.; Claude N. Williams, Jr., and Michael A. Palecki (2010). "On the reliability of the U.S. surface temperature record" (PDF). *J. Geophys. Res.* **115**: D11108. Bibcode:2010JGRD..11511108M. doi:10.1029/2009JD013094. In summary, we find no evidence that the CONUS average temperature trends are inflated due to poor station siting...The reason why station exposure does not play an obvious role in temperature trends probably warrants further investigation.

[116] Cook, John (27 January 2010). "Climate sceptics distract us from the scientific realities of global warming". London: The Guardian. Retrieved 5 February 2010.

[117] Jeff Tollefson (20 October 2011). "Different method, same result: global warming is real". *Nature News*. doi:10.1038/news.2011.607. Archived from the original on 14 January 2012. Retrieved 22 October 2011.

[118] "Cooling the Warming Debate: Major New Analysis Confirms That Global Warming Is Real". *Science Daily*. 21 October 2011. Retrieved 22 October 2011.

[119] Ian Sample (20 October 2011). "Global warming study finds no grounds for climate sceptics' concerns". *The Guardian* (London). Retrieved 22 October 2011.

[120] "Climate change: The heat is on". *The Economist*. 22 October 2011. Retrieved 22 October 2011.

[121] Santer, B. D.; Thorne, P. W.; Haimberger, L.; K. E. Taylor, T. M. L. Wigley, J. R. Lanzante, S. Solomon, M. Free, P. J. Gleckler, P. D. Jones, T. R. Karl, S. A. Klein, C. Mears, D. Nychka, G. A. Schmidt, S. C. Sherwood, and F. J. Wentz (2008). "Consistency of modelled and observed temperature trends in the tropical troposphere" (PDF). *International Journal of Climatology* **28** (13): 1703–22. Bibcode:2008IJCli..28.1703S. doi:10.1002/joc.1756.

[122] IPCC. "Summary for Policymakers". Direct Observations of Recent Climate Change Missing or empty |title= (help), in IPCC AR4 WG1 2007.

[123] Davidson, Keay (4 February 2002). "Media goofed on Antarctic data / Global warming interpretation irks scientists". *The San Francisco Chronicle*. Retrieved 13 April 2013.

[124] Peter N. Spotts (18 January 2002). "Guess what? Antarctica's getting colder, not warmer". *The Christian Science Monitor*. Retrieved 13 April 2013.

[125] Chang, Kenneth (3 May 2002). "Ozone Hole Is Now Seen as a Cause for Antarctic Cooling". *The New York Times*. Retrieved 13 April 2013.

[126] "America Reacts To Speech Debunking Media Global Warming Alarmism". U.S. Senate Committee on Environment and Public Works. 28 September 2006. Retrieved 13 April 2013.

[127] Bijal P. Trivedi (25 January 2002). "Antarctica Gives Mixed Signals on Warming". *National Geographic*. Retrieved 13 April 2013.

[128] Derbyshire, David (14 January 2002). "Antarctic cools in warmer world". *The Daily Telegraph* (London). Retrieved 13 April 2013.

[129] "Scientific winds blow hot and cold in Antarctica". *CNN*. 25 January 2002. Archived from the original on 9 June 2012. Retrieved 13 April 2013.

[130] Chang, Kenneth (2 April 2002). "The Melting (Freezing) of Antarctica; Deciphering Contradictory Climate Patterns Is Largely a Matter of Ice". *The New York Times*. Retrieved 13 April 2013.

[131] Chapman WL, Walsh JE (2007). "A Synthesis of Antarctic Temperatures". *Journal of Climate* **20** (16): 4096–4117. Bibcode:2007JCli...20.409 doi:10.1175/JCLI4236.1.

[132] Kenneth Chang (21 January 2009). "Warming in Antarctica Looks Certain". The New York Times. Archived from the original on 24 January 2009. Retrieved 13 April 2013.

[133] Ding, Qinghua; Eric J. Steig, David S. Battisti & Marcel Küttel (10 April 2011). "Winter warming in West Antarctica caused by central tropical Pacific warming". *Nature Geoscience* **4** (6): 398–403. Bibcode:2011NatGe...4..398D. doi:10.1038/ngeo1129. Retrieved 12 January 2012.

[134] "Antarctic cooling pushing life closer to the edge". *USA Today*. 16 January 2002. Retrieved 13 April 2013.

[135] Doran PT; Priscu JC; Lyons WB; et al. (January 2002). "Antarctic climate cooling and terrestrial ecosystem response" (PDF). *Nature* **415** (6871): 517–20. doi:10.1038/nature710. PMID 11793010. Archived from the original (PDF) on 11 December 2004.

[136] Doran; et al. (13 January 2002). "Antarctic climate cooling and terrestrial ecosystem response" (PDF). University of Illinois at Chicago. Retrieved 13 April 2013. *PDF version: advance online publication Letters to Science* (archived original)

[137] Crichton, Michael (2004). *State of Fear*. HarperCollins, New York. p. 109. ISBN 0-06-621413-0. First Edition

[138] Michael Crichton (25 January 2005). "The Case for Skepticism in Global Warming" (PDF). Michael Crichton The official site. Retrieved 13 April 2013. *Speech at the National Press Club, Washington, D.C.* (restored from archived copy)

[139] Michael Crichton (28 September 2005). "Statement of Michael Crichton, M.D. – The Role of Science in Environmental Policy-Making". U.S. Senate Committee on Environment and Public Works. Retrieved 13 April 2013. Testimony before the Committee on Environment and Public Works, Washington, D.C.

[140] Peter Doran (27 July 2006). "Cold, Hard Facts". The New York Times. Retrieved 13 August 2013.

[141] IPCC, Glossary A-D: "Climate Sensitivity", in IPCC AR4 SYR 2007.

[142] SYR 2.3: Climate sensitivity and feedbacks

[143] Response to Comments on "Heat capacity, time constant, and sensitivity of Earth's climate system". Accepted for publication in Journal of Geophysical Research

[144] Schwartz, Stephen E. (2007). "Heat Capacity, Time Constant and Sensitivity of Earth's Climate System" (PDF). *Journal of Geophysical Research* **112**: D24S05. Bibcode:2007JGRD..11224S05S. doi:10.1029/2007JD008746.

[145] Comment on 'Heat Capacity, Time Constant, and Sensitivity of Earth's Climate System,' *Schwartz et al.* Journal of Geophysical Research DRAFT September 2007

[146] Climate Insensitivity RealClimate September 2007

[147] "Aerosol Optical Depth, Climate Sensitivity and Global Warming". Agu.org. Retrieved 29 August 2010.

[148] "A sensitive matter". *The Economist*. 30 March 2013. Retrieved 7 April 2013.

[149] The Research Council of Norway (25 January 2013). "Global Warming Less Extreme Than Feared? New Estimates from a Norwegian Project On Climate Calculations". Science Daily. Retrieved 7 April 2013.

[150] Revkin, Andrew (28 January 2013). "When Publicity Precedes Peer Review in Climate Science". NYTimes.com. Retrieved 22 July 2013.

[151] "Unpublished estimates of climate sensitivity". CICERO. 28 January 2013. Retrieved 22 July 2013.

[152] Richard S. Lindzen, Ming-Dah Chou, and Arthur Y. Hou (March 2001). "Does the Earth Have an Adaptive Infrared Iris?" (PDF). *Bulletin of the American Meteorological Society* **82** (3): 417–432. Bibcode:2001BAMS...82..417L. doi:10.1175/1520-0477(2001)082<0417:DTEHAA>2.3.CO;2.

[153] Ari Jokimäki, 2009, List of Papers on the iris hypothesis of Lindzen (Retrieved 26 March 2012)

[154] Spencer, Roy W., Braswell, William D., Christy, John R. & Hnilo, Justin (2007). "Cloud and radiation budget changes associated with tropical intraseasonal oscillations" (PDF). *Geophysical Research Letters* **34** (15): L15707. Bibcode:2007GeoRL..3415707S. doi:10.1029/2007GL029698.

[155] Bing Lin, Bruce A. Wielicki, Lin H. Chambers, Yongxiang Hu, Kuan-Man Xu (2002). "The iris hypothesis: a negative or positive cloud feedback?". *Journal of Climate* **15** (1): 3–7. Bibcode:2002JCli...15....3L. doi:10.1175/1520-0442(2002)015<0003:TIHANO>2.0.CO;2. ISSN 1520-0442.

[156] Roy W. Spencer (15 April 2008). "Internal Radiative Forcing And The Illusion Of A Sensitive Climate System By Roy Spencer". climatescience.org. Archived from the original on 28 November 2009.

[157] Roy Spencer, "Global Warming and Nature's Thermostat" (the essay, no longer available on Sperser's site weatherquestions.com, can be still be found on the web, for example, at http://petesplace-peter.blogspot.com/2008/01/global-warming-and-natures-thermostat. html (Retrieved 26 March 2012)

[158] http://www.realclimate.org/index.php/archives/2012/02/2011-updates-to-model-data-comparisons/

[159] http://www.columbia.edu/~{}mhs119/Temperature/T_moreFigs/PNAS_GTCh_Fig2.pdf

[160] Steve Connor (16 September 2005). "Global Warming 'Past the Point of No Return'". The Independent. Retrieved 7 September 2007.

[161] Science News, 9 May 2009

[162] Lawrence Solomon. "The limits of predictability". Retrieved 23 July 2013.

[163] David Orrell. "Frequently asked questions on Apollo's Arrow/The Future of Everything, by David Orrell". Retrieved 11 September 2007.

[164] "New Study Increases Concerns About Climate Model Reliability". Sciencedaily.com. 20 December 2007. Retrieved 4 April 2008.

[165] J. Hansen, M. Sato, R. Ruedy (15 January 2013). "Global Temperature Update Through 2012" (PDF). NASA. Retrieved 7 April 2013.

[166] Stott, P., Good, P., Jones, G., Gillett, N. and Hawkins, E (2013). "The upper end of climate model temperature projections is inconsistent with past warming". *Environmental Research Letters* **8** (1): 014024. Bibcode:2013ERL.....8a4024S. doi:10.1088/1748-9326/8/1/014024. Available on line here , see Figure 4.

[167] *What to Make of a Warming Plateau* 10 June 2013 New York Times

[168] "Climate Models and Their Evaluation" (PDF). Retrieved 29 August 2010.

[169] "Skeptic: The Magazine: A Climate of Belief".

[170] Climate Science: Roger Pielke Sr. Research Group News » Comment On Real Climate's Post On The Relevance Of The Sensitivity Of Initial Conditions In The IPCC Models

[171] "On the credibility of climate predictions" (PDF). Retrieved 29 December 2008.

[172] Kesten C. Greene and J. Scott Armstrong (2007). "Global Warming: Forecasts by Scientists Versus Scientific Forecasts" (PDF). *Energy & Environment* (Multi-Science Publishing Co. Ltd) **18** (7): 997–1021. doi:10.1260/095830507782616887. Archived from the original (PDF) on 20 June 2010.

[173] William Chapman (9 August 2007). "New historic sea ice minimum". The Cryosphere Today. Retrieved 11 September 2007.

[174] David Adam (4 September 2007). "Loss of Arctic ice leaves experts stunned". London: Guardian Unlimited. Retrieved 7 September 2007.

[175] Vidal, John (17 September 2012). "Climate change (Environment),Environment,Sea ice (environment),Polar regions (Environment),Arctic (News),World news, Geoengineering (environment),Carbon emissions (Environment),Climate change (Science),Science". *The Guardian* (London).

[176] http://nsidc.org/arcticseaicenews/2012/01/

[177] Richard Monastersky (1 July 2005). "Congressman Demands Complete Records on Climate Research by 3 Scientists Who Support Theory of Global Warming — Archives". The Chronicle of Higher Education. Retrieved 4 March 2011.

[178] "The Committee on Energy and Commerce, Joe Barton, Chairman". *Letters Requesting Information Regarding Global Warming Studies*. U.S. House of Representatives. letters dated 23 June 2005. Archived from the original on 10 April 2011. Retrieved 4 March 2011. Check date values in: |date= (help)

[179] Joe Barton; Ed Whitfield (23 June 2005). "letter to Dr. Michael Mann" (PDF). United States House Committee on Energy and Commerce. Archived from the original (PDF) on 7 February 2012. Retrieved 4 March 2011.

[180] editorial (23 July 2005). "Hunting Witches – washingtonpost.com". *Washington Post*. Retrieved 4 March 2011.

[181] Juliet Eilperin (18 July 2005). "GOP Chairmen Face Off on Global Warming – washingtonpost.com". Washington Post. Retrieved 4 March 2011.

[182] Henry A. Waxman (1 July 2005). "Letter to Chairman Barton" (PDF). henrywaxman.house.gov. Archived from the original (PDF) on 14 March 2012. Retrieved 4 March 2011.

[183] Michael E. Mann (15 July 2005). "Letter to Chairman Barton and Chairman Whitfield" (PDF). RealClimate. Retrieved 4 March 2011.
Gavin Schmidt; Stefan Rahmstorf (18 July 2005). "Scientists respond to Barton". RealClimate. Retrieved 4 March 2011.

[184] 20 scientists as listed (15 July 2005). "letter to Chairman Barton and Chairman Whitfield" (PDF). RealClimate. Retrieved 4 March 2011.

[185] Roland Pease (18 July 2005). "Science/Nature | Politics plays climate 'hockey'". BBC News. Retrieved 4 March 2011.

[186] Alan I. Leshner (13 July 2005). "www.aaas.org" (PDF). American Association for the Advancement of Science. Retrieved 4 March 2011.

[187] Steven Milloy (31 July 2005). "Tree Ring Circus – Opinion – FOXNews.com". *Fox News.* Retrieved 9 March 2011.

[188] "The Weekly Closer from U.S. Senate, September 23, 2005." (PDF). Archived from the original (PDF) on 28 November 2007. Retrieved 29 December 2008.

[189] Mascaro, Lisa (12 February 2007). "GOP still cool on global warming". Las Vegas Sun. Archived from the original on 2 January 2010. Retrieved 14 April 2007.

[190] Waxman, Henry (20 March 2007). "The Safe Climate Act of 2007". Rep. Henry Waxman. Archived from the original on 29 March 2007. Retrieved 14 April 2007. H.R. 1590

[191] Trenberth, Kevin (2001). "The IPCC Assessment of global warming 2001". *Journal of the Forum for Environmental Law, Science, Engineering, and Finance* (8–26). Archived from the original (– Scholar search) on 6 December 2006. Retrieved 19 February 2014.

[192] "What's up with the weather: the debate: Stephen H. Schneider". PBS Nova & Frontline. Retrieved 13 April 2007.

[193] "Global Warming, the Anatomy of a Debate: A speech by Jerry Taylor of the Cato Institute". Archived from the original on 24 January 2012.

[194] "What's up with the weather: the debate: Fred Palmer". PBS Nova & Frontline. Retrieved 13 April 2007.

[195] Nicholas Stern. (2006). "7. Projecting the Growth of Greenhouse-Gas Emissions" (PDF). In Stern, Nicolas. *Stern Review: The Economics of Climate Change* (– Scholar search). HM Treasury, Cambridge University Press. ISBN 978-0-521-70080-1. Retrieved 19 February 2014.

[196] Palmer, Brian (30 January 2012). "Global warming would harm the Earth, but some areas might find it beneficial". *The Washington Post.*

[197] Will, George, "When Bambi becomes Godzilla", *Denver Post*, 5 September 2010.

[198] Darragh, Ian (1998). "A Guide to Kyoto: Climate Change and What it Means to Canadians: Does the Kyoto treaty go far enough... or too far?" (PDF). International Institute for Sustainable Development. Retrieved 14 April 2007.

[199] "Kyoto protocol status" (PDF). UNFCCC. Retrieved 7 November 2006. (Niue,The Cook Islands, Nauru consider reductions "inadequate")

[200] Catherine Brahic (25 April 2007). "China's emissions may surpass the US in 2007". New Scientist. Archived from the original on 27 April 2007. Retrieved 20 May 2007.

[201] Saeed Shah (8 November 2006). "China to pass US greenhouse gas levels by 2010". London: The Independent. Archived from the original on 30 September 2007. Retrieved 20 May 2007.

[202] "China fears disasters, grain cut from global warming". Reuters AlertNet. 27 December 2006. Retrieved 20 May 2007.

[203] China now no. 1 in CO_2 emissions; US in second position Netherlands Environmental Assessment Agency. Retrieved 20 June 2007.

[204] Vidal, John; Adam, David (19 June 2007). "China overtakes US as world's biggest CO_2 emitter". *The Guardian* (London). Retrieved 9 February 2010.

[205] Singer, S. Fred (24 May 2000). *Climate Policy –From Rio to Kyoto: A Political Issue for 2000—and Beyond.* Essays in Public Policy, No. 102. Stanford University: Hoover Institution. p. 49. ISBN 978-0-8179-4372-1. Retrieved 13 April 2007.

[206] Prins, Gwyn et. al (May 2010). "The Hartwell Paper – A new direction for climate policy after the crash of 2009" (PDF). London School of Economics. Retrieved 12 May 2010.

[207] "Oblique strategies". The Economist. 11 May 2010. Retrieved 12 May 2010.

[208] "Do You Heart 'The Hartwell Paper'?". Science Insider. 12 May 2010. Retrieved 12 May 2010.

[209] "Canvassing Works". Canvassing Works. Retrieved 19 July 2013.

[210] Bradsher, Keith (7 December 1999). "Ford Announces Its Withdrawal From Global Climate Coalition". New York Times. Retrieved 21 July 2013.

[211] "GCC Suffers Technical Knockout, Industry defections decimate Global Climate Coalition".

[212] "globalclimate.org". Internet Archive. 19 April 2003. Archived from the original on 19 April 2003.

[213] Vidal, John (27 June 2011). "Climate sceptic Willie Soon received $1m from oil companies, papers show". London: The Guardian.

[214] Brahic, Catherine (February 25, 2015). "Climate change sceptic's work called into question". *New Scientist*. Retrieved March 17, 2015.

[215] McCoy, Terrence (February 23, 2015). "Things just got very hot for climate deniers' favorite scientist". *Washington Post*. Retrieved March 17, 2015.

[216] Gillis, Justin; Schwartz, John (February 21, 2015). "Deeper Ties to Corporate Cash for Doubtful Climate Researcher". *New York times*. Retrieved 2015-02-21.

[217] Yuhas, Alan (13 Mar 2015). "Koch Industries refuses to comply with US senators' climate investigation". *the Guardian*. Retrieved 17 Apr 2015.

[218] Naomi Oreskes; Erik Conway (2010). *Merchants of Doubt: How a Handful of Scientists Obscured the Truth on Issues from Tobacco Smoke to Global Warming*. US: Bloomsbury. ISBN 978-1-59691-610-4.

[219] Clive Hamilton (2010). *Requiem for a Species: Why We Resist the Truth about Climate Change*. Allen & Unwin. pp. 103–105. ISBN 978-1-74237-210-5. Retrieved 19 February 2014.

[220] Monbiot, George (19 September 2006). "The denial industry". London: Guardian. Retrieved 11 August 2007. "By May 1993, as another memo from APCO to Philip Morris shows, the fake citizens' group had a name: the Advancement of Sound Science Coalition".

[221] Adam, David (27 January 2005). "Oil firms fund climate change 'denial'". London: Guardian. Retrieved 14 April 2007.

[222] Sample, Ian (2 February 2007). "Scientists offered cash to dispute climate study". London: Guardian. Retrieved 14 April 2007.

[223] "Climate Controversy and AEI: Facts and Fictions". American Enterprise Institute for Public Policy Research. 9 February 2007. Archived from the original on 13 April 2007. Retrieved 14 April 2007.

[224] Hayward, Steven F.; Kenneth Green (5 July 2006). "AEI Letter to Pf. Schroeder" (PDF). Archived from the original (PDF) on 8 February 2007. Retrieved 14 April 2007.

[225] ABC News Reporting Cited As Evidence In Congressional Hearing On Global Warming ABC August 2006

[226] "Lewandowski memo" (PDF). Retrieved 29 December 2008.

[227] FEATURE-Carbon backlash: coal divides corporations *James, Steve* Reuters, July 2007

[228] "Smoke, Mirrors & Hot Air – How ExxonMobil Uses Big Tobacco's Tactics to Manufacture Uncertainty on Climate Science". Union of Concerned Scientists. January 2007. Retrieved 14 April 2007.

[229] Exxon cuts ties to global warming skeptics MSNBC January 2007

[230] Exxon Still Funding Climate Change Deniers Greenpeace May 2007

[231] "Links". Western Fuels. Archived from the original on 15 January 2006. Retrieved 13 April 2007.

[232] Borenstein, Seth (27 July 2006). "Utilities Paying Global Warming Skeptic". CBS News from Associated Press. Archived from the original on 3 March 2007. Retrieved 14 April 2007.

[233] Real Clear Politics: Hooey Denier Deniers. 24 June 2007.

[234] "Must-See Global Warming TV". Fox News. March 2007. Retrieved 14 May 2007.

[235] Trulock, Notra, "Science for Sale: the Global Warming Scam" Accuracy in Media, 26 August 2002

[236] "Climate of Fear". Wall Street Journal. April 2006. Retrieved 14 May 2007.

[237] Gelbspan, Ross (December 1995). "The Heat Is On: The warming of the world's climate sparks a blaze of denial". Harper's Magazine. Retrieved 8 February 2008.

[238] Lindzen, Richard S.; Constantine Giannitsis (2002). "Reconciling observations of global temperature change" (PDF). *Geophysical research letters* **29** (12): 24–26. Bibcode:2002GeoRL..29l..24L. doi:10.1029/2001GL014074. Retrieved 10 September 2007.

[239] Ronald Bailey (11 August 2005). "We're All Global Warmers Now". Reason Online. Archived from the original on 9 April 2008. Retrieved 27 April 2008.

[240] Bailey, Ronald (2 February 2007). "Global Warming—Not Worse Than We Thought, But Bad Enough". Reason (magazine). Retrieved 13 April 2007.

[241] Hayward, Steven F. (15 May 2006). "Acclimatizing – How to Think Sensibly, or Ridiculously, about Global Warming". American Enterprise Institute. Archived from the original on 4 February 2007. Retrieved 13 April 2007.

[242] "How Dangerous Is Global Warming?". Los Angeles Times. 17 June 2001. Archived from the original on 17 June 2001. Retrieved 14 April 2007.

[243] Keller, Michelle (15 February 2005). "World to celebrate Kyoto Protocol start". The Stanford Daily. Archived from the original on 27 September 2007. Retrieved 14 April 2007.

[244] Harrison, Paul; Pearce, Fred (2000). "Foreword by Peter H. Raven". In Victoria Dompka Markham. *AAAS Atlas of Population & Environment*. American Association for the Advancement of Science & University of California Press. p. 215. ISBN 978-0-520-23081-1. Retrieved 14 April 2007.

[245] "Environmental Task Force". National Center for Policy Analysis. Archived from the original on 6 February 2007. Retrieved 14 April 2007.

[246] Burnett, H. Sterling (19 September 2005). "Climate Change: Consensus Forming around Adaptation". National Center for Policy Analysis. Archived from the original on 29 September 2007. Retrieved 14 April 2007.

[247] Logan, Andrew; Grossman, David (May 2006). "ExxonMobil's Corporate Governance on Climate Change" (PDF). Ceres & Investor Network on Climate Risk. Archived from the original (PDF) on 23 September 2006. Retrieved 14 April 2007.

[248] "Letter to Michael J. Boskin, Secretary Exxon Mobil Corporation" (PDF). Investor Network on Climate Risk. 15 May 2006. Archived from the original (PDF) on 23 September 2006. Retrieved 14 April 2007.

[249] Easterbrook, Gregg (24 May 2006). "Finally Feeling the Heat". New York Times. Retrieved 23 November 2009.

[250] Revkin, Andrew C. (3 June 2002). "Bush climate plan says adapt to inevitable Cutting gas emissions not recommended". San Francisco Chronicle. Retrieved 14 April 2007.

[251] "Climate Compendium: International Negotiations: Vulnerability & Adaptation". Climate Change Knowledge Network & International Institute for Sustainable Development. 2007. Retrieved 14 April 2007.

[252] Revkin, Andrew C. (23 October 2002). "US Pullout Forces Kyoto Talks To Focus on Adaptation – Climate Talks Will Shift Focus From Emissions". New York Times (reprinted by heatisonline.org). Retrieved 14 April 2007.

[253] Eilperin, Juliet (7 April 2007). "U.S., China Got Climate Warnings Toned Down". Washingtonpost.com. pp. A05. Retrieved 30 December 2008.

[254] "Letter to The Honorable George W. Bush — State Attorneys General – A Communication From the Chief Legal Officers of the Following States: Alaska, California, Connecticut, Maine, Maryland, Massachusetts, New Hampshire, New Jersey, New York, Rhode Island, Vermont". 17 July 2002. Retrieved 14 April 2007.

[255] Monbiot, George (December 2006). "Costing Climate Change". New Internationalist. Retrieved 14 April 2007.

[256] Schwartz, Peter; Randall, Doug (February 2004). "An Abrupt Climate Change Scenario and Its Implications for United States National Security". Global Business Network for the Department of Defense. Archived from the original on 18 February 2007. Retrieved 14 April 2007.

[257] Engineering, and Public Policy (U. S.) Panel on Policy Implications of Greenhouse Warming Committee on Science (1992). *Policy Implications of Greenhouse Warming: Mitigation, Adaptation, and the Science Base*. National Academies Press. p. 944. ISBN 978-0-309-04386-1. Retrieved 14 April 2007.

[258]

[259] Adapt or die

[260] US climate scientists pressured on climate change, NewScientist, 31 January 2007

[261] Goddard, Jacqui (4 June 2008). "Nasa 'played down' global warming to protect Bush". *The Scotsman* (Edinburgh). Archived from the original on 24 November 2010. Retrieved 12 February 2010.

[262] Campbell, D. (20 June 2003) "White House cuts global warming from report" *Guardian Unlimited*

[263] Donaghy, T., *et al.* (2007) "Atmosphere of Pressure:" a report of the Government Accountability Project (Cambridge, Massachusetts: UCS Publications)

[264] Rule, E. (2005) "Possible media attention" Email to NOAA staff, 27 July. Obtained via FOIA request on 31 July 2006. and Teet, J. (2005) "DOC Interview Policy" Email to NOAA staff, 29 September. Originally published by Alexandrovna, L. (2005) "Commerce Department tells National Weather Service media contacts must be pre-approved" *The Raw Story*, 4 October. Retrieved 22 December 2006.

[265] Zabarenko, D. (2007) "'Don't discuss polar bears:' memo to scientists" *Reuters*

[266] Revkin, Andrew C. (29 January 2006). "Climate Expert Says NASA Tried to Silence Him". New York Times. Retrieved 14 April 2007.

[267] Eilperin, J. (6 April 2006) "Climate Researchers Feeling Heat From White House" *Washington Post*

[268] "Climate chaos: Bush's climate of fear". BBC Panorama. 1 June 2006. Retrieved 14 April 2007.

[269] Hulme, Mike (4 November 2006). "Chaotic world of climate truth". BBC News. Retrieved 14 April 2007.

[270] "Groups Say Scientists Pressured On Warming". CBC and Associated Press. 30 January 2007. Retrieved 14 April 2007.

[271] Donaghy, Timothy; Freeman, Jennifer; Grifo, Francesca; Kaufman, Karly; Maassarani, Tarek; Shultz, Lexi (February 2007). "Appendix A: UCS Climate Scientist Survey Text and Responses (Federal)". *Atmosphere of Pressure – Political Interference in Federal Climate Science* (PDF). Union of Concerned Scientists & Government Accountability Project.

[272] Taranto, James (1 February 2007). "They Call This Science?". OpinionJournal.com. Retrieved 14 April 2007.

[273] "ABC World News Sunday". ABC News. 23 May 2010.

[274] "Statement of the AAAS Board Of Directors Concerning the Virginia Attorney General's Investigation of Prof. Michael Mann's Work While on the Faculty of University of Virginia" (PDF). AAAS. 18 May 2010. Retrieved 30 July 2010.

[275] Gentile, Sal. "Climate scientist calls Va. attorney general's fraud probe 'harassment'". *PBS.org* (PBS). Retrieved 7 September 2010.

[276] Judge Dismisses Ken Cuccinelli's Misguided Investigation of Michael Mann | Union of Concerned Scientists

[277] Kumar, Anita (2 March 2012). "Va. Supreme Court tosses Cuccinelli's case against former U-Va. climate change researcher – Virginia Politics". The Washington Post. Retrieved 2 March 2012.

[278] Goldenberg, Suzanne (2 March 2012). "Virginia court rejects sceptic's bid for climate science emails : Environment". London: The Guardian. Retrieved 2 March 2012.

[279] Reddy, Sudhakara (2009). "The Great Climate Debate". *Energy Policy* **37** (8): 2997–3008. doi:10.1016/j.enpol.2009.03.064.

[280] de Granados, Oriana Zill. "The Doubters of Global Warming". PBS. Retrieved 19 February 2014.

[281] Lifsher, Marc (18 September 2007). "Global warming lawsuit dismissed". LA Times. Archived from the original on 4 October 2009.

[282] Tanner, Adam (18 September 2007). "Calif. suit on car greenhouse gases dismissed". Reuters. Archived from the original on 15 February 2013.

[283] Pidot, Justin R. (2006). "Global Warming in the Courts – An Overview of Current Litigation and Common Legal Issues" (PDF). Georgetown University Law Center. Archived from the original (PDF) on 4 June 2007. Retrieved 13 April 2007.

[284] http://www.ca5.uscourts.gov/opinions/pub/07/07-60756-CV0.wpd.pdf

[285] "Proposed Settlement Agreement, Clean Air Act Citizen Suit". United States Environmental Protection Agency. 12 August 2005. Retrieved 13 April 2007.

[286] *The Sierra Club vs. Stephen L. Johnson (United States Environmental Protection Agency)*, 03-10262 (United States Court of Appeals for the Eleventh Circuit 20 January 2006).

1.2.6   References

- Ipcc ar4 syr (2007). Core Writing Team; Pachauri, R.K; and Reisinger, A., eds. *Climate Change 2007: Synthesis Report*. Contribution of Working Groups I, II and III to the Fourth Assessment Report of the Intergovernmental Panel on Climate Change. IPCC. ISBN 92-9169-122-4

- Ipcc ar4 wg1 (2007). Solomon, S.; Qin, D.; Manning, M.; Chen, Z.; Marquis, M.; Averyt, K.B.; Tignor, M.; and Miller, H.L., ed. *Climate Change 2007: The Physical Science Basis*. Contribution of Working Group I to the Fourth Assessment Report of the Intergovernmental Panel on Climate Change. Cambridge University Press. ISBN 978-0-521-88009-1 (pb: 978-0-521-70596-7)

- Ipcc tar syr (2001). Watson, R. T.; and the Core Writing Team, ed. *Climate Change 2001: Synthesis Report*. Contribution of Working Groups I, II, and III to the Third Assessment Report of the Intergovernmental Panel on Climate Change. Cambridge University Press. ISBN 0 521 80770 0 (pb: 0-521-01507-3)

- Ipcc tar wg1 (2001). Houghton, J.T.; Ding, Y.; Griggs, D.J.; Noguer, M.; van der Linden, P.J.; Dai, X.; Maskell, K.; and Johnson, C.A., ed. *Climate Change 2001: The Scientific Basis*. Contribution of Working Group I to the Third Assessment Report of the Intergovernmental Panel on Climate Change. Cambridge University Press. ISBN 0-521-80767-0 (pb: 0-521-01495-6)

- Ipcc sar wg3 (1996). Bruce, J.P.; Lee, H.; and Haites, E.F., ed. *Climate Change 1995: Economic and Social Dimensions of Climate Change*. Contribution of Working Group III to the Second Assessment Report of the Intergovernmental Panel on Climate Change. Cambridge University Press. ISBN 0-521-56051-9 (pb: 0-521-56854-4) pdf.

1.2.7   Further reading

- Hulme, Mike (2009). *Why we disagree about climate change: understanding controversy, inaction and opportunity*. Cambridge, UK: Cambridge University Press. ISBN 978-0-521-72732-7.

1.2.8   External links

- ResearchChannel — The American Public's Views of Global Climate Change. A video of a lecture given by Jon A. Krosnick, social scientist, Stanford University. Produced by the National Science Foundation, 25 October 2007

- Spirit that Freed South Africa Must Now Rescue the Planet by Desmond Tutu

- American Petroleum Institute (A.K.A. Energy Citizens). Mother Jones article about astroturfing by petroleum industry trade group American Petroleum Institute

- Skeptical Science: Examining Global Warming Skepticism

- Climate Change Deniers vs The Consensus

- It's Global Warming, Stupid Bloomberg BusinessWeek, 2 November 2012

- Global Warming and other Fictions LifeIvy Magazine, 15 April 2013

The Washington Monument illuminated with a message from Greenpeace criticizing American environmental policy

Chapter 2

Related Articles

2.1 American Petroleum Institute

The **American Petroleum Institute (API)** is the largest U.S trade association for the oil and natural gas industry. It claims to represent about 400 corporations involved in production, refinement, distribution, and many other aspects of the petroleum industry.

The association's chief functions on behalf of the industry include advocacy and negotiation with governmental, legal, and regulatory agencies; research into economic, toxicological, and environmental effects; establishment and certification of industry standards; and education outreach.[2] API both funds and conducts research related to many aspects of the petroleum industry.[2] The current CEO is Jack Gerard.

It has many front groups, including the NH Energy Forum that in August 2011 hosted a New Hampshire event for Republican presidential candidate Rick Perry[3][4]

2.1.1 Standards and certification

API distributes more than 200,000 copies of its publications each year. The publications, technical standards, and electronic and online products are designed, according to API itself, to help users improve the efficiency and cost-effectiveness of their operations, comply with legislative and regulatory requirements, and safeguard health, ensure safety, and protect the environment. Each publication is overseen by a committee of industry professionals, mostly member company engineers.

These technical standards tend to be uncontroversial. For example, API 610 is the specification for centrifugal pumps, API 675 is the specification for controlled volume positive displacement pumps, both packed-plunger and diaphragm types are included. Diaphragm pumps that use direct mechanical actuation are excluded. API 677 is the standard for gear units and API 682 governs mechanical seals.

API also defines the industry standard for the energy conservation of motor oil. **API SN** is the latest specification to which motor oils intended for spark-ignited engines should adhere since 2010. It supersedes **API SM**.[5] Different specifications exist for compression-ignited engines.

API provides vessel codes and standards for the design and fabrication of pressure vessels that help safeguard the lives of people and environments all over the world.

API also defines and drafts standards for measurement for manufactured products such as:

- Precision thread gauges
- Plain plug and ring gauges
- Thread measuring systems

- Metrology and industrial supplies

- Measuring instruments

- Custom gauges

- Precision machining and grinding

- ISO 17025 registered calibration

API RP 500 and RP 505 classify the locations for electrical equipment in hazardous areas. [6] [7]

API has entered petroleum industry nomenclature in a number of areas:

- API gravity, a measure of the density of petroleum.

- API number, a unique identifier applied to each petroleum exploration or production well drilled in the United States.

- API unit, a standard measure of natural gamma radiation measured in a borehole.

2.1.2 Educator intervention

In addition to training industry workers and conducting seminars, workshops, and conferences on public policy, API develops and distributes materials and curricula for schoolchildren and educators. The association also maintains a website, Classroom Energy. These materials take a boldly pro-oil-industry view of various major controversies including oil spills, pipelines, global warming, and ocean acidity.

2.1.3 Public advocacy

In the second half of 2008, as the US presidential election neared, API began airing a series of television ads where spokeswoman Brooke Alexander encourages people to visit their new website, EnergyTomorrow.org API does not use their own name in the ads but does call themselves "The People of America's Oil and Natural Gas Industry."

In January 2012, the American Petroleum Institute launched the voter education campaign - Vote 4 Energy. The campaign claims that increased domestic energy production can create jobs, increase government revenue, and provide U.S. energy security. The Vote 4 Energy campaign does not promote any specific candidate or party, but rather provides voters with energy information to equip them to evaluate candidates on the federal and local levels and make decisions in favor of domestic energy on Election Day. The main components of the Vote 4 Energy campaign include the website - Vote4Energy.org - and social media communities, along with a series of advertisements and events around the country.

2.1.4 Lobbying

API spent more than $3 million annually during the period 2005 to 2009 on lobbying; $3.6 million in 2009.[8] As of 2009, according to API's quarterly "Lobbying Report" submitted to the US Senate, the organization had 16 lobbyists lobbying various Congressional activities.[9]

API conducts lobbying and organizes its member employees' attendance at public events to communicate the industry's position on various issues. A leaked summer 2009 memo from API President Jack Gerard asked its member companies to urge their employees to participate in planned protests (designed to appear independently organized) against the cap-and-trade legislation the House passed that same summer. "The objective of these rallies is to put a human face on the impacts of unsound energy policy and to aim a loud message at [20 different] states," including Florida, Georgia, and Pennsylvania. Gerard went on to assure recipients of the memo that API will cover all organizational costs and handling of logistics. In response to the memo, an API spokesman told media that participants will be there (at protests) because of their own concerns, and that API is just helping them assemble.[10]

To help fight climate control legislation that has been approved by the US House, API supports the Energy Citizens group, which is holding public events.[11][12] API encouraged energy company employees to attend one of its first Energy Citizen events held in Houston in August 2009, but turned away Texas residents who were not employed by the energy industry. *Fast Company* reported that some attendees had no idea of the purpose of the event, and called it "astroturfing at its finest."[13][14] In December 2009, *Mother Jones* magazine said API and Energy Citizens were promulgating climate disinformation.[15]

2.1.5   See also

- United States Oil and Gas Association, formerly the **Mid-Continent Oil and Gas Association**

2.1.6   References

[1] "Jack N. Gerard - President and Chief Executive Officer, American Petroleum Institute - Biography". Congressional Coalition on Adoption Institute. Retrieved January 20, 2011.

[2] "About API". American Petroleum Institute. Retrieved March 29, 2012.

[3] Johnson, Brad (August 15, 2011). "Rick Perry's First Stop In New Hampshire Is Funded By Big Oil". ThinkProgress. Retrieved March 29, 2012.

[4] "Rick Perry stumps Manchester - next stop Iowa", New Hampshire Public Radio, 14 August 2011.

[5] "Engine Oil Guide" (PDF). American Petroleum Institute. March 2010.

[6] API RP 505 Recommended Practice for Classification of Locations for Electrical Installation at Petroleum Facilities Classified as Class I, Zone 0, Zone 1 and Zone 2 (2002).

[7] API RP 500 Recommended Practice for Classification of Locations for electrical Installation at Petroleum Facilities Classified as Class I, Division 1 and Division 2.

[8] "Lobbying: American Petroleum Institute". Center for Responsive Politics. Retrieved March 29, 2012.

[9] "Second Quarter Lobbying Form, 2009, Secretary of the Senate". Retrieved March 29, 2012.

[10] Stone, Daniel (August 20, 2009). "The Browning of Grassroots". Newsweek. Retrieved March 29, 2012.

[11] New York Times, "Oil industry backs protests of emissions bill," August 19, 2009

[12] McNulty, Sheila (August 20, 2009). "The big oil backlash?". Financial Times. Retrieved March 29, 2012.

[13] Schwartz, Ariel (August 21, 2009). "American Petroleum Institute Demonstrates How to Screw Up a Grassroots Event". Fast Company. Retrieved March 29, 2012.

[14] Talley, Ian (August 11, 2009). "Lobby Groups to Use Town Hall Tactics to Oppose Climate Bill". *The Wall Street Journal*.

[15] Harkinson, Josh (December 4, 2009). "The Dirty Dozen of Climate Change Denial". *Mother Jones*. Retrieved August 17, 2015.

2.1.7   External links

- API Website

- Organizational Profile – National Center for Charitable Statistics (Urban Institute)

- Center for Biological Diversity v Dept of the Interior DC Appellate Decision stopping offshore Alaska Oil Leases. April 17, 2009

- Sourcewatch profile

- Center for Responsive Politics profile

- Energy Citizens, API-sponsored organization

- Vote 4 Energy, API-sponsored voter education campaign

- API code list at Piping-Designer.com

- American Petroleum Institute Internal Revenue Service filings archived at the ProPublica Nonprofit Explorer

2.2 Business action on climate change

Business action on climate change includes a range of activities relating to global warming, and to influencing political decisions on global-warming-related regulation, such as the Kyoto Protocol. Major multinationals have played and to some extent continue to play a significant role in the politics of global warming, especially in the United States, through lobbying of government and funding of global warming skeptics. Business also plays a key role in the mitigation of global warming, through decisions to invest in researching and implementing new energy technologies and energy efficiency measures. (See also individual and political action on climate change.)

2.2.1 Overview

In 1989 in the US, the petroleum and automotive industries and the National Association of Manufacturers created the Global Climate Coalition (GCC) to oppose mandatory actions to address global warming. In 1997, when the US Senate overwhelmingly passed a resolution against ratifying the Kyoto Protocol, the industry funded a $13 million industry advertising blitz in the run-up to the vote.[1]

In 1998 the *New York Times* published an American Petroleum Institute (API) memo outlining a strategy aiming to make "recognition of uncertainty ... part of the 'conventional wisdom.'"[2] The memo has been compared to a late 1960s memo by tobacco company Brown and Williamson, which observed: "Doubt is our product since it is the best means of competing with the 'body of fact' that exists in the mind of the general public. It is also the means of establishing a controversy."[3] Those involved in the memo included Jeffrey Salmon, then executive director of the George C. Marshall Institute, Steven Milloy, a prominent skeptic commentator, and the Competitive Enterprise Institute's Myron Ebell.[3] In June 2005 a former API lawyer, Philip Cooney, resigned his White House post after accusations of politically motivated tampering with scientific reports.[4]

In 2002 the GCC considered its work in the US against regulation on global warming to have been so successful that it "deactivated" itself, although the loss of some leading members may also have been a factor.

At the same time, since 1989 many previously skeptical petroleum and automobile industry corporations have changed their position as the political and scientific consensus has grown, with the creation of the Kyoto Protocol and the publication of the International Panel on Climate Change's Second and Third Assessment Reports. These corporations include major petroleum companies like Royal Dutch Shell, Texaco, and BP, as well as automobile manufacturers like Ford, General Motors, and DaimlerChrysler. Some of these have joined with the Center for Climate and Energy Solutions (formerly the Pew Center on Global Climate Change), a non-profit organization aiming to support efforts to address global climate change.[5]

Since 2000, the Carbon Disclosure Project has been working with major corporations and investors to disclose the emissions of the largest companies. By 2007, the CDP published the emissions data for 2400 of the largest corporations in the world, and represented major institutional investors with $41 trillion combined assets under management.[6] The pressure from these investors had had some success in working with companies to reduce emissions.

The World Business Council for Sustainable Development, a CEO-led association of some 200 multinational companies, has called on governments to agree on a global targets, and suggests that it is necessary to cut emissions by 60-80 percent from current levels by 2050.[7]

2.2.2   Global Climate Coalition

A central organization in climate skepticism was the Global Climate Coalition (1989–2002), a group of mainly United States businesses opposing immediate action to reduce greenhouse gas emissions. The coalition funded skeptical scientists to be public spokespeople, provided industry a voice on climate change, and fought the Kyoto Protocol. The *New York Times* reported that "even as the coalition worked to sway opinion [towards skepticism], its own scientific and technical experts were advising that the science backing the role of greenhouse gases in global warming could not be refuted."[8]

In the year 2000, the rate of corporate members leaving accelerated when they became the target of a national divestiture campaign run by John Passacantando and Phil Radford with the organization Ozone Action. According to the New York Times, when Ford Motor Company was the first company to leave the coalition, it was "the latest sign of divisions within heavy industry over how to respond to global warming."[9][10] After that, between December, 1999 and early March, 2000, the GCC was deserted by Daimler-Chrysler, Texaco, the Southern Company and General Motors.[11]

The organization closed in 2002, or in their own words, 'deactivated'.

2.2.3   U.S. Climate Action Partnership

The U.S. Climate Action Partnership (USCAP) was formed in January 2007 with the primary goal of influencing the US government's regulation of greenhouse gas emissions. Original members included General Electric, Alcoa, Natural Resources Defense Council, etc., but they were joined in April, 2007 by ConocoPhilips and AIG.

2.2.4   Energy industry

ExxonMobil

ExxonMobil has been a leading figure in the business world's position on climate change, providing substantial funding to a range of global-warming-skeptical organizations. *Mother Jones* counted some 40 ExxonMobil-funded organization that "either have sought to undermine mainstream scientific findings on global climate change or have maintained affiliations with a small group of "skeptic" scientists who continue to do so." Between 2000 and 2003 these organizations received more than $8m in funding.[3]

It has also had a key influence in the Bush administration's energy policy, including on the Kyoto Protocol,[12] supported by both $55m spent on lobbying since 1999,[3] and direct contacts between the company and leading politicians. It was a leading member of the Global Climate Coalition. It encouraged (and may have been instrumental in) the replacement in 2002 of the head of the IPCC, Robert Watson. .[13] It has also invested $100m into the Global Climate and Energy Project, with Stanford University, and other programs at institutions such as the Massachusetts Institute of Technology, Carnegie Mellon University and the International Energy Agency Greenhouse Gas Research and Development Program.

Some of Exxon's activities on climate change produced strong criticism from environmental groups, including reactions such as a leaflet produced by the Stop Esso campaign, saying 'Don't buy E$$o', and featuring a tiger hand setting fire to the Earth. The company's carbon dioxide emissions are more than 50% higher than those of British rival BP, despite the US firm's oil and gas production being only slightly larger.[14]

According to a 2004 study commissioned by Friends of the Earth, ExxonMobil and its predecessors caused 4.7 to 5.3 percent of the world's man-made carbon dioxide emissions between 1882 and 2002. The group suggested that such studies could form the basis for eventual legal action.[15]

BP

BP left the Global Climate Coalition in 1997 and said that global warming was a problem that had to be dealt with, although it subsequently joined others in lobbying the Australian government not to sign the Kyoto Protocol unless the US did.[16] In March 2002 BP's chief executive, Lord Browne, declared in a speech that global warming was real and that urgent action was needed, saying that "Companies composed of highly skilled and trained people can't live in denial of mounting evidence gathered by hundreds of the most reputable scientists in the world.".[17] In 2005 BP was considering testing carbon sequestration in one of its North Sea oil fields, by pumping carbon dioxide into them (and thereby also increasing yields).[18] Throughout 2006 BP, led by their CEO Lord John Browne, has continued to take a leadership stance on climate change. It has cut its own operational emissions of CO_2 by 10%. It is investing $8 billion in renewable energy over the next 10 years. And most recently it has launched a 'target zero' campaign in the UK to encourage its customers to offset their vehicle emissions when they fill up at the petrol station.

BP's American division is a member of the U.S. Climate Action Partnership (USCAP) (see above).

Koch Industries

Main article: Political activities of the Koch family

From 2005 to 2008, Koch Industries donated $5.7 million on political campaigns and $37 million on direct lobbying to support fossil fuel industries. Between 1997 and 2008, Koch Industries donated a total of nearly $48 million to climate opposition groups.[19] According to Greenpeace, Koch Industries is the major source of funds of what Greenpeace calls "climate denial".[20][21] Koch Industries and its subsidiaries spent more than $20 million on lobbying in 2008 and $12.3 million in 2009, according to the Center for Responsive Politics, a nonpartisan research group.[22][23]

Others

American Electric Power, the world's largest private producer of carbon dioxide, said in 2005 that targets for carbon reduction "represent a common-sense approach that can begin the process of lowering emissions along a gradual, cost-effective path." The company complained that "uncertainties over the cost of carbon" made it very difficult to make decisions about capital investment.[24]

DuPont has cut its greenhouse gas emissions by 65% since 1990, saving hundreds of millions of dollars in the process. "Give us a date, tell us how much we need to cut, give us the flexibility to meet the goals, and we'll get it done", Xcel Energy CEO Wayne Brunetti told *Business Week* in 2004.[25]

Duke Energy, FPL Group, and PG&E Corporation are members of the U.S. Climate Action Partnership (USCAP) (see above).

2.2.5 Transportation

A large proportion of carbon dioxide emissions occur because of transportation. Several companies have formed or invested in electric substitutes for standard automobiles. The Tesla Roadster is an all-electric sports car, available now (production ended). Tesla produces a sedan, Tesla Model S now. Vectrix produces and sells an electric scooter rated for 100 km/h (60 MPH).

There has also been greatly increased interest in personal rapid transit, which applies system engineering principles to reduce energy use, eliminate traffic jams, and produce an acceptable substitute to replace cars, all at the same time. Most systems fully meet Kyoto Treaty carbon emission goals now, 60 years ahead of schedule. Korean steel maker POSCO and its partner Vectus Ltd. have produced a working safety case, including test track and vehicles, that remains fully functional in Swedish winters. Vectus and Suncheon S. Korea signed a memorandum of understanding to install a system.[26] Advanced Transportation Systems' ULTra passed safety certification by the UK Rail Inspectorate in 2003, and won a demonstration project at Heathrow Airport due to be in service in early 2010. ATS Ltd. estimates its ULTra PRT will consume 839 BTU per passenger mile (0.55 MJ per passenger km).[27][28] By comparison, automobiles consume 3,496 BTU, and personal trucks consume 4,329 BTU per passenger mile.[29] 2getthere Inc. sells automated electric freight handling and transit vehicles designed to share existing rights of way with normal traffic.[30] The company recently won the personal rapid transit competition for Masdar.[31]

2.2.6 Insurance industry

In 2004 Swiss Re, the world's second largest reinsurance company, warned that the economic costs of climate-related disasters threatened to reach $150 billion a year within ten years.[32]

In 2006 Lloyd's of London, published a report highlighting the latest science and implications for the insurance industry.[33]

Swiss Re, has said that if the shore communities of four Gulf Coast states choose not to implement adaptation strategies, they could see annual climate-change related damages jump 65 percent a year to $23 billion by 2030. "Society needs to reduce its vulnerability to climate risks, and as long as they remain manageable, they remain insurable, which is our interest as well," said Mark D. Way, head of Swiss Re's sustainable development for the Americas.[34][35]

AIG is a member of the U.S. Climate Action Partnership (USCAP) (see above).

2.2.7 Media

Main article: Media coverage of climate change

In the UK, some newspapers (*Daily Mail, Daily Telegraph*) are significantly skeptical, while most others (with varying enthusiasm, *The Independent* giving it most prominence) support action on global warming. Overall, British newspapers have given the issue three times more coverage than US newspapers.[36] In 2006 (*British Sky Broadcasting* (Sky) became the world's first media company to go 'climate neutral' by purchasing enough carbon offsets. The CEO of the company James Murdoch (son of Rupert Murdoch and heir apparent for the News International empire) is a strong advocate of action on climate change and is thought to be influential on the issue within the wider group of companies, *The Sun* announced it was "going green" and now covers the global warming issue extensively.[37] In June 2006, to much industry interest, Rupert Murdoch invited Al Gore to make his climate change presentation at the annual News Corp (including the Fox Network) gathering at the Pebble Beach golf resort, (USA).[38] In August 2007, Rupert Murdoch announced plans for News Corp. to be carbon neutral by 2010.[39]

2.2.8 More on business action

Businesses take action on climate change for several reasons. Action improves corporate image and better aligns corporate actions with the environmental interests of owners, employees, suppliers, and customers. Action also occurs to reduce costs, increase return on investments, and to reduce dependency on uncontrollable costs.

Increased energy efficiency

For many companies, looking at more efficient energy use can pay off in the medium to long term; unfortunately, shareholders need to be satisfied in the short term, so regulatory intervention is often required, to encourage prudent conservation measures. However, as carbon intensity starts to show up on balance books through organizations such as the Carbon Disclosure Project, voluntary action is starting to take place.

Recently there has been a spate of companies acting to improve their energy efficiency. Possibly the most prominent of these companies is Wal-Mart. Wal-Mart, the largest retailer in the US, has announced specific environmental goals to reduce energy use in its stores and pressure its 60,000 suppliers in its worldwide supply chain to follow its lead. On energy efficiency, Wal-Mart wants to increase the fuel efficiency of its truck fleet by 25% over the next three years and double it within ten years, moving from 6.5 mpg. This seems an attainable goal, and by 2020, it is expected to save the company $494 million a year. The company also wants to build a store that is at least 25% more energy efficient within four years.

Use of renewable energies

Main article: Renewable energy commercialization

In August 2002, the largest gathering of ministers in the history of the world met at the World Summit on Sustainable Development WSSD in Johannesburg. The global environmental community discussed the role of renewables and energy efficiency in lowering carbon emissions, mitigating poverty reduction (energy access) and improving energy security. One result from WSSD was the formation of Partnerships for Sustainable Development to carry forward the international dialogue on sustainable energy and its role in the energy mix.

Partnerships formed include the Renewable Energy and Energy Efficiency Partnership REEEP, the Global Village Energy Partnership GVEP, the Johannesburg Renewable Energy Coalition (JREC), and the Global Network on Energy for Sustainable Development GNESD.

Renewable energies and renewable energy technologies have many advantages over their fossil fuel counterparts. These advantages include the absence of local pollution such as particulates, sulphur oxides (SOX's) and nitrous oxides (NOX's). For the business community, the economic advantages are also becoming clearer. Numerous studies have shown that the working environment has a significant effect on workforce morale. Renewable energy solutions are a part of this, wind turbines in particular being seen by many as a potent symbol of a new modernity, where environmental considerations are taken seriously. A workforce seeing a forward-looking and responsible company is more likely to feel good about working for such a company. A happier workforce is a more productive workforce.

More directly, the high petroleum (oil) and gas prices of 2005 have only added to the attraction of renewable energy sources. Although most renewable energies are more expensive at current fuel prices, the difference is narrowing, and uncertainty in oil and gas markets is a factor worth considering for highly energy-intensive businesses.

Another factor affecting the uptake of renewable energies in Europe is the EU Energy Trading Scheme (ETS or EUTS). Many large businesses are fined for increases in emissions, but can sell any "excess" reductions they make.

Companies with high-profile renewable energy portfolios include an aluminium smelter (Alcan), a cement company (Lafarge), and a microchip manufacturer (Intel). Many examples of corporate leadership in this area can be found on the website of The Climate Group, an independent organization set up for promoting such action by business and government.

Carbon offsets

The principle of carbon offset is fairly simple: a business decides that it doesn't want to contribute further to global warming, and it has already made efforts to reduce its carbon (dioxide) emissions, so it decides to pay someone else to further reduce its net emissions by planting trees or by taking up low-carbon technologies. Every unit of carbon that is absorbed by trees—or not emitted due to funding of renewable energy deployment—offsets the emissions from fossil fuel use. In many cases, funding of renewable energy, energy efficiency, or tree planting—particularly in developing nations—can be a relatively cheap way of making an event, project, or business "carbon neutral". Many carbon offset providers—some as inexpensive as \$0.10 per ton of carbon dioxide—are referenced in the Carbon Offset article of this encyclopedia.

Many businesses are now looking to carbon offset all their work. An example of a business going carbon neutral is FIFA: their 2006 World Cup Final will be carbon neutral. FIFA estimate they are offsetting one hundred thousand tons of carbon dioxide created by the event, largely as a result of people travelling there. Other carbon neutral companies include the bank HSBC, the consumer staples manufacturer Annie's Homegrown, world leading society publisher Blackwell Publishing, and the publishing house New Society Publishers. The Guardian newspaper also offsets its carbon emissions resulting from international air travel.

2.2.9 See also

- Avoiding dangerous climate change

- Business for Innovative Climate and Energy Policy

- Clean Tech Nation

- Climate change denial

- Ecology summit

- Economics of global warming

- Global warming controversy

- Green Globe Lite

- Greenwashing

- Individual and political action on climate change

- List of scientists opposing the mainstream scientific assessment of global warming

- Low-carbon economy

- Mitigation of global warming

- Politics of global warming

- Religious action on climate change

- Renewable energy commercialization

- The Cool War

- Sustainable business

2.2.10 Notes

[1] "Snowed". Mother Jones. Retrieved 2009-12-03.

[2] http://www.environmentaldefense.org/documents/3860_GlobalClimateSciencePlanMemo.pdf

[3] "Some Like It Hot". Mother Jones. Retrieved 2009-12-03.

[4] Borger, Julian (2005-06-09). "Ex-oil lobbyist watered down US climate research | Environment". London: The Guardian. Retrieved 2009-12-03.

[5]

[6] Carbon Disclosure Project: Homepage

[7] WBCSD (2007-09-05). "Climate Change Debate Needs Revolution, Financial Times, 5 September 2007". Wbcsd.org. Retrieved 2009-12-03.

[8] Revkin, Andrew C. Industry Ignored Its Scientists on Climate, *New York Times.* April 23, 2009.

[9] "Canvassing Works". Canvassing Works. Retrieved 2013-07-19.

[10] "Ford Announces Its Withdrawal From Global Climate Coalition". New York Times. Retrieved 2013-07-21.

[11] "GCC Suffers Technical Knockout, Industry defections decimate Global Climate Coalition".

[12] "Revealed: how oil giant influenced Bush | World news". London: The Guardian. 2005-06-08. Retrieved 2009-12-03.

[13]

[14] MacAlister, Terry (2004-10-07). "Exxon admits greenhouse gas increase | Environment". London: The Guardian. Retrieved 2009-12-03.

[15] "Friends of the Earth: Archived press release: Exxonmobil's contribution to global warming revealed". Friends of the Earth (EWNI) foe.co.uk. 2004-01-29. Retrieved 2009-12-03.

[16] "Environmentalists and corporate reputation management". Uow.edu.au. Retrieved 2009-12-03.

[17] mindfully.org (2001-02-13). "How Green Is BP? DARCY FREY / NY Times 8dec02". Mindfully.org. Retrieved 2009-12-03.

[18] McKie, Robin (2005-04-25). "Seabed supplies a cure for global warming crisis | Science | The Observer". London: Observer.guardian.co.uk. Retrieved 2009-12-03.

[19] Vidal, John (30 March 2010). "US oil company donated millions to climate sceptic groups, says Greenpeace". *The Guardian* (London).

[20] "Secretly Funding the Climate Denial Machine". *Global Warming*. Washington: Greenpeace. 2010-03-29. Retrieved 2010-04-01.

[21] DeMelle, Brendan (2010-03-30). "Greenpeace Unmasks Koch Industries' Funding of Climate Denial Industry". Los Angeles: Huffington Post.

[22] http://www.opensecrets.org/lobby/clientsum.php?year=2008&lname=Koch+Industries&id= Center For Responsive Politics, www.opensecrets.org

[23] http://www.opensecrets.org/lobby/clientsum.php?year=2009&lname=Koch+Industries&id= Center For Responsive Politics, www.opensecrets.org

[24] "Global Inc. Goes Green". Mother Jones. 2007-08-10. Retrieved 2009-12-03.

[25] "BW Online | August 16, 2004 | Global Warming". Businessweek.com. 2004-08-16. Retrieved 2009-12-03.

[26] "POSCO signs MOU with Suncheon City Government for Asia's first Personal Rapid Transit System". Vectusprt.com. Retrieved 2009-12-03.

[27] Lowson, Martin (2004). "(doc) A New Approach to Sustainable Transport Systems" (DOC). Retrieved 2007.

[28] The conversion is: 0.55 MJ = 521.6 BTU; 1.609 km = 1 mi; therefore, 521.6 x 1.609 = 839

[29] "Transportation Energy Databook, 26th Edition, Ch. 2, Table 2-12". U.S. Dept. of Energy. 2004.

[30] "2getthere web site" (in Dutch). 2getthere.eu. Retrieved 2009-12-03.

[31] Mogge, John, *The Technology of Personal Transit*, "Figure 6. MASDAR Phase 1A Prototype Passenger PRT." Paper delivered at the World Future Energy Summit, January 20, 2009. Available in WFES online media center.

[32] "911 in New Orleans". Mother Jones. 2005-09-01. Retrieved 2009-12-03.

[33] "Climate Change - Adapt or Bust"

[34] US Gulf Coast could face average annual losses of up to US$23 billion by 2030 and cumulative economic damages of $350 billion from climate risks, says Swiss Re research; 20 October 2010

[35] A City Prepares for a Warm Long-Term Forecast in the New York Times May 2011

[36] Ross Gelbspan (May 2005). "Snowed". Mother Jones. Retrieved 2007-09-12.

[37] "Go Green". *The Sun*. Retrieved 2007-09-12.

[38] Emiko Terazono (6 October 2006). "Gore and Murdoch join forces in TV deal". Financial Times. Retrieved 2007-09-12.

[39] Marc Gunther (27 August 2007). "Rupert Murdoch's climate crusade". CNN. Retrieved 2007-09-12.

2.2.11 References

- Ross Gelbspan, *Boiling Point: How Politicians, Big Oil and Coal, Journalists and Activists Are Fueling the Climate Crisis—And What We Can Do to Avert Disaster*, Basic Books, (August 1, 2004) ISBN 0-465-02761-X

- Lowe, EA and Harris, RJ (1998), " Taking Climate Change Seriously: British Petroleum's business strategy", *Corporate Environmental Strategy*, Winter 1998

2.2.12   External links

- Grant Thornton International Business Report Energy & Environment survey

- History Commons, Timeline of industry influence on climate change science.

- *Business Week*, 16 August 2004, "Global Warming"

- *The Guardian*, 8 December 2005, Oil industry targets EU climate policy

- Articles on the role of business organizations in sustainable development - by Frank Dixon, MBA (Harvard) - includes actions on climate change.

- "Some Like It Hot" by Chris Mooney, *Mother Jones*, May/June 2005

2.3   Climate Audit

Climate Audit is a blog which was founded on 31 January 2005[2][3] by Steve Mcintyre.

The *New York Times* has called it "a popular skeptics' blog".[4]

2.3.1   Founding

In 2004 Stephen McIntyre blogged on his website climate2003.com about his efforts with Ross McKitrick to get an extended analysis of the hockey stick graph into the journal *Nature*.[5][6]

On 25 October 2004 McIntyre posted comments on climate2003.com about a piece by William Connolley circulated on various blogs, and on 26 October wrote, "Maybe I'll start blogging some odds and ends that I'm working on. I'm going to post up some more observations on some of the blog criticisms."[7] On 1 December Mann and nine other scientists launched the RealClimate website as "a resource where the public can go to see what actual scientists working in the field have to say about the latest issues."[8] On climate2003.com McIntyre noted this development in a blog post on 10 December, where he wrote "Mann and some of his colleagues have set up a blog at the above address. A couple of Mann's first postings have been arguments against our papers. I'll post up a two quick comments below."[7] On 2 February 2005 McIntyre set up his Climate Audit blog, having found difficulties with posting comments on the climate2003.com layout.[9]

Dr Judith Curry of the Georgia Institute of Technology has said "McIntyre started the blog climateaudit.org so that he could defend himself against claims being made at the blog Realclimate with regards to his critique of the "hockey stick" since he was unable to post his comments there". She has also referred to this site as one of several "Climate Auditor" websites.[10]

2.3.2   Climatic Research Unit email controversy

See also: Climatic Research Unit email controversy

On 12 August 2009, Olive Heffernan wrote in *naturenews* that "Since 2002, Steve McIntyre, the editor of Climate Audit, a blog that investigates the statistical methods used in climate science, has repeatedly asked Phil Jones, director of the Climatic Research Unit (CRU) at the University of East Anglia, UK, for access to monthly global surface temperature data held by the institute. But in recent weeks, Jones has been swamped by a sudden surge in demands for data". She described how the Climatic Research Unit of the University of East Anglia had received 58 FOIA requests between 24 and 29 July 1999 from McIntyre or others associated with the blog, for raw climate data used in processed temperature datasets which were freely available. The raw data was restricted to academics, and the unit's director Phil Jones said that the data was subject to confidentiality agreements with various governments, but he was seeking agreement to get the raw data available online. He said that "Data release needs to be done in a systematic way."[11]

The site was one of the first to receive word of the e-mails which had been leaked[12][13][14] from the University of East Anglia with Jonathan Leake of *The Times* writing, "The storm began with just four cryptic words. 'A miracle has happened,' announced a contributor to Climate Audit, a website devoted to criticising the science of climate change."[15] Louise Gray wrote in The Daily Telegraph, "Climate Audit was one of the first to post up the stolen emails from the University of East Anglia that led to the 'climategate' scandal".[16]

Bloomberg said of the controversy, "Web sites and blogs including the Climate Audit Mirror Site have carried copies of e-mails, correspondence between climatologists and commentary. In one e-mail cited widely on blogs including Climate Audit, Phil Jones writes about completing "Mike's nature trick of adding in the real temps" in order to hide the decline."[17] According to *Science Magazine*, Jones wrote e-mails stating that he convinced Freedom of Information Act officers to not release data to "greenhouse skeptics" because Jones believed that they planned to harm the UAE or setback climate science.[18] "Think I've managed to persuade UEA to ignore all further FOIA requests if the people have anything to do with Climate Audit," Jones wrote.[18] The House of Commons' Science and Technology Committee largely vindicated the scientists involved in the scandal, but left consideration of the quality of the science and the conduct of the research to committees chaired by Lord Oxburgh and Sir Muir Russell. Fox News said that McIntyre "who also worked at the IPCC and submitted notes to the Science and Technology Committee for its investigation, wrote a lengthy rebuttal of the decision on his blog", and disputed the committee's conclusion that the word *trick* "appears to be a colloquialism for a 'neat' method of handling data".[19]

2.3.3 Reception

James Hansen, the former director of NASA's Goddard Institute, has dismissed McIntyre as a "court jester.[20]

"If a single person can be credited with setting the stage for Climategate, it's Stephen McIntyre, the retired mining consultant behind the popular skeptic blog Climate Audit," wrote Kate Sheppard at *Mother Jones* in 2011. [21] "Emails from this period show the scientists lashing out against McIntyre. He is referred to as a "bozo " and "a playground bully ." McIntyre clearly gets a rise out of irking scientists, whom he frequently refers to as 'the Team'—another play on the hockey-stick metaphor. He likes to 'tease these guys and kind of make fun of them,' he says, and their evident aggravation at his inquiries only egged him on. 'I think it was a mistake for them to in effect adopt a fatwa against Climate Audit,' says McIntyre." [21]

Patrick J. Michaels, a former contributor to the IPCC, named Climate Audit as part of "a new "parallel universe" of emerging online publications, manned by serious scientists critical of world governments approach to climate change. "A parallel universe is assembling itself parallel to the IPCC. This universe has become very technical -- very proficient at taking apart the U.N.'s findings."[22]

The website won the 2007 Best Science Blog award[23] and was a runner up in the same category in 2008.[24]

2.3.4 See also

- Global warming controversy

- Hockey stick controversy

- *The Hockey Stick Illusion*

- Tree ring

2.3.5 References

[1] "Climateaudit.org Site Info". Alexa Internet. Retrieved 2014-04-01.

[2] .net, Whois (31 Jan 2005). "Whois". Whois. Retrieved 4 May 2010.

[3] "Alexa Ranking". Amazon. p. 1. Retrieved 26 August 2010.

[4] Revkin, Andrew C. (11,27,2009). "Hacked E-Mail Data Prompts Calls for Changes in Climate Research". *New York Times* (New York Times). p. 1. Retrieved 27 May 2010. Check date values in: |date= (help)

[5] Pearce, Fred, *The Climate Files: The Battle for the Truth about Global Warming*, (2010) Guardian Books, ISBN 978-0-85265-229-9, pp. 93–96.

[6] McIntyre, Stephen (2004), *Welcome to Climate2003*, archived from the original on 2 September 2004, retrieved 10 September 2012, McIntyre, Stephen; McKitrick, Ross (1 July 2004), *M&M03 Page*, archived from the original on 12 September 2004, retrieved 10 September 2012

[7] McIntyre, Stephen (5 January 2005), *Welcome to Climate2003*, archived from the original on 24 January 2005, retrieved 10 September 2012

[8] David Appell (February 21, 2005). "Behind the Hockey Stick". Scientific American. Retrieved 2011-03-07.

[9] McIntyre, Stephen (2 February 2005), *Climate Audit*, archived from the original on 4 February 2005, retrieved 10 September 2012

[10] Curry, Judith (February 24, 2010). "Can scientists rebuild the public trust in climate science?". Physics Today. Retrieved 4 May 2010.

[11] Heffernan, Olive (2009-08-12). "Climate data spat intensifies Growing demands for access to information swamp scientist.". Nature. Archived from the original on 2010-08-31. Retrieved 2010-05-04. Since 2002, Steve McIntyre, the editor of Climate Audit, a blog that investigates the statistical methods used in climate science, has repeatedly asked Phil Jones, director of the Climatic Research Unit (CRU) at the University of East Anglia, UK, for access to monthly global surface temperature data held by the institute.

[12] "UN climate panel ordered to make fundamental reforms". *AFP*. Google News. 2010-08-30. Archived from the original on 2010-08-31. Retrieved 2010-08-31. the IPCC was rocked by a scandal involving leaked emails which critics say showed that they skewed data.

[13] Spotts, Pete (2010-08-30). "IPCC climate change panel needs transparency, review panel finds". *csmonitor.com*. The Christian Science Monitor. Archived from the original on 2010-08-31. Retrieved 2010-08-31. Many of these controversies came to light within the past 10 months. Emails leaked from the University of East Anglia revealed a handful of influential climate scientists displaying a circle-the-wagons mentality as some analysts tried to gain access to their data and analysis methods. Critics alleged that the emails also held evidence of fudged results.

[14] Helderman, Rosalind S. (2010-08-31). "Judge rejects Ken Cuccinelli's probe of U-Va. global warming records". *washingtonpost.com*. The Washington Post. Archived from the original on 2010-08-31. Retrieved 2010-08-31. has long been under attack by those who doubt global warming, particularly after his work was referenced in a series of leaked e-mails from the University of East Anglia's Climatic Research Unit.

[15] Leake, Jonathan (November 29, 2009) "The great climate change science scandal". The Sunday Times. Retrieved 4 May 2010. IT was against this background that the emails were leaked last week

[16] Gray, Louise (9 Apr 2010). "Climate change: Key influencers in the debate". The Telegraph. Retrieved 4 May 2010.

[17] Efstathiou Jr, Jim; Alex Morales (12/02 /2009). "U.K. Climate Scientist Steps Down After E-Mail Flap (Update4". Bloomberg.com. p. 1. Retrieved 27 May 2010. Check date values in: |date= (help)

[18] Antonio Regalado (2009-11-23). "In Climate Hack Story, Could Talk of Cover-Up Be as Serious as Crime?". Retrieved 2010-09-08.

[19] "Scientists Cleared -- After One-Day Probe". Fox News. 2010-03-31. Retrieved 2010-09-08.

[20] Global warming's most dangerous apostate speaks out about the state of climate change science. by Anne Jolis, Wall Street Journal, Nov. 18, 2009

[21] Climategate: What Really Happened?, Mother Jones, Apr. 21, 2011

[22] Koprowski, Gene J (April 28, 2010). "EXCLUSIVE: Citizen's Group Plans Extensive Audit of U.N. Climate Report". Fox News. Retrieved 3 May 2010.

[23] Aylward, Kevin (November 1, 2007). "Best Science Blog". Web Blog Awards. Retrieved 4 May 2010.

[24] Aylward, Kevin (December 31, 2008). "Best Science Blog". Web Blog Awards. Retrieved 4 May 2010.

2.3.6 External links

- ClimateAudit

2.4 Climate change policy of the George W. Bush administration

This article is about the climate change policy of the United States under the George W. Bush administration.

2.4.1 Kyoto Protocol

In March 2001, the Bush Administration announced that it would not implement the Kyoto Protocol, an international treaty signed in 1997 in Kyoto, Japan that would require nations to reduce their greenhouse gas emissions, claiming that ratifying the treaty would create economic setbacks in the U.S. and does not put enough pressure to limit emissions from developing nations.[1] In February 2002, Bush announced his alternative to the Kyoto Protocol, by bringing forth a plan to reduce the intensity of greenhouse gasses by 18 percent over 10 years. The intensity of greenhouse gasses specifically is the ratio of greenhouse gas emissions and economic output, meaning that under this plan, emissions would still continue to grow, but at a slower pace. Bush stated that this plan would prevent the release of 500 million metric tons of greenhouse gases, which is about the equivalent of 70 million cars from the road. This target would achieve this goal by providing tax credits to businesses that use renewable energy sources.[2]

2.4.2 Influence of industry groups

In June 2005, US State Department papers showed the Bush administration thanking Exxon executives for the company's "active involvement" in helping to determine climate change policy, including the U.S. stance on Kyoto. Input from the business lobby group Global Climate Coalition was also a factor.[3]

The Bush administration implemented an industry-formulated disinformation campaign designed to actively mislead the American public on global warming and to forestall limits on "climate polluters," according to a report in *Rolling Stone* magazine which reviews hundreds of internal government documents and former government officials.[4]

"'They've got a political clientele that does not want to be regulated,' says Rick S. Piltz, a former Bush climate official who blew the whistle on White House censorship of global-warming documents in 2005. 'Any honest discussion of the science would stimulate public pressure for a stronger policy. They're not stupid.'

"Bush's do-nothing policy on global warming began almost as soon as he took office. By pursuing a carefully orchestrated policy of delay, the White House blocked even the most modest reforms and replaced them with token investments in futuristic solutions like hydrogen cars. 'It's a charade,' says Jeremy Symons, who represented the EPA on Dick Cheney's energy task force, the industry-studded group that met in secret to craft the administration's energy policy. 'They have a single-minded determination to do nothing—while making it look like they are doing something.' . . .

"The CEQ became Cheney's shadow EPA, with industry calling the shots. To head up the council, Cheney installed James Connaughton, a former lobbyist for industrial polluters, who once worked to help General Electric and ARCO skirt responsibility for their Superfund waste sites. "two weeks after Bush took office - ExxonMobil's top lobbyist, Randy Randol, demanded a housecleaning of the scientists in charge of studying global warming. . . .Exxon's wish was the CEQ's command.[5]

2.4.3 Political pressure on scientists

See also: Politics and science in the United States

Also according to testimony taken by the U.S. House of Representatives, the Bush White House pressured American scientists to suppress discussion of global warming[6][7]

"High-quality science" was "struggling to get out," as the Bush administration pressured scientists to tailor their writings on global warming to fit the Bush administration's skepticism, in some cases at the behest of an ex-oil industry lobbyist. "Nearly half of all respondents perceived or personally experienced pressure to eliminate the words 'climate change,' 'global warming' or other similar terms from a variety of communications."

Similarly, according to the testimony of senior officers of the Government Accountability Project, the White House attempted to bury the report "National Assessment of the Potential Consequences of Climate Variability and Change," produced by U.S. scientists pursuant to U.S. law.[8] Some U.S. scientists resigned their jobs rather than give in to White House pressure to underreport global warming.[6]

Also, the White House removed key portions of a Centers for Disease Control and Prevention (CDC) report given to the U.S. Senate Environment and Public Works Committee about the dangers to human health of global warming.[9] According to one CDC official familiar with both the CDC version and the version given to the Senate, the version given to the Senate was "eviscerated." The White House prevented the Senate and thus the public from receiving key CDC estimates in the report about diseases likely to flourish in a warmer climate, increased injuries and deaths from severe weather such as hurricanes, more respiratory problems from drought-driven air pollution, an increase in waterborne diseases including cholera, increases in vector-borne diseases including malaria and hantavirus, mental health problems such as depression and post-traumatic stress, and how many people might be adversely affected because of increased warming.

US officials, such as Philip Cooney, have repeatedly edited scientific reports from US government scientists, [10] many of whom, such as Thomas Knutson, have been ordered to refrain from discussing climate change and related topics.[11][12][13]

Climate scientist James E. Hansen, director of NASA's Goddard Institute for Space Studies, claimed in a widely cited *New York Times* article [14] in 2006 that his superiors at the agency were trying to "censor" information "going out to the public." NASA denied this, saying that it was merely requiring that scientists make a distinction between personal, and official government, views in interviews conducted as part of work done at the agency. Several scientists working at the National Oceanic and Atmospheric Administration have made similar complaints;[15] once again, government officials said they were enforcing long-standing policies requiring government scientists to clearly identify personal opinions as such when participating in public interviews and forums.

The BBC's long-running current affairs series *Panorama* recently investigated the issue, and was told that "scientific reports about global warming have been systematically changed and suppressed."[16]

According to an Associated Press release on January 30, 2007,

> "Climate scientists at seven government agencies say they have been subjected to political pressure aimed at downplaying the threat of global warming.

> "The groups presented a survey that shows two in five of the 279 climate scientists who responded to a questionnaire complained that some of their scientific papers had been edited in a way that changed their meaning. Nearly half of the 279 said in response to another question that at some point they had been told to delete reference to "global warming" or "climate change" from a report."[17]

Critics writing in the *Wall Street Journal* editorial page claim that the survey [18] was itself unscientific.[19]

Attempts to suppress scientific information on global warming and other issues have been described by Chris Mooney as constituting a Republican War on Science.

2.4.4 Allegations of attempts to mislead the public

The book *Hell and High Water* asserts that there has been a disingenuous, concerted and effective campaign to convince Americans that the science is not proven, or that global warming is the result of natural cycles, and that there needs to be more research. The book claims that, to delay action, industry and government spokesmen suggest falsely that "technology breakthroughs" will eventually save us with hydrogen cars and other fixes. It calls on voters to demand immediate government action to curb emissions. Tyler Hamilton, in his review of the book for The Toronto Star, wrote that the book offers "alarming detail on how the U.S. public was being misled by [the Bush administration] (backed by conservative political forces) that is intent on inaction, and that's also on a mission to derail international efforts to curb emissions."[20]

Papers presented at an International Scientific Congress on Climate Change, held in 2009 under the sponsorship of the University of Copenhagen in cooperation with nine other universities in the International Alliance of Research Universities (IARU), maintained that the climate-change skepticism which is so prevalent in the USA[21] "was largely generated and kept alive by a small number of conservative think tanks, often with direct funding from industries having special interests in delaying or avoiding the regulation of greenhouse gas emissions".[22]

In *Merchants of Doubt* (2010) Naomi Oreskes and Erik M. Conway, both American science historians, argue that Fred Seitz, Fred Singer, and a few other contrarian scientists joined forces with conservative think tanks and private corporations to challenge the scientific consensus on climate change, by spreading doubt and confusion.[23][24]

In *Requiem for a Species* (2010), Clive Hamilton suggests that the roots of climate change denial lie in the reaction of American conservatism to the collapse of the Soviet Union in 1991. He argues that as the "red menace" receded, conservatives who had put energy into opposing communism sought other outlets. Hamilton contends that the conservative backlash against climate science was led by three prominent physicists -- Frederick Seitz, Robert Jastrow, and William Nierenberg.[25]

2.4.5 Attempts to undermine U.S. and state efforts

The Bush Administration worked to undermine state efforts to mitigate global warming. Mary Peters, the Transportation Secretary at that time, personally directed US efforts to urge governors and dozens of members of the House of Representatives to block California's first-in-the-nation limits on greenhouse gases from cars and trucks, according to e-mails obtained by Congress.[26]

2.4.6 References

[1] Alex Kirby, US blow to Kyoto hopes, 2001-03-28, BBC News (online).

[2] Bush unveils voluntary plan to reduce global warming, CNN.com, 2002-02-14.

[3] Vidal, John (2005-06-08). "Revealed: how oil giant influenced Bush". *The Guardian* (London). Retrieved 2010-01-24.

[4] Dickinson, Tim (2007-06-08). "The Secret Campaign of President Bush's Administration To Deny Global Warming". *Rolling Stone*. Retrieved 2010-01-24.

[5] The Washington Post, June 21, 2007 "http://www.washingtonpost.com/wp-dyn/content/blog/2007/06/21/BL2007062101075_2.html?nav=hcmodule , citing the Rolling Stone investigative report published 2007/6/13

[6] Reuters, January 30, 2007, free archived version at http://www.commondreams.org/headlines07/0130-10.htm, last visited Jan. 30, '07

[7] Written testimony of Dr. Grifo before the Committee on Oversight and Government Reform of the U.S. House of Representatives on January 30, 2007, archived at http://oversight.house.gov/Documents/20070130113153-55829.pdf

[8] written testimony of Rick Piltz before the Committee on Oversight and Government Reform of the U.S. House of Representatives on January 30, 2007, archived at http://oversight.house.gov/Documents/20070130113813-92288.pdf last visited Jan. 30, 07

[9] Associated Press, Oct. 24, 2007, http://hosted.ap.org/dynamic/stories/G/GLOBAL_WARMING_HEALTH?SITE=NVREN&SECTION=HOME&TEMPLATE=DEFAULT; also archived at http://www.commondreams.org/archive/2007/10/24/4772/

[10] Campbell, D. (June 20, 2003) "White House cuts global warming from report" *Guardian Unlimited*

[11] Donaghy, T., *et al.* (2007) "Atmosphere of Pressure:" a report of the Government Accountability Project (Cambridge, Mass.: UCS Publications)

[12] Rule, E. (2005) "Possible media attention" Email to NOAA staff, July 27. Obtained via FOIA request on July 31, 2006. and Teet, J. (2005) "DOC Interview Policy" Email to NOAA staff, September 29. Originally published by Alexandrovna, L. (2005) "Commerce Department tells National Weather Service media contacts must be pre-approved" *The Raw Story,* October 4. Accessed December 22, 2006

[13] Zabarenko, D. (2007) "'Don't discuss polar bears:' memo to scientists" *Reuters*

[14] Revkin, Andrew C. (January 29, 2006). "Climate Expert Says NASA Tried to Silence Him". The New York Times. Retrieved 2007-04-14.

[15] Eilperin, Julie (2006-04-06). "Climate Researchers Feeling Heat From White House". *The Washington Post.* Retrieved 2010-01-24.

[16] "Climate chaos: Bush's climate of fear". BBC Panorama. June 1, 2006. Retrieved 2007-04-14.

[17] "Groups Say Scientists Pressured On Warming". CBS News and Associated Press. January 30, 2007. Retrieved 2007-04-14.

[18] Donaghy, Timothy; Jennifer Freeman; Francesca Grifo; Karly Kaufman; Tarek Maassarani; Lexi Shultz (February 2007). "Appendix A: UCS Climate Scientist Survey Text and Responses (Federal)" (PDF). *Atmosphere of Pressure – Political Interference in Federal Climate Science* (PDF). Union of Concerned Scientists & Government Accountability Project. Retrieved 2007-04-14.

[19] Taranto, James (February 1, 2007). "They Call This Science?". OpinionJournal.com. Retrieved 2007-04-14.

[20] Hamilton, Tyler (2007-01-01). "Fresh alarm over global warming". *The Toronto Star.* Retrieved 2010-01-30.

[21] Riley Dunlap, "Why climate-change skepticism is so prevalent in the USA: the success of conservative think tanks in promoting skepticism via the media," *Climate Change: Global Risks, Challenges and Decisions, IOP Conf. Series: Earth and Environmental Science* 6 (2009) 532010 doi:10.1088/1755-1307/6/3/532010

[22] William Freudenburg, "The effects of journalistic imbalance on scientific imbalance: special interests, scientific consensus and global climate disruption," *Climate Change: Global Risks, Challenges and Decisions, IOP Conf. Series: Earth and Environmental Science* 6 (2009) 532011 doi:10.1088/1755-1307/6/3/532011

[23] Mike Steketee. Some sceptics make it a habit to be wrong *The Australian*, November 20, 2010.

[24] Naomi Oreskes and Erik M. Conway (2010). *Merchants of Doubt*, Bloomsbury Press, p. 6.

[25] Clive Hamilton (2010). *Requiem for a Species* , pp. 98-103.

[26] "How the White House Worked to Scuttle California's Climate Law", San Francisco Chronicle, September 25, 2007 http://www.commondreams.org/archive/2007/09/25/4099/

2.5   Environmental skepticism

Environmental skepticism is the belief that claims by environmentalists, and the environmental scientists who support them, are false or exaggerated. The term is also applied to those who are critical of environmentalism in general. Environmental skepticism is closely linked with anti-environmentalism.

Environmental skeptics have argued that the extent of harm coming from human activities is less certain than some scientists and scientific bodies claim, or that it is too soon to be introducing curbs in these activities on the basis of existing evidence, or that further discussion is needed regarding who should pay for such environmental initiatives.[1] One of the focus themes in the environmental skeptics movement is the idea that environmentalism is a growing threat to social and economic progress and the civil liberties.

The popularity of the term was enhanced by Bjørn Lomborg's book *The Skeptical Environmentalist.*[2] Lomborg approached environmental claims from a statistical and economic standpoint, and concluded that often the claims made by environmentalists were overstated. Lomborg argued, on the basis of cost–benefit analysis, that few environmentalist claims warranted serious concern. However, in 2010, Lomborg reversed his position and he now agrees with "tens of billions of dollars a year to be invested in tackling climate change" and declared global warming to be "undoubtedly one of the chief concerns facing the world today" and "a challenge humanity must confront".[3][4] He summarized his position, saying "Global warming is real - it is man-made and it is an important problem. But it is not the end of the world."

2.5.1 Criticism

Environmentalist organizations and lobbies argue that such widespread skeptical doubts have not developed independently, but have been "encouraged by lobbying and PR campaigns financed by the polluting industries". Supporters of environmentalists argue that "skepticism" implies a form of denialism, and that, in the US particularly, "large donations [have been made] to Senators and Congressmen and [have] sponsored neoliberal think tanks and contrarian scientific research. ExxonMobil, the oil major, has been accused by Friends of the Earth and others of giving millions of dollars to a long list of think-tanks and lobbyists opposed to Kyoto."[1]

A recent study from progressive authors about the environmental skepticism movement claim that the overwhelming majority of environmentally skeptical books published since the 1970s were either written or published by authors or institutions affiliated with right-wing think tanks. They "conclude that scepticism is a tactic of an elite-driven counter-movement designed to combat environmentalism, and that the successful use of this tactic has contributed to the weakening of US commitment to environmental protection."[5] The skeptical environmental counter-movement is a problem for many grant writers with connections in Washington, as well as for statists who cannot bear disagreement, and in dealing with the propositions for the counter-movement people who are not skeptics are forced to reach down the bedrock issues of epistemology, identities, and other core work for politics. [6]

2.5.2 See also

- Anti-environmentalism

- Climate change denial

- Ecofascism

- Global warming controversy

- Media coverage of climate change

2.5.3 References

[1] "'Denial lobby' turns up the heat". London: The Observer. 2005-03-06. Retrieved 2008-02-07.

[2] Lomborg, Bjrn (2004). *Global crises, global solutions.* Cambridge, UK: Cambridge University Press. ISBN 0-521-60614-4.

[3] Jowit, Juliette (30 August 2010). "Bjørn Lomborg: $100bn a year needed to fight climate change". *guardian.co.uk home Location* (London). Retrieved 30 August 2010.

[4] Brett Michael Dykes, "Noted anti-global-warming scientist reverses course", Yahoo News (August 31, 2010)

[5] Jacques, P.J.; Dunlap, R.E.; Freeman, M. (June 2008). "The organisation of denial: Conservative think tanks and environmental scepticism". *Environmental Politics* **17** (3): 349–385. doi:10.1080/09644010802055576.

[6] Jacques, Peter. Environmental Skepticism: Ecology, Power and Public Life. Farnham, England: Ashgate, 2009. Print.

- "Environmental sceptics overwhelmingly politicised, says study". Carbon News. June 17, 2008. Retrieved December 12, 2011.

2.5.4 Selected works and analyses

- Horner, Christopher C., *The Politically Incorrect Guide to Global Warming: (and Environmentalism)*

- Bethell, Tom (2005). *The Politically Incorrect Guide to Science.* Regnery Pub. ISBN 978-0-89526-031-4.

- Huber, Peter (1999). *Hard Green: Saving the Environment from the Environmentalists : a Conservative Manifesto.* Basic Books. ISBN 978-0-465-03113-9.

- Lomborg, Bjørn (2001). *The Skeptical Environmentalist: Measuring the Real State of the World*. ISBN 978-0-521-01068-9.

- Mooney, Chris (2006). *The Republican War on Science*. Basic Books. ISBN 978-0-465-04676-8.

- de Steiguer, J.E. 2006. *The Origins of Modern Environmental Thought*. The University of Arizona Press. Tucson. 246 pp.

- Michaels, David (2008). *Doubt is Their Product: How Industry's Assault on Science Threatens Your Health*. Oxford University Press. ISBN 978-0-19-530067-3.

- *In a Dark Wood: The Fight Over Forests and the Myths of Nature*. Transaction Pub. 1995. ISBN 978-0-7658-0752-6.

- Driessen, Paul (2003). *Eco-Imperialism: Green Power, Black Death*. Free Enterprise Press. ISBN 0-939571-23-4. External link in |title= (help)

- Arun Kumar Shrivastava (2007). *Global Warming*. ISBN 978-1-55263-212-3.

- Patrick J. Michaels; Robert C. Balling (2000). *The satanic gases: clearing the air about global warming*. Cato Institute. ISBN 978-1-882577-92-7.

- José Ortega y Gasset; Datus Proper (1995). *Meditations on Hunting*. ISBN 978-1-885106-18-6.

- Reisman, George, *The Toxicity of Environmentalism*, Laguna Hills, CA, The Jefferson School of Philosophy, Economics & Psychology, 1990 ISBN 1-931089-01-9

- Shearer, Christine (2011). *Kivalina: A Climate Change Story*. Haymarket Books. ISBN 978-1-60846-128-8.

- James A. Swan (1995). *In defense of hunting*. Harpercollins. ISBN 0-06-251237-4.

2.6 Global warming conspiracy theory

A **global warming conspiracy theory** invokes claims that scientific consensus on global warming is based on conspiracies to produce false data or suppress dissent. It is one of a number of tactics used in climate change denial to legitimise political global warming controversy disputing this consensus.[1] Global warming conspiracy theorists typically allege that, through worldwide acts of professional and criminal misconduct, the science behind global warming has been invented or distorted for ideological or financial reasons, or both.[2][3]

2.6.1 Background

As stated by the Intergovernmental Panel on Climate Change (IPCC), the largest contributor to global warming is the increase in atmospheric carbon dioxide (CO_2) since 1750, particularly from fossil fuel combustion, cement production, and land use changes such as deforestation.[4] The IPPC's Fifth Assessment Report (AR5) states

> *Human influence has been detected in warming of the atmosphere and the ocean, in changes in the global water cycle, in reductions in snow and ice, in global mean sea level rise, and in changes in some climate extremes. This evidence for human influence has grown since AR4. It is extremely likely (95–100%) that human influence has been the dominant cause of the observed warming since the mid-20th century.* - IPCC AR5 WG1 Summary for Policymakers[5][6]

The evidence for global warming due to human influence has been recognized by the national science academies of all the major industrialized countries.[7] No scientific body of national or international standing maintains a formal opinion dissenting from the summary conclusions of the IPCC.[8]

Despite this scientific consensus on climate change, allegations have been made that scientists and institutions involved in global warming research are part of a global scientific conspiracy or engaged in a manipulative hoax.[9] There have been allegations of malpractice, most notably in the Climatic Research Unit email controversy. Eight committees investigated these allegations and published reports, each finding no evidence of fraud or scientific misconduct.[10] The Muir Russell report stated that the scientists' "rigour and honesty as scientists are not in doubt," that the investigators "did not find any evidence of behaviour that might undermine the conclusions of the IPCC assessments," but that there had been "a consistent pattern of failing to display the proper degree of openness."[11][12] The scientific consensus that global warming is occurring as a result of human activity remained unchanged at the end of the investigations.[13]

2.6.2   Claims

In a speech given to the US Senate Committee on the Environment and Public Works on July 28, 2003, entitled "The Science of Climate Change",[14] Senator James Inhofe (Republican, for Oklahoma) concluded by asking the following question: "With all of the hysteria, all of the fear, all of the phony science, could it be that man-made global warming is the greatest hoax ever perpetrated on the American people?" He further stated, "some parts of the IPCC process resembled a Soviet-style trial, in which the facts are predetermined, and ideological purity trumps technical and scientific rigor."[15] Inhofe has suggested that supporters of the Kyoto Protocol such as Jacques Chirac are aiming at global governance.[16]

Commenting on criticism of the Lavoisier Group by Clive Hamilton, the Cooler Heads Coalition notes that "Hamilton accuses the Lavoisier Group of painting the UN's global warming negotiations as "an elaborate conspiracy in which hundreds of climate scientists have twisted their results to support the climate change theory in order to protect their research funding" and adds, "Sounds plausible to us."[17]

William M. Gray said in 2006 that global warming became a political cause because of the lack of any other enemy following the end of the Cold War. He went on to say that its purpose was to exercise political influence, to try to introduce world government, and to control people, adding, "I have a demonic view on this."[3] The TV documentary *The Great Global Warming Swindle* was made by Martin Durkin, who called global warming "a multibillion-dollar worldwide industry, created by fanatically anti-industrial environmentalists." In the *Washington Times* in 2007 he said that his film would change history, and predicted that "in five years the idea that the greenhouse effect is the main reason behind global warming will be seen as total bunk."[18]

Climate change has also been called the "greatest scam in history" by John Coleman, who co-founded the Weather Channel.[19] When questioned by the IPCC regarding his claims, he responded "The polar ice is increasing, not melting away. Polar Bears are increasing in number."[19]

2.6.3   Criticism

Steve Connor links the terms "hoax" and "conspiracy," saying, "Reading through the technical summary of this draft (IPCC) report, it is clear that no one could go away with the impression that climate change is some conspiratorial hoax by the science establishment, as some would have us believe."[20]

The documentary *The Great Global Warming Swindle* received criticism from several experts. George Monbiot described it as "the same old conspiracy theory that we've been hearing from the denial industry for the past ten years".[21] Similarly, in response to James Delingpole, Monbiot stated that his Spectator article was "the usual conspiracy theories [...] working to suppress the truth, which presumably now includes virtually the entire scientific community and everyone from Shell to Greenpeace and The Sun to Science."[22] Some Australian meteorologists also weighed in, saying that the film made no attempt to offer a "critical deconstruction of climate science orthodoxies", but instead used various other means to suggest that climate scientists are guilty of lying or are seriously misguided. Although the film's publicist's asserted that "global warming is 'the biggest scam of modern times'", these meteorologists concluded that the film was "not scientifically sound and presents a flawed and very misleading interpretation of the science".[23]

Former UK Secretary of State for Environment, Food and Rural Affairs David Miliband presented a rebuttal of the main points of the film and stated "There will always be people with conspiracy theories trying to do down the scientific consensus, and that is part of scientific and democratic debate, but the science of climate change looks like fact to me."[24]

National Geographic fact checked 6 persistent scientific conspiracy theories. Regarding the persistent belief in a global warming hoax they note that the Earth is continuing to warm and the rate of warming is increasing as documented in numerous scientific studies. The rise in global temperature and its rate of increase coincides with the rise of greenhouse gases in the atmosphere due to human activity. Moreover, global warming is causing Arctic sea ice to thaw at historic rates, many species of plants are blooming earlier than expected, and the migration routes of many birds, fish, mammals, and insects are changing.[25]

2.6.4   Funding

See also: Climate change denial § Lobbying and ExxonMobil § Funding of global warming disinformation and denial

There is evidence that some of those alleging such conspiracies are part of well-funded misinformation campaigns designed to manufacture controversy, undermine the scientific consensus on climate change and downplay the projected effects of global warming.[26][27] Individuals and organisations kept the global warming debate alive long after most scientists had reached their conclusions. These doubts have influenced policymakers in both Canada and the US, and have helped to form government policies.[27]

> Since the late 1980s, this well-coordinated, well-funded campaign by contrarian scientists, free-market think tanks and industry has created a paralyzing fog of doubt around climate change. Through advertisements, op-eds, lobbying and media attention, greenhouse doubters (they hate being called deniers) argued first that the world is not warming; measurements indicating otherwise are flawed, they said. Then they claimed that any warming is natural, not caused by human activities. Now they contend that the looming warming will be minuscule and harmless. "They patterned what they did after the tobacco industry," says former senator Tim Wirth, who spearheaded environmental issues as an under secretary of State in the Clinton administration. "Both figured, sow enough doubt, call the science uncertain and in dispute. That's had a huge impact on both the public and Congress."
>
> — The truth about denial, S. Begley, Newsweek[28]

Greenpeace presented evidence of the energy industry funding climate change denial in their 'Exxon Secrets' project.[29][30] An analysis conducted by *The Carbon Brief* in 2011 found that 9 out of 10 of the most prolific authors who cast doubt on climate change or speak against it had ties to ExxonMobil. Greenpeace have said that Koch industries invested more than US$50 million in the past 50 years on spreading doubts about climate change.[31][32][33] ExxonMobil announced in 2008 that it would cut its funding to many of the groups that "divert attention" from the need to find new sources of clean energy, although in 2008 still funded over "two dozen other organisations who question the science of global warming or attack policies to solve the crisis."[34] A survey carried out by the UK Royal Society found that in 2005 ExxonMobil distributed US$2.9 million to 39 groups that "misrepresented the science of climate change by outright denial of the evidence".[34]

2.6.5   Fictional representations

The novel *State of Fear* by Michael Crichton, published in December 2004, describes a conspiracy by scientists and others to create public panic about global warming. The novel includes 20 pages of footnotes, described by Crichton as providing a factual basis for the non-plotline elements of the story.[35] In a Senate speech on 4 January 2005, Inhofe mistakenly described Crichton as a "scientist", and said the book's fictional depiction of environmental organizations primarily "focused on raising money, principally by scaring potential contributors with bogus scientific claims and predictions of a global apocalypse" was an example of "art imitating life."[36]

In a piece headed *Crichton's conspiracy theory*, Harold Evans described Crichton's theory as being "in the paranoid political style identified by the renowned historian Richard Hofstadter," and went on to suggest that "if you happen to be in the market for a conspiracy theory today, there's a rather more credible one documented by the pressure group Greenpeace," namely the funding by ExxonMobil of groups opposed to the theory of global warming.[37]

2.6.6 See also

- Global warming controversy

- List of scientists opposing the mainstream scientific assessment of global warming

- *Merchants of Doubt* - Analysis of denialist movements in the United States by Erik M. Conway and Naomi Oreskes

2.6.7 Notes

[1] Pascal Diethelm & Martin McKee (January 2009). "Denialism: what is it and how should scientists respond?" (PDF). *European Journal of Public Health* **19** (1): 2–4. doi:10.1093/eurpub/ckn139. PMID 19158101.

[2] Goldenberg, Suzanne (1 March 2010). "US Senate's top climate sceptic accused of waging 'McCarthyite witch-hunt'". *The Guardian*. Retrieved 7 July 2015.

[3] Achenbach, Joel. "The Tempest". *The Washington Post*. Retrieved 2010-03-31.

[4] "Total radiative forcing is positive, and has led to an uptake of energy by the climate system. The largest contribution to total radiative forcing is caused by the increase in the atmospheric concentration of CO_2 since 1750." (p 11) "From 1750 to 2011, CO_2 emissions from fossil fuel combustion and cement production have released 375 [345 to 405] GtC to the atmosphere, while deforestation and other land use change are estimated to have released 180 [100 to 260] GtC." (p 10), IPCC, Climate Change 2013: The Physical Science Basis - Summary for Policymakers, Observed Changes in the Climate System, p. 10&11, in IPCC AR5 WG1 2013.

[5] IPCC, Climate Change 2013: The Physical Science Basis - Summary for Policymakers, Observed Changes in the Climate System, p. 15, in IPCC AR5 WG1 2013. "Extremely likely" is defined as a 95–100% likelihood on p 2.

[6] [Notes-SciPanel] America's Climate Choices: Panel on Advancing the Science of Climate Change; National Research Council (2010). *Advancing the Science of Climate Change*. Washington, D.C.: The National Academies Press. ISBN 0-309-14588-0. (p1) ... there is a strong, credible body of evidence, based on multiple lines of research, documenting that climate is changing and that these changes are in large part caused by human activities. While much remains to be learned, the core phenomenon, scientific questions, and hypotheses have been examined and found wanting, lacking consistent support in the face of serious scientific debate and careful evaluation of alternative explanations. * * * (p21-22) Some scientific conclusions or theories have been so thoroughly examined and tested, and supported by so many independent observations and results, that their likelihood of subsequently being found to be wrong is vanishingly small. Such conclusions and theories are then regarded as settled facts. This is the case for the conclusions that the Earth system is warming and that much of this warming is very likely due to human activities.

[7] [Notes-SciAcademy Statement] "Joint Science Academies' Statement" (PDF). 2005. Retrieved 2014-04-20. It is likely that most of the warming in recent decades can be attributed to human activities (IPCC 2001). This warming has already led to changes in the Earth's climate.

[8] Julie Brigham-Grette (September 2006). "Petroleum Geologists' Award to Novelist Crichton Is Inappropriate" (PDF). *Eos* **87** (36). Bibcode:2006EOSTr..87..364B. doi:10.1029/2006EO360008. Retrieved 2007-01-23. The AAPG stands alone among scientific societies in its denial of human-induced effects on global warming.

[9] Goertzel, Ted (June 2010). "Conspiracy theories in science". *EMBO Reports* **11** (7): 493–499. doi:10.1038/embor.2010.84. PMC 2897118. PMID 20539311. Retrieved 30 December 2013.

[10] Six of the major investigations covered by secondary sources include: 1233/uk-climategate-inquiry-largely-clears.html House of Commons Science and Technology Committee (UK); Independent Climate Change Review (UK); International Science Assessment Panel (UK); Pennsylvania State University (US); United States Environmental Protection Agency (US); Department of Commerce (US).

[11] Jonsson, Patrik (7 July 2010). "Climate scientists exonerated in 'climategate' but public trust damaged". Christian Science Monitor. p. 2. Retrieved 17 Aug 2011.

[12] Russell, Sir Muir (July 2010). "The Independent Climate Change E-mails Review" (PDF). p. 11. Retrieved 17 Aug 2011.

[13] Biello, David (Feb., 2010). "Negating 'Climategate'". *Scientific American.* **(302)**:2. 16. ISSN 00368733. "In fact, nothing in the stolen material undermines the scientific consensus that climate change is happening and that humans are to blame"; See also: Lubchenco, Jane (2 December 2009) House Select Committee on Energy Independence and Global Warming (House Select Committee). "The Administration's View on the State of Climate Science". House Hearing, 111 Congress. U.S. Government Printing Office. "...the e-mails really do nothing to undermine the very strong scientific consensus and the independent scientific analyses of thousands of scientists around the world that tell us that the Earth is warming and that the warming is largely a result of human activities." As quoted in the report published by Office of Inspector General.

[14] James M. Inhofe - U.S. Senator - Oklahoma

[15] Senator James Inhofe, Chairman of Committee on Environment and Public Works, U.S. Senate.The Facts and Science of Climate Change

[16] "Senate Environment And Public Works Committee".

[17] "Antarctic Cooling Down; The Antarctic Ice Sheet is Growing; Hansen Downgrades Warming Threat". Cooler Heads Coalition.

[18] "Global warming labeled a 'scam' - Washington Times". washingtontimes.com. Retrieved 2010-03-15.

[19] Taylor, Jason. "express news". Retrieved 3 January 2015.

[20] "Steve Connor: Global warming is not some conspiratorial hoax - Independent Online Edition > Commentators". *The Independent* (London). 2007-01-29. Retrieved 2007-11-16.

[21] "Another Species of Denial". Retrieved 2014-01-02.

[22] George Monbiot, *Spectator recycles climate rubbish published by sceptic*, 2009-07-09

[23] Jones, D; Watkins, A.; Braganza, K.; Coughlan, M (2007). ""The Great Global Warming Swindle": a critique". *Bulletin of the Australian Meteorologi cal and Oceanographic Society* **20** (3): 63–72.

[24] "The Great Climate Change Swindle?".

[25] Than, Ker. "Fact Checking 6 Persistent Science Conspiracy Theories". National Geographic. Retrieved 22 May 2013.

[26] Griffiths, Jenny; Mala Rao; Fiona Adshead (2009). *The health practitioner's guide to climate change: diagnosis and cure.* Earthscan. p. 228. ISBN 1-84407-729-2.

[27] "*The Denial Machine* - synopsis". CBC/Radio-Canada. 24 October 2007. Retrieved 3 September 2011.

[28] Begley, Sharon; Eve Conant; Sam Stein; Eleanor Clift; Matthew Philips (13 August 2007). "The Truth About Denial" (PDF). Newsweek. p. 20. Retrieved 3 September 2011.

[29] "Exxon Secrets". Retrieved 2008-12-23.

[30] Monbiot, George (2006-09-19). "The denial industry". *The Guardian* (London). Retrieved 2008-12-23.

[31] "9 out of 10 top climate change deniers linked with Exxon Mobil".

[32] "Analysing the '900 papers supporting climate scepticism': 9 out of top 10 authors linked to ExxonMobil".

[33] "Exposing the dirty money behind fake climate science".

[34] Adam, David (2008-05-28). "Exxon to cut funding to climate change denial groups". *The Guardian* (London). Retrieved 2008-12-23.

[35] Mooney, Chris (2005-02-06). "Checking Crichton's Footnotes". Boston Globe.

[36] Inhofe, James M. (4 January 2005), *Climate Change Update Senate Floor Statement*, U.S. Senator James M. Inhofe, archived from the original on 12 January 2005, retrieved 2011-03-07.
Mooney, Chris (11 January 2005), *Warmed Over*, CBS News, retrieved 2011-03-07. Reprinted from The American Prospect, 10 January 2005.

[37] Evans, Harold (2005-10-07). "Crichton's conspiracy theory". *BBC News* (London). Retrieved 2007-11-16.

2.6.8 Further reading

- Lahsen, M. (1999). The Detection and Attribution of Conspiracies: The Controversy Over Chapter 8. In G. E. Marcus (Ed.), *Paranoia Within Reason: A Casebook on Conspiracy as Explanation* (pp. 111–136). Chicago, IL: University of Chicago Press. ISBN 0-226-50458-1.

2.7 Information Council on the Environment

The **Information Council on the Environment** (ICE), was a U.S. organization created by the National Coal Association, the Western Fuels Association, and Edison Electrical Institute.

2.7.1 History

ICE launched a $500,000 advertising and public relations campaign to, in ICE's words, "reposition global warming as theory (not fact)." Patrick Michaels, Robert Balling and Sherwood B. Idso all lent their names in 1991 to its scientific advisory panel.

Its publicity plan called for placing these three scientists, along with fellow global warming skeptic S. Fred Singer, in broadcast appearances, op-ed pages, and newspaper interviews. Bracy Williams & Co., a Washington D.C.-based PR firm, did the advance publicity work for the interviews. Another company was contracted to conduct opinion polls, which identified "older, less-educated males from larger households who are not typically active information-seekers" and "younger, lower-income women" as "good targets for radio advertisements" that would "directly attack the proponents of global warming . . . through comparison of global warming to historical or mythical instances of gloom and doom."

One print advertisement prepared for the ICE campaign showed a sailing ship about to drop off the edge of a flat world into the jaws of a waiting dragon. The headline read: "Some say the earth is warming. Some also said the earth was flat." Another featured a cowering chicken under the headline "Who Told You the Earth Was Warming . . . Chicken Little?" Another ad was targeted at Minneapolis readers and asked, "If the earth is getting warmer, why is Minneapolis getting colder?"

Collapse

The ICE campaign collapsed after embarrassing internal memoranda related to the PR campaign were leaked to the press. An embarrassed Michaels hastily disassociated himself from ICE, citing what he called its "blatant dishonesty."

Following the collapse of the organization, Michaels, Balling, Idso, and Singer have continued to express their skepticism about the theory of global warming. Singer has been the most visible and vocal of the group.

2.7.2 See also

- Politicization of science

2.7.3 External links

- The Coal Industry's "ICE" Campaign (1999)

2.8 Leipzig Declaration

The **Leipzig Declaration on Global Climate Change** is a statement made in 1995, seeking to refute the claim there is a scientific consensus on the global warming issue.[1] It was issued in an updated form in 1997 and revised again in 2005,[2] claiming to have been signed by 80 scientists and 25 television news meteorologists while the posting of 33 additional signatories was pending verification that those 33 additional scientists still agreed with the statement.[3] All versions of the declaration, which opposes the global warming hypothesis and the Kyoto Protocol, were penned by Fred Singer's Science and Environmental Policy Project (SEPP).

The first declaration was based on a November 9–10, 1995 conference, organized by Helmut Metzner in Leipzig, Germany.[4] The second declaration was additionally based on a successor conference in Bonn, Germany on November 10–11, 1997. The conferences were cosponsored by SEPP and the European Academy for Environmental Affairs and titled *International Symposium on the Greenhouse Controversy*.

2.8.1   Versions

The 1995 Declaration

The 1995 declaration asserts: "There does not exist today a general scientific consensus about the importance of greenhouse warming from rising levels of carbon dioxide. On the contrary, most scientists now accept the fact that actual observations from earth satellites show no climate warming whatsoever." The latter statement was broadly accurate at the time, but with additional data and correction of errors, all analyses of satellite temperature measurements now show statistically significant warming.

The declaration also criticised the United Nations Framework Convention on Climate Change, saying: "Energy is essential for all economic growth, and fossil fuels provide today's principal global energy source. In a world in which poverty is the greatest social pollutant, any restriction on energy use that inhibits economic growth should be viewed with caution. For this reason, we consider 'carbon taxes' and other drastic control policies ... to be ill-advised, premature, wrought with economic danger, and likely to be counterproductive."

Signatures   According to the SEPP website, there were 79 signatures to the 1995 declaration, including Frederick Seitz: the current SEPP chair. Perhaps the most prominent signatory to the declaration was Dr. Robert E. Stevenson, a former research scientist for NASA and the Scripps Institution of Oceanography.[5] The signature list was last updated on July 16, 1996. Of these 79, 33 failed to respond when SEPP asked them to sign the 1997 declaration. SEPP calls the signatories "nearly 100 climate experts".

The signatures to the 1995 declaration were disputed by David Olinger of the *St. Petersburg Times*. In an article on July 29, 1996, he revealed that many signers, including Chauncey Starr, Robert Balling, and Patrick Michaels, have received funding from the oil industry, while others had no scientific training or could not be identified.[6]

The 1995 declarations begins: "As scientists, we are intensely interested in the possibility that human activities may affect the global climate". However, those identified as scientists and climate experts include at least ten weather presenters, including Dick Groeber of Dick's Weather Service in Springfield, Ohio. Groeber, who had not completed a university degree, labelled himself a scientist by virtue of his thirty to forty years of self-study.[6]

In any case, it is difficult to accurately evaluate the list of signatures of the 1995 declaration, as the SEPP website provides no additional details about them except for their university, if they are professors.

The 1997 Declaration

The 1997 declaration updated the 1995 declaration in a number of ways. The most obvious difference was its focus on the Kyoto Protocol, as the Kyoto conference was in the process of being finalised. The declaration says:

> "We believe the Kyoto Protocol -- to curtail carbon dioxide emissions from only part of the world community -- is dangerously simplistic, quite ineffective, and economically destructive to jobs and standards-of-living. ... We consider the drastic emission control policies deriving from the Kyoto conference -- lacking credible support from the underlying science -- to be ill-advised and premature."

The 1997 declaration also updated its citations of evidence that appeared to run contrary to the consensus on global warming. For example, the 1995 declaration cites "observations from earth satellites," where the 1997 declaration cites "observations from both weather satellites and balloon-borne radiosondes." As with satellite data, subsequent analysis of radiosondes has shown a statistically significant warming trend.

Signatures   The declaration begins: "As independent scientists concerned with atmospheric and climate problems, we...". As with the 1995 declaration, questions have been raised about the scientific background of the signers, and others have questioned the degree to which they can be deemed to be independent. Because many of those who signed the 1997 declaration also signed the 1995 declaration, the concerns raised by David Olinger and others after the 1995 declaration are still relevant.

The signers are generally described by Fred Singer and his supporters as climate scientists, although the current signers also include 25 weather presenters. One key report opposing the scientific credentials of the signers was a Danish Broadcasting Company TV special by Øjvind Hesselager.[7] Hesselager attempted to contact the declaration's 33 European signers and found that four of them could not be located, twelve denied ever having signed, and some had not even heard of the Leipzig Declaration. Those who verified signing included a medical doctor, a nuclear scientist, and an entomologist. After discounting the signers whose credentials were inflated, irrelevant, false, or unverifiable, Hesselager claimed that only 20 of the names on the list had any scientific connection with the study of climate change, and some of those names were known to have obtained grants from the oil and fuel industry, including the German coal industry and the government of Kuwait (a major oil exporter). As a result of Hesselager's report, Singer removed some, but not all, of the discredited signatures. The number of signatures on the document, according to SEPP's own press releases, has declined from 140 (according to a December 1997 press release) to 105 (as of February 2003).

SEPP's position is that "a few of the original signers did not have the 'proper' academic credentials - even though they understand the scientific climate issues quite well. To avoid this kind of smear, we want to restrict the Leipzig Declaration to signers with impeccable qualifications." To address the signer credibility issue, SEPP has provided considerably more information about each signer on their website and lists the weather presenters separately from the other signers.

2005, revised

As of 2010, Singer's SEPP website listed the "2005, revised" declaration (which still spoke of the 1997 Kyoto conference as a future event).[2] This version included the claim: "In fact, many climate specialists now agree that actual observations from weather satellites show no global warming whatsoever".

2.8.2   Use of the declarations

The declarations have been widely cited by some in the "sound science" movement. It has been cited by Fred Singer in editorial columns appearing in hundreds of websites and major publications, including *The Wall Street Journal*, *Miami Herald*, *Detroit News*, *Chicago Tribune*, *The Plain Dealer*, *Memphis Commercial Appeal*, *The Seattle Times*, and *The Orange County Register*. Jeff Jacoby, a columnist with the *Boston Globe*, described the signers of the Leipzig Declaration as "climate scientists" that "include prominent scholars." Think tanks such as the Heritage Foundation, The Heartland Institute, and Australia's Institute for Public Affairs called them "noted scientists." Both the Leipzig Declaration and Frederick Seitz's Oregon Petition have been quoted as authoritative sources during deliberations in the U.S. Senate and House of Representatives.

Although the key data on which the Leipzig declaration relied (such as satellite temperature measurements) has been invalidated by subsequent research, and much new evidence has accumulated,[8] the declaration continues to be cited, along with the Oregon Petition, as evidence of the current views of scientists on climate change. Moreover, the organizers have not changed their stated position of rejecting anthropogenic global warming.

2.8.3   Original texts

1995 declaration:

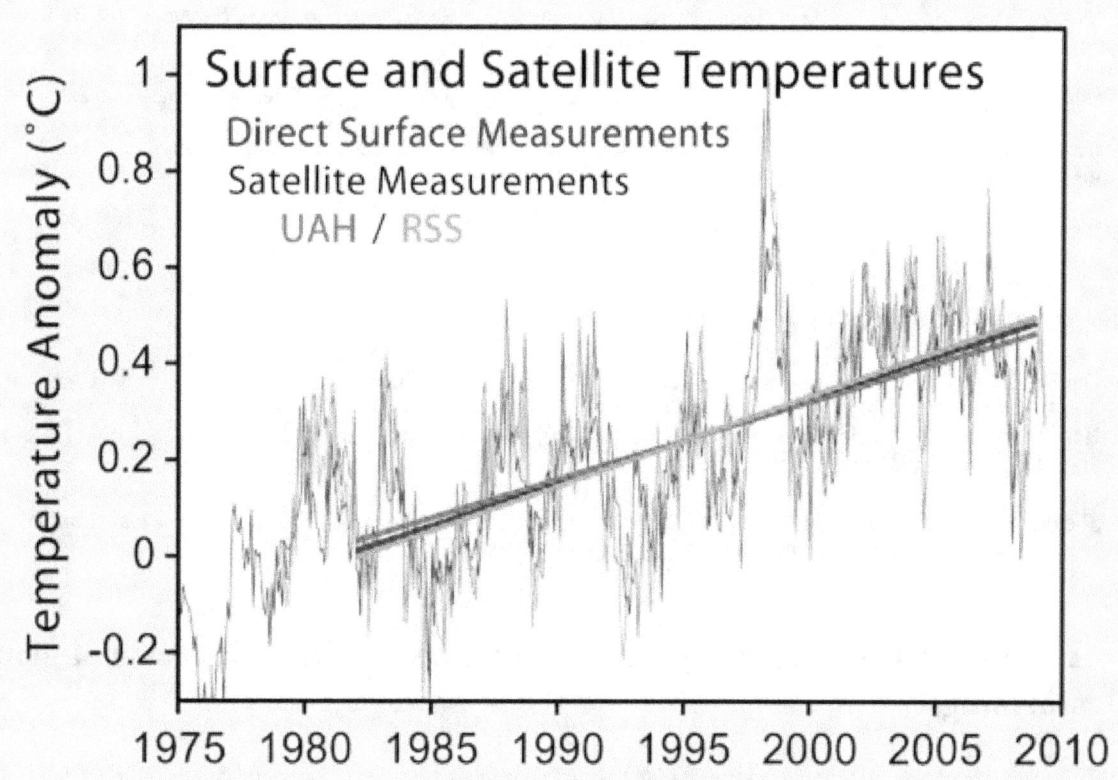

Comparison of ground-based (blue) and satellite-based (red: UAH; green: RSS) records of temperature variations since 1979. Trends plotted since January 1982. (Source: Publicly available data, figure by Robert A. Rohde)

- "The Leipzig Declaration On Global Climate Change". SEPP. Archived from the original on 1998-12-06.

1997 declaration:

- "The Leipzig Declaration On Global Climate Change". SEPP. Archived from the original on 2006-10-29.

2005 declaration (revised):

- "The Leipzig Declaration on Global Climate Change (2005, revised)". SEPP. Archived from the original on 2006-08-27.

2.8.4   See also

- Scientific opinion on climate change

2.8.5   References

[1] "The Leipzig Declaration On Global Climate Change". SEPP. Archived from the original on 1998-12-06.

[2] "The Leipzig Declaration on Global Climate Change (2005, revised)". SEPP. Archived from the original on 2006-08-27.

[3] "Signatories to the Leipzig Declaration". SEPP. Archived from the original on 2006-09-28.

[4] "Obituary in "The Week That Was". SEPP. 25 Dec 1999. Archived from the original on 2002-12-18.

[5] "NASA Johnson Space Center Oral History Project, Biographical Data Sheet, Name: Robert E. Stevenson" (PDF). Retrieved 2014-08-16.

[6] Olinger, David (July 29, 1996). "Cool to the warnings of global warming's dangers". St. Petersburg Times. (paywall)

[7] Jensen, Christian (11 February 1998). "How many climate researchers support the 'Leipzig Declaration'?". naturalSCIENCE. Retrieved 2014-08-16.

[8] "Summary for Policy Makers" (PDF). IPCC Fourth Assessment Report (4AR).

2.8.6 Related resources

- Danish Broadcasting Corporation (DR1) report, TV report cited by Christian Jensen.

- Hans Bulow and Poul-Eric Heilburth, "The Energy Conspiracy" (video documentary), Filmakers Library, 124 East 40th Street, New York, NY 10016.

- Sheldon Rampton and John Stauber, *Trust Us, We're Experts: How Industry Manipulates Science and Gambles With Your Future* (New York, NY: Tarcher Putnam, 2002).

2.9 List of scientists opposing the mainstream scientific assessment of global warming

This is an incomplete list that may never be able to satisfy particular standards for completeness. You can help by expanding it with reliably sourced entries.

This is a list of scientists who have made statements that conflict with the scientific consensus on global warming as summarized by the Intergovernmental Panel on Climate Change and endorsed by other scientific bodies.

The scientific consensus is that the global average surface temperature has risen over the last century. The scientific consensus and scientific opinion on climate change were summarized in the 2001 Third Assessment Report of the Intergovernmental Panel on Climate Change (IPCC). The main conclusions on global warming at that time were as follows:

1. The global average surface temperature has risen 0.6 ± 0.2 °C since the late 19th century, and 0.17 °C per decade in the last 30 years.[3]

2. "There is new and stronger evidence that most of the warming observed over the last 50 years is attributable to human activities", in particular emissions of the greenhouse gases carbon dioxide and methane.[4]

3. If greenhouse gas emissions continue the warming will also continue, with temperatures projected to increase by 1.4 °C to 5.8 °C between 1990 and 2100.[A] Accompanying this temperature increase will be increases in some types of extreme weather and a projected sea level rise.[5] The balance of impacts of global warming become significantly negative at larger values of warming.[6]

These findings are recognized by the national science academies of all the major industrialized nations.[7]

There have been several efforts to compile lists of dissenting scientists, including a 2008 US senate minority report,[8] the Oregon Petition,[9] and a 2007 list by the Heartland Institute,[10] all three of which have been criticized on a number of grounds.[11][12][13]

For the purpose of this list, a "scientist" is defined as an individual who has published at least one peer-reviewed article in the broad field of natural sciences, although not necessarily in a field relevant to climatology. Since the publication of the IPCC Third Assessment Report, each has made a clear statement in his or her own words (as opposed to the name being found on a petition, etc.) disagreeing with one or more of the report's three main conclusions. Their views on climate

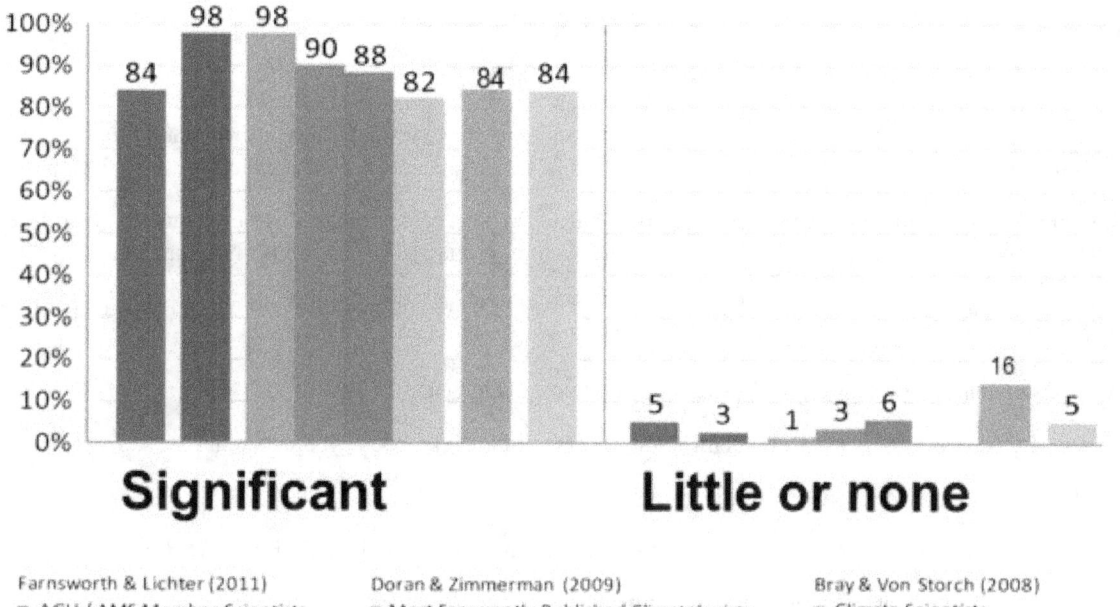

A majority of earth and climate scientists are convinced by the evidence that humans are significantly contributing to global warming.[1][2]

change are usually described in more detail in their biographical articles. Few of the statements in the references for this list are part of the peer-reviewed scientific literature; most are from other sources such as interviews, opinion pieces, online essays and presentations.

NB: Only scientists who have their own Wikipedia article may be included in the list.

2.9.1   Scientists questioning the accuracy of IPCC climate projections

These scientists have said that it is not possible to project global climate accurately enough to justify the ranges projected for temperature and sea-level rise over the next century. They may not conclude specifically that the current IPCC projections are either too high or too low, but that the projections are likely to be inaccurate due to inadequacies of current global climate modeling.

- David Bellamy, botanist.[14][15][16][17]

- Lennart Bengtsson, meteorologist, Reading University.[18][19]

- Judith Curry, Professor and former chair of the School of Earth and Atmospheric Sciences at the Georgia Institute of Technology.[20][21][22][23]

- Freeman Dyson, professor emeritus of the School of Natural Sciences, Institute for Advanced Study; Fellow of the Royal Society [24][25]

- Steven E. Koonin, theoretical physicist and director of the Center for Urban Science and Progress at New York University[26][27]

- Richard Lindzen, Alfred P. Sloan emeritus professor of atmospheric science at the Massachusetts Institute of Technology and member of the National Academy of Sciences[28][29][30][31]

- Craig Loehle, ecologist and chief scientist at the National Council for Air and Stream Improvement.[32][33][34][35][36][37][38]

- Patrick Moore, former president of Greenpeace Canada[39][40][41]

- Nils-Axel Mörner, retired head of the Paleogeophysics and Geodynamics Department at Stockholm University, former chairman of the INQUA Commission on Sea Level Changes and Coastal Evolution (1999–2003)[42][43]

- Garth Paltridge, retired chief research scientist, CSIRO Division of Atmospheric Research and retired director of the Institute of the Antarctic Cooperative Research Centre, visiting fellow Australian National University[44][45]

- Denis Rancourt, former professor of physics at University of Ottawa, research scientist in condensed matter physics, and in environmental and soil science[46][47][48][49]

- Harrison Schmitt, geologist, Apollo 17 Astronaut, former U.S. Senator.[50]

- Peter Stilbs, professor of physical chemistry at Royal Institute of Technology, Stockholm[51][52]

- Philip Stott, professor emeritus of biogeography at the University of London[53][54]

- Hendrik Tennekes, retired director of research, Royal Netherlands Meteorological Institute [55][56]

- Anastasios Tsonis, distinguished professor at the University of Wisconsin-Milwaukee[57][58]

- Fritz Vahrenholt, German politician and energy executive with a doctorate in chemistry[59][60]

2.9.2 Scientists arguing that global warming is primarily caused by natural processes

These scientists have said that the observed warming is more likely to be attributable to natural causes than to human activities. Their views on climate change are usually described in more detail in their biographical articles.

- Khabibullo Abdusamatov, astrophysicist at Pulkovo Observatory of the Russian Academy of Sciences[62][63]

- Sallie Baliunas, retired astrophysicist, Harvard-Smithsonian Center for Astrophysics[64][65][66]

- Timothy Ball, historical climatologist, and retired professor of geography at the University of Winnipeg[67][68][69]

- Robert M. Carter, former head of the school of earth sciences at James Cook University[70][71]

- Ian Clark, hydrogeologist, professor, Department of Earth Sciences, University of Ottawa[72][73]

- Chris de Freitas, associate professor, School of Geography, Geology and Environmental Science, University of Auckland[74][75]

- David Douglass, solid-state physicist, professor, Department of Physics and Astronomy, University of Rochester[76][77]

- Don Easterbrook, emeritus professor of geology, Western Washington University[78][79]

- William M. Gray, professor emeritus and head of the Tropical Meteorology Project, Department of Atmospheric Science, Colorado State University[80][81]

- William Happer, physicist specializing in optics and spectroscopy; emeritus professor, Princeton University[82][83]

- Ole Humlum, professor of geology at the University of Oslo[84][85]

- Wibjörn Karlén, professor emeritus of geography and geology at the University of Stockholm.[86][87]

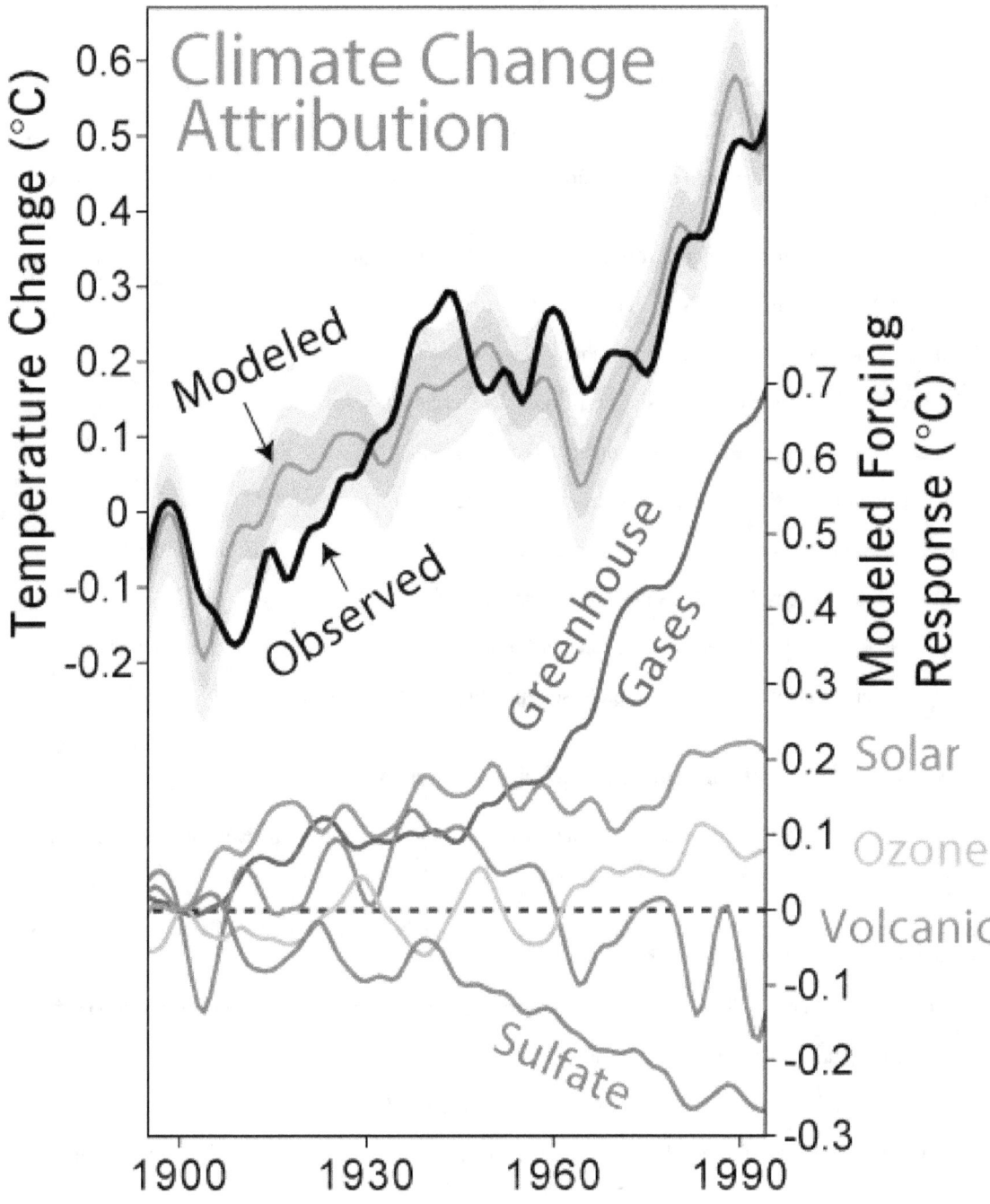

Graph showing the ability with which a global climate model is able to reconstruct the historical temperature record, and the degree to which those temperature changes can be decomposed into various forcing factors. It shows the effects of five forcing factors: greenhouse gases, man-made sulfate emissions, solar variability, ozone changes, and volcanic emissions.[61]

- William Kininmonth, meteorologist, former Australian delegate to World Meteorological Organization Commission for Climatology[88][89]

- David Legates, associate professor of geography and director of the Center for Climatic Research, University of Delaware[90][91]

- Anthony Lupo, professor of atmospheric science at the University of Missouri[92][93]

- Tad Murty, oceanographer; adjunct professor, Departments of Civil Engineering and Earth Sciences, University of Ottawa[94][95]

- Tim Patterson, paleoclimatologist and professor of geology at Carleton University in Canada.[96][97][98]

- Ian Plimer, professor emeritus of mining geology, the University of Adelaide.[99][100]

- Arthur B. Robinson, American politician, biochemist and former faculty member at the University of California, San Diego[101][102]

- Murry Salby, atmospheric scientist, former professor at Macquarie University and University of Colorado[103][104]

- Nicola Scafetta, research scientist in the physics department at Duke University[105][106][107]

- Tom Segalstad, geologist; associate professor at University of Oslo[108][109]

- Nir Shaviv, professor of physics focusing on astrophysics and climate science at the Hebrew University of Jerusalem[110][111]

- Fred Singer, professor emeritus of environmental sciences at the University of Virginia[112][113][114][115]

- Willie Soon, astrophysicist, Harvard-Smithsonian Center for Astrophysics[116][117]

- Roy Spencer, meteorologist; principal research scientist, University of Alabama in Huntsville[118][119]

- Henrik Svensmark, physicist, Danish National Space Center[120][121]

- George H. Taylor, retired director of the Oregon Climate Service at Oregon State University[122][123]

- Jan Veizer, environmental geochemist, professor emeritus from University of Ottawa[124][125]

2.9.3 Scientists arguing that the cause of global warming is unknown

These scientists have said that no principal cause can be ascribed to the observed rising temperatures, whether man-made or natural.

- Syun-Ichi Akasofu, retired professor of geophysics and founding director of the International Arctic Research Center of the University of Alaska Fairbanks.[126][127]

- Claude Allègre, French politician; geochemist, emeritus professor at Institute of Geophysics (Paris).[128][129]

- Robert Balling, a professor of geography at Arizona State University.[130][131]

- Pål Brekke, solar astrophycisist, senior advisor Norwegian Space Centre.[132][133]

- John Christy, professor of atmospheric science and director of the Earth System Science Center at the University of Alabama in Huntsville, contributor to several IPCC reports.[134][135][136]

- Petr Chylek, space and remote sensing sciences researcher, Los Alamos National Laboratory.[137][138]

- David Deming, geology professor at the University of Oklahoma.[139][140]

- Ivar Giaever, professor emeritus of physics at the Rensselaer Polytechnic Institute and a Nobel laureate.[141][142]

- Vincent R. Gray, New Zealand physical chemist with expertise in coal ashes[143][144]

- Keith E. Idso, botanist, former adjunct professor of biology at Maricopa County Community College District and the vice president of the Center for the Study of Carbon Dioxide and Global Change[145][146]

- Antonino Zichichi, emeritus professor of nuclear physics at the University of Bologna and president of the World Federation of Scientists.[147][148]

2.9.4 Scientists arguing that global warming will have few negative consequences

These scientists have said that projected rising temperatures will be of little impact or a net positive for society or the environment.

- Indur M. Goklany, science and technology policy analyst for the United States Department of the Interior[149][150][151]

- Craig D. Idso, faculty researcher, Office of Climatology, Arizona State University and founder of the Center for the Study of Carbon Dioxide and Global Change [152][153]

- Sherwood B. Idso, former research physicist, USDA Water Conservation Laboratory, and adjunct professor, Arizona State University[154][155]

- Patrick Michaels, senior fellow at the Cato Institute and retired research professor of environmental science at the University of Virginia[156][157]

2.9.5 Dead scientists

This section includes deceased scientists who would otherwise be listed in the prior sections.

- August H. "Augie" Auer Jr. (1940–2007), retired New Zealand MetService Meteorologist and past professor of atmospheric science at the University of Wyoming[158]

- Reid Bryson (1920–2008), Emeritus Professor of Atmospheric and Oceanic Sciences, University of Wisconsin-Madison, said in a 2007 magazine interview that he believed global warming was primarily caused by natural processes:[159]

- Robert Jastrow (1925–2008), American astronomer, physicist and cosmologist. He was a leading NASA scientist. Together with Fred Seitz and William Nierenberg he established the George C. Marshall Institute[159] to counter the scientists who were arguing against Reagan's Starwars Initiative, arguing for equal time in the media. This institute later took the view that tobacco was having no effect, that acid rain was not caused by human emissions, that ozone was not depleted by CFCs, that pesticides were not environmentally harmful and it was also critical of the consensus view of anthropogenic global warming.[160] Jastrow acknowledged the Earth was experiencing a warming trend, but claimed that the cause was likely to be natural variation.[161]

- Harold ("Hal") Warren Lewis (1923-2011), Emeritus Professor of Physics and former department chairman at the University of California, Santa Barbara. In 2010, after 67 years of membership, Lewis resigned from the American Physical Society, writing in a letter about the "corruption" from "the money flood" of government grants.[162]

- Frederick Seitz (1911–2008), solid-state physicist and former president of the National Academy of Sciences and co-founder of the George C. Marshall Institute in 1984.[159][163]

2.9.6 See also

- Environmental skepticism

- Global warming controversy

  - Climate change denial
  - Global warming conspiracy theory
  - Hockey stick controversy

- List of authors from the IPCC AR4 WGI report

- List of climate scientists

- Merchants of Doubt

- Oregon Petition

2.9.7   Notes

1. ^ In its 2007 assessment report, IPCC projected likely temperature rise for various hypothetical levels of future greenhouse gas emissions, known as "emissions scenarios". They reported that during the 21st century the global surface temperature is likely to rise a further 1.1 to 2.9 °C (2.0 to 5.2 °F) for the lowest emissions scenario used in the report, and 2.4 to 6.4 °C (4.3 to 11.5 °F) for the highest.[164]

2. ^ The compilation criteria for including scientists in the list is that they are relevant enough to have their own Wikipedia article, according to Wikipedia's notability guidelines.

2.9.8   References

[1] Anderegg, William R L; James W. Prall; Jacob Harold; Stephen H. Schneider (2010). "Expert credibility in climate change" (PDF). *Proc. Natl. Acad. Sci. USA* **107** (27): 12107–9. Bibcode:2010PNAS..10712107A. doi:10.1073/pnas.1003187107. PMC 2901439. PMID 20566872.

[2] Doran, Peter T.; Maggie Kendall Zimmerman (January 20, 2009). "Examining the Scientific Consensus on Climate Change" (PDF). *EOS* **90** (3): 22–23. Bibcode:2009EOSTr..90...22D. doi:10.1029/2009EO030002.

[3] Climate Change 2001: Working Group I: The Scientific Basis p.5 – IPCC

[4] Climate Change 2001: Working Group I: The Scientific Basis p.7 – IPCC

[5] Climate Change 2001: Working Group I: The Scientific Basis p.8 – IPCC

[6] Working Group II: Impacts, Adaptation and Vulnerability p.958 – IPCC

[7] "Joint Science Academies' Statement" (PDF). Retrieved August 9, 2010.

[8] Morano, Marc (11 December 2008). "U. S. Senate Minority Report: More Than 700 International Scientists Dissent Over Man-Made Global Warming Claims". Senate Committee on Environment and Public Works. Retrieved 1 September 2013.

[9] "Global Warming Petition Project". Retrieved 2 March 2014.

[10] 500 Scientists Whose Research Contradicts Man-Made Global Warming Scares, by Dennis T. Avery. From the Heartland Institute website; published September 14, 2007, accessed June 20, 2008.

[11] Kaufman, Leslie (April 9, 2009). "Dissenter on Warming Expands His Campaign". New York Times.

[12] McKnight, David (August 2, 2008). "The climate change smokescreen". *Sydney Morning Herald*. Retrieved December 28, 2009.

[13] Grandia, Kevin (22 July 2009). "The 30,000 Global Warming Petition Is Easily-Debunked Propaganda". *Huffington Post*. Retrieved 2 March 2014.

[14] Global Warming? What a load of poppycock! by David Bellamy, Daily Mail 9 July 2004.

[15] http://www.newscientist.com/article/mg18624950.100-glaciers-are-cool.html

[16] Monbiot, George (2005-05-10), *Junk Science*, London: Guardian.co.uk, retrieved 2008-11-07

[17] Bellamy, David (2005-05-29), *In an Adverse Climate*, London: Times Online, retrieved 2008-11-07

[18] Wills Robinson (15 May 2014).Climate change scientist claims he has been forced from new job. Daily Mail.

[19] Axel Bojanowski (May 12, 2014). Climate Change Debate: A Famous Scientist Becomes a Skeptic. Spiegel Online International.

[20] Curry, Judith A.; Webster, Peter J. (2011). "Climate Science and the Uncertainty Monster". *Bureau of the American Meteorological Society* **175**: 1667–1682. doi:10.1175/2011BAMS3139.1. http://journals.ametsoc.org/doi/pdf/10.1175/2011BAMS3139.1

[21] Curry, Judith A. (2010-11-17). "Statement to the Subcommittee on Energy and Environment of the United States House of Representatives" (PDF). Retrieved 2014-10-25.

[22] Curry, Judith A. (2013-04-25). "Statement to the Subcommittee on Environment of the United States House of Representatives" (PDF). Retrieved 2014-10-25. [The IPCC AR4] estimate of equilibrium climate sensitivity is not easily reconciled with recent forcing estimates and observational data. There is increasing support for values of climate sensitivity around or below 2°C.

[23] Mann, Michael E. (2014-10-17). "Curry Advocates Against Action on Climate Change". Retrieved 2014-10-25.

[24] Freeman Dyson, in correspondence with editor Steve Connor (February 25, 2011), "Letters to a heretic: An email conversation with climate change sceptic Professor Freeman Dyson", *The Independent*, First, the computer models are very good at solving the equations of fluid dynamics but very bad at describing the real world. [...] Sixth, summing up the other five reasons, the climate of the earth is an immensely complicated system and nobody is close to understanding it.

[25] Epstein, Ethan (13 January 2014). "What Catastrophe?". The Weekly Standard. Retrieved 31 October 2014. Nor, of course, is he the only skeptic with serious scientific credentials... famed physicist Freeman Dyson are among dozens of scientists who have gone on record questioning various aspects of the IPCC's line on climate change

[26] Koonin, Steven (September 2014). "Climate Science Is Not Settled". The Wall Street Journal. Retrieved 13 October 2014. [Many open questions] are not "minor" issues to be "cleaned up" by further research. Rather, they are deficiencies that erode confidence in the computer projections. [...They are] fundamental challenges to our understanding of human impacts on the climate, and they should not be dismissed with the mantra that 'climate science is settled.' While the past two decades have seen progress in climate science, the field is not yet mature enough to usefully answer the difficult and important questions being asked of it.

[27] "Turn down heat on climate debate". *Tampa Tribune*. 23 September 2014.

[28] "The Climate Science Isn't Settled", *The Wall Street Journal online*, November 30, 2009, Claims that climate change is accelerating are bizarre. [...] The quality of the data is poor [...] The general support for warming is based not so much on the quality of the data, but rather on the fact that there was a little ice age from about the 15th to the 19th century.

[29] "What Catastrophe?", *The Weekly Standard*, January 13, 2014

[30] Zedillo, Ernesto, ed. (2008). *Global Warming: Looking Beyond Kyoto*. Brookings Institution Press. pp. 21–. ISBN 0-8157-9716-8.

[31] Epstein, Ethan (13 January 2014). "What Catastrophe?". The Weekly Standard. Retrieved 31 October 2014. But Lindzen, plainly, is different. He can't be dismissed. Nor, of course, is he the only skeptic with serious scientific credentials.

[32] Loehle, Craig. "Climate change: detection and attribution of trends from long-term geologic data". *Ecological Modelling* **171** (4): 433–450. doi:10.1016/j.ecolmodel.2003.08.013.

[33] Carter, Bob (20 January 2009). "Facts debunk global warming alarmism". *The Australian*. Retrieved 11 May 2014.

[34] Loehle, Craig (2007). "A 2000 Year Global Temperature Reconstruction based on Non-Tree ring Proxies" (PDF). *Energy & Environment* **18** (7 & 8): 1049–1058.

[35] Milloy, Steven (21 November 2007). "U.N. Climate Distractions". *Fox News Channel*. Retrieved 13 February 2014. A new temperature reconstruction for the past 2,000 years created by Craig Loehle of the National Council for Air and Stream Improvement indicates that, 1,000 years ago, globally averaged temperature was about 0.3 degrees Celsius warmer than the current temperature.

[36] Singer, S. Fred (9 January 2010). "Index of Editorials Global Warming Junkscience". *Science and Environmental Policy Project*. Retrieved 1 February 2014.

[37] "Skeptics, unite!". *National Post*. Retrieved 1 February 2014.

[38] Loehle, Craig; McCulloch, J. Huston (2008). "Correction to: A 2000-Year Global Temperature Reconstruction Based on Non-Tree Ring Proxies". *Energy & Environment* (Multi-Science Publishing Co. Ltd.) **19** (1): 93–100. doi:10.1260/095830508783563109. Retrieved 18 September 2015.

[39] A.R. Lucas & P.A. Moore, The Utah Controversy: A Case Study of Public Participation in Pollu-tion Control, *Natural Resources Journal* Vol. 36, Pages 46-47, 1973

[40] Jacqueline Echevarria, Greenpeace Co-founder: CO2 does not cause global warming, *Energy live News*, October 15, 2015

[41] Patrick Moore, Obama's Half-Baked Alaska: Yes, the glacier of Glacier Bay is receding—as it has from time to time for centuries, *The Wall Street Journal*, September 3, 2015

[42] Nils-Axel Mörner (Mar 30, 2005), *Economics of Climate Change: 12-ii Session 2005–06 Evidence to Select Committee on Economic Affairs* **II**, The Stationery Office, p. 269, In conclusion, observational data do not support the sea level rise scenario. On the contrary, they seriously contradict it.

[43] Kelly, Jack. "The facts don't add up for human-caused global warming". Pittsburgh Post-Gazette. prominent skeptics...Nils-Axel Mörner

[44] Paltridge, Garth (2009). *the Climate Caper*. Connor Court Publishing. ISBN 978-1-921421-25-9. There are good and straightforward scientific reasons to believe that the burning of fossil fuel and consequent increase in atmospheric carbon dioxide will lead to an increase in the average temperature of the world above that which would otherwise be the case. Whether the increase will be large enough to be noticeable is still an unanswered question.

[45] Bolt, Andrew (12 February 2010). "Warmists are hot under the collar as scepticism rules". Herald Sun. Retrieved 1 November 2014.

[46] "Denis Rancourt on climate science and on climate politics", *Climate Guy blog*, February 23, 2014

[47] Cockburn, A., "Dissidents Against Dogma", *The Nation*, 25 June 2007.

[48] "Inhofe reveals how scientists and activists believe global warming has 'co-opted' the environmental movement," US Senate Committee on Environment and Public Works, 26 October 2007

[49] Newton, David E. *Science and Political Controversy: A Reference Handbook*. ABC-CLIO. p. 156. ISBN 1610693191.

[50] No Need to Panic About Global Warming. Wall Street Journal, January 27, 2012.

[51] Peter Stilbs and Åke Ortmark (12 January 2014), *Expressen, gå inte på klimatbluffen*, IPCC gör ingen egen forskning, utan söker som grupp stöd för en given hypotes - att koldioxiden har en avgörande betydelse för jordens framtida klimat. Detta är egentligen ogörligt, då ingen ännu har klarlagt klimatsystemets naturliga variationer. Enligt de vetenskapliga principer som växt fram under hundratals år tyder de senaste 20 årens observationer snarare på att hypotesen är falsk. (Own translation to English: The IPCC does not make its own research, but is a group searching for a given hypothesis – that carbon dioxide is crucial for the earth's future climate. This is actually impossible since nobody has yet clarifed the climate system's natural variability. According to the scientific principles that have developed over hundreds of years, the last 20 years of observations rather indicate that the hypothesis is false.)

[52] "Enögt fokus på CO2-utsläpp leder fel" (in Swedish). Göteborgs-Posten. 1 February 2012. Retrieved 1 November 2014.

[53] *Global Warming Is Not A Crisis*, It is claimed, on the basis of computer models, that this should lead to 1.1 – 6.4 C warming. What is rarely noted is that we are already three-quarters of the way into this in terms of radiative forcing, but we have only witnessed a 0.6 (+/−0.2) C rise, and there is no reason to suppose that all of this is due to humans.

[54] Glasse, Jennifer (5 December 2009). "UN Panel to Investigate Claims Climate Change Scientists Suppressed Data". Voice of America. Retrieved 1 November 2014. Philip Stott, a professor in biogeography at the University of London and a climate change skeptic

[55] Tennekes, Hendrik. "A Skeptical View of Climate Models" (PDF). The blind adherence to the harebrained idea that climate models can generate 'realistic' simulations of climate is the principal reason why I remain a climate skeptic.

[56] Timmer, Edwin (13 February 2010). "Het Gelijk Van Henk Tennekes" (in Dutch). De Telegraaf. Retrieved 1 November 2014.

[57] Roberts, John (30 September 2013). "UN's massive new climate report adds little explanation for 'pause' in warming". *Fox News Channel*. Retrieved 10 October 2014. "I know that the models are not adequate," Tsonis told Fox News. "There are a lot of climate models out there. They don't agree with each other – and they don't agree with reality."

[58] Lee Bergquist and Thomas Content (25 March 2009). "Natural forces stalling global warming, UWM pair say". Retrieved 1 November 2014. The findings of mathematicians Kyle L. Swanson and Anastasios Tsonis contradict the assumptions of many climate scientists... who say the planet is currently warming.

[59] "Breaking Global Warming Taboos: 'I Feel Duped on Climate Change'". *Spiegel Online*. 8 February 2012. Retrieved 19 February 2014. CO2 alone will never cause a warming of more than 2 degrees Celsius (3.6 degrees Fahrenheit) by the end of the century. Only with the help of supposed amplification effects, especially water vapor, do the computers arrive at a drastic temperature increase.

[60] Delingpole, James (16 June 2012). "It's no wonder the world's cooling on climate change". Daily Mail. Retrieved 1 November 2014. leading German green - former activist and Hamburg state environment senator Prof Fritz Vahrenholt. The evidence for man-made global warming is looking shakier by the day, Germany's answer to Jonathon Porritt or George Monbiot admitted. Far more likely a culprit is the sun.

[61] Meehl, G.A.; W.M. Washington; C.A. Ammann; J.M. Arblaster; T.M.L. Wigleym; C. Tebaldi (2004). "Combinations of Natural and Anthropogenic Forcings in Twentieth-Century Climate". Journal of Climate 17: 3721–3727. Bibcode:2004JCli...17.3721M. doi:10.1175/1520-0442(2004)017<3721:CONAAF>2.0.CO;2.

[62] "Russian academic says CO2 not to blame for global warming". Russian International News Agency. 15 January 2007. Retrieved 24 August 2012. Global warming results not from the emission of greenhouse gases [...], but from an unusually high level of solar radiation and [...] growth in its intensity.

[63] "Change climate change!". Hindustan Times. 19 January 2010. Retrieved 1 November 2014. A Russian astronomer named Khabibullo Abdusamatov from St Petersburg has predicted the next ice age will start between 2035 and 2045 due to a decline in solar activity

[64] Baliunas, Sallie (August 2002). "Warming Up to the Truth". The Heritage Foundation. Retrieved 31 August 2012.

[65] Baliunas, Sallie; Willie Soon (22 August 2002). "Global Warming Science vs. Computer Model Speculation: Just Ask the Experts". Capitalism Magazine. Retrieved 31 August 2012. [T]he recent warming trend in the surface temperature record cannot be caused by the increase of human-made greenhouse gases in the air.

[66] Rowland, Christopher (5 November 2013). "Researcher helps sow climate-change doubt". Boston Globe. Retrieved 1 November 2014. prominent climate-change doubter, Sallie Baliunas, who was studying variations in solar radiation

[67] Coren, Michael (13 February 2010). "Climatology expert threatened for climate change views". Toronto Sun. Retrieved 6 March 2014. There has always been and always will be climate change, but it has very little to do with human activity and has nothing at all to do with pollution of course.

[68] Plumer, Bradford (7 February 2007). "The dire global cooling problem". The Guardian. Retrieved 1 November 2014. global warming-skeptic Timothy Ball

[69] Dyck, M.J. et al. (with Timothy F. Ball), Polar bears of western Hudson Bay and climate change: Are warming spring air temperatures the "ultimate" survival control factor?, Ecological Complexity, Volume 4, Issue 3, September 2007, Pages 73–84. PDF of full article

[70] McLean, J. D.; de Freitas, C. R.; Carter, R. M. (2009). "Influence of the Southern Oscillation on tropospheric temperature" (PDF). Journal of Geophysical Research 114: D14104. Bibcode:2009JGRD..11414104M. doi:10.1029/2008JD011637.

[71] "A little warming, a lot of hysteria". Washington Times. 11 April 2006. Retrieved 1 November 2014. professor [Robert Carter], writing in the London Daily Telegraph, does not dispute the evidence that we're in an era of rising temperatures. Who does? But he suggests that man exhibits considerable hubris — insolence, even — if he imagines that he's responsible.

[72] Ian Clark (March 22, 2004). "Letter to the editor of The Hill Times". Natural Resources Stewardship Project. Archived from the original on February 10, 2009. Retrieved August 26, 2011. We know that [the sun] was responsible for climate change in the past, and so is clearly going to play the lead role in present and future climate change. And interestingly... solar activity has recently begun a downward cycle.

[73] Huberman, Joel A. (6 October 2007). "Skeptics need to be held to scientific standards, too". The Buffalo News. Retrieved 1 November 2014. Harris and [Ian] Clark are global warming "skeptics."

[74] Chris de Freitas (May 9, 2006). "Chris de Freitas: Evidence must prevail" (PDF). The New Zealand Herald. Archived from the original (PDF) on May 23, 2006. Retrieved August 26, 2011. To support the argument that carbon dioxide is causing [global warming], the evidence would have to distinguish between human-caused and natural warming. This has not been done.

[75] "Half of Kiwis doubt global warming: poll". New Zealand Herald. 18 January 2010. Retrieved 1 November 2014. climate sceptic Chris de Freitas

[76] Phillip V Brennan (December 10, 2007). "New Study Explodes Human-Global Warming Story". Newsmax.com. Archived from the original on May 11, 2008. Retrieved August 26, 2011. [...]observed increases in carbon dioxide and other greenhouse gases make only a negligible contribution to climate warming.

[77] Asten, Michael (29 December 2009). "More evidence CO2 not culprit". The Australian. Retrieved 1 November 2014.

[78] Easterbrook, Don (22–25 October 2006). "THE CAUSE OF GLOBAL WARMING AND PREDICTIONS FOR THE COMING CENTURY". *Philadelphia Annual Meeting*. Retrieved 31 August 2012. Because the warming periods in these oscillations [of glaciers] occurred well before atmospheric CO_2 began to rise rapidly in the 1940s, they could not have been caused by increased atmospheric CO_2, and global warming since 1900 could well have happened without any effect of CO_2. If the cycles continue as in the past, the current warm cycle should end soon[...]

[79] "The views of retired geology professor Don Easterbrook are considered in the minority.". 26 March 2013. Retrieved 1 November 2014. ...global warming skeptic who argued that federal scientists have been manipulating climate data to inflate temperatures. The views of retired geology professor Don Easterbrook are considered in the minority.

[80] Achenbach, Joel (28 May 2006). "The Tempest". *The Washington Post* (Washington DC: WPC). ISSN 0190-8286. Retrieved 1 September 2012. I am of the opinion that [global warming] is one of the greatest hoaxes ever perpetrated on the American people.

[81] "Hurricane predictor will update forecast Wednesday". The Washington Post. 27 September 2010. Retrieved 1 November 2014. Gray - who has gotten attention lately for calling global warming a hoax

[82] Raymond Brusca (January 12, 2009). "Professor denies global warming theory". [Global warming] probably has little to do with carbon dioxide, just like past warmings had little to do with carbon dioxide

[83] Epstein, Ethan (13 January 2014). "What Catastrophe?". The Weekly Standard. Retrieved 31 October 2014. Nor, of course, is he the only skeptic with serious scientific credentials... William Happer, professor of physics at Princeton... among dozens of scientists who have gone on record questioning various aspects of the IPCC's line on climate change

[84] Halfdan Carstens (2013). "Klimatolog i hardt vær". Retrieved January 9, 2014. Based on my own observations of how the climate varies naturally, I am skeptical of the CO2 hypothesis (own translation from Norwegian)

[85] Prestrud, Pal (18 October 2011). "Questionable climate debate" (in Norwegian). Aftenposten. Retrieved 31 October 2014.

[86] Wibjörn Karlén (January 7, 2010). "Lilla istiden kan redan vara här". Retrieved January 16, 2014. After a long time of studying climate variations, I have come to the conclusion that the space weather suggests that we are more likely heading towards a colder period than a warmer. (own translation from Swedish)

[87] "Skeptics from a range of scientific disciplines get louder in their opposition to doomsday claims". Orange County Register. 1 January 2008. Retrieved 31 October 2014.

[88] William Kininmonth, *Climate Change: A Natural Hazard* (PDF), archived from the original (PDF) on August 28, 2007, retrieved August 26, 2011, Natural variability of the climate system has been underestimated by IPCC and has, to now, dominated human influences.

[89] Shand, Adam (10 January 2013). "Heat likely to return despite southerlies - NATURE'S FURY -". *The Australian*. Retrieved 31 October 2014. William Kininmonth, a noted climate change sceptic (Subscription required.)

[90] Legates, David (May 2006). "Climate Science: Climate Change and Its Impacts". National Center for Policy Analysis. Retrieved 31 August 2012. About half of the warming during the 20th century occurred prior to the 1940s, and natural variability accounts for all or nearly all of the warming.

[91] Montgomery, Jeff (19 March 2013). "Climate change skeptics say 'sick' science distorts facts". USA Today. Retrieved 31 October 2014. Professor David Legates of the University of Delaware, a former climatologist for the state, bluntly rejected leading climate change claims

[92] Silvey, Janese (5 March 2012). "Professor details role as climate consultant". *Columbia Tribune*. Retrieved 15 April 2014. There's no doubt the climate is changing; that's a given," he said. "But the question is: What's causing it. Is it mankind alone, which a lot of people say? Is it some mix of man and nature? Or is it nature? I would say nature is mostly responsible. There may be a role for man in there somewhere, but how much, I don't know.

[93] Gerken, James (28 August 2014). "Utility-Sponsored Teacher Training At Mizzou Brings Climate Skepticism And Anti-EPA Message". The Huffington Post. Retrieved 31 October 2014.

[94] Robinson, Cindy (Spring 2005). "**Global warning?** Controversy heats up in the scientific community". Carleton University Magazine. Retrieved 31 August 2012. There is no global warming due to human anthropogenic activities.

[95] Montgomery, Charles (12 August 2006). "Nurturing doubt about climate change is big business". The Globe and Mail. Retrieved 31 October 2014. Canada's most vocal climate skeptics...University of Ottawa lecturer Tad Murty

[96] Tom, Harris (June 12, 2006). "Global warming, Scientists, Al Gore climate change". Canada Free Press. Retrieved 31 August 2012. There is no meaningful correlation between CO_2 levels and Earth's temperature over this [geologic] time frame.

[97] Patterson, Timothy (June 2007). "Read the Sunspots". Financial Post.

[98] "He's in the hot seat". Edmonton Journal. 23 September 2007. Retrieved 31 October 2014. The main driver of climate change, [Tim Patterson] believes, is a combination of solar changes (well-known cycles of the sun's intensity) as well as cosmic rays.

[99] "Wild weather ignites climate change debate". Australian Broadcasting Company - Lateline. 31 August 2012. Natural climate changes occur unrelated to carbon dioxide contents.

[100] Manning, Paddy (26 November 2012). "Roy Hill to push through pain". Sydney Morning Herald. Retrieved 31 October 2014. Mr Plimer, a noted climate sceptic

[101] Robinson, Arthur B. (1997). "Science Has Spoken: Global Warming is a Myth" (PDF). Dow Jones & Company. Retrieved 18 February 2014. we needn't worry about human use of hydrocarbons warming the Earth. We also needn't worry about environmental calamities, even if the current, natural warming trend continues: After all the Earth has been much warmer during the past 3,000 years without ill effects.

[102] Gaston, Christian (10 August 2013). "Former Peter DeFazio opponent Art Robinson elected to lead Oregon Republican Party". The Oregonian. Retrieved 31 October 2014. Robinson, a chemist and outspoken skeptic of human-caused global warming

[103] Bolt, Andrew (3 August 2011). "New research: warmth produces these carbon dioxide concentrations". *Herald Sun*. Retrieved 2 April 2014. Salby...suggests that its warmth which tends to produce more CO2, rather than vice versa - which, incidentally is the story of the past recoveries from ice ages.

[104] Darwall, Rupert (Summer 2014). "An Unsettling Climate". City Journal (New York City). Retrieved 30 October 2014. Another dissenter, the American atmospheric physicist Murry Salby...

[105] "I cambi climatici e le loro cause, una discussione su alcuni punti chiave (Climate Change and Its Causes, A Discussion About Some Key Issues)" (PDF). *La Chimica e l'Industria*. 2010. pp. 70–75. Retrieved 31 August 2012. At least 60% of the warming of the Earth observed since 1970 appears to be induced by natural cycles which are present in the solar system. A climatic stabilization or cooling until 2030–2040 is forecast by the phenomenological model.

[106] "Scafetta webpage".

[107] Taylor, James (30 May 2013). "Global Warming Alarmists Caught Doctoring '97-Percent Consensus' Claims". Forbes. Retrieved 30 October 2014. prominent, vigorous skeptic... Nicola Scafetta

[108] Segalstad, Tom. "What is CO_2 – friend or foe?" (PDF). Retrieved July 4, 2009. The IPCC's temperature curve (the so-called 'hockey stick' curve) must be in error [...] All measurements of solar luminosity and 14C isotopes show that there is at present an increasing solar radiation which gives a warmer climate

[109] Stratton, Allegra (20 November 2009). "Climate change denial MEP attacks church". The Guardian. Retrieved 30 October 2014. Tom Segalstad, a Norwegian geologist who says human-released CO2 would not have a large effect on the climate

[110] Shaviv, Nir. "Prof Nir Shaviv: The influence of cosmic radiation on the climate!". European Institute for Climate and Energy. Retrieved 24 October 2014. The story we hear from the IPCC is faulty in many respects

[111] Taylor, James (30 May 2013). "Global Warming Alarmists Caught Doctoring '97-Percent Consensus' Claims". Forbes. Retrieved 30 October 2014. prominent, vigorous skeptic... Nir Shaviv

[112] Singer, S. Fred (April 22, 2005). "'Flat Earth Award' nominee's challenge to Chicken Littles". Christian Science Monitor. The greenhouse effect is real. However, the effect is minute, insignificant, and very difficult to detect.

[113] "The Denial Machine (ABC Interview)". 2008.

[114] "Climate of Doubt". *PBS Frontline*. October 23, 2012.

[115] Mook, Dean (8 February 2014). "Connecting the dots for climate skeptics". The Roanoke Times. Retrieved 30 October 2014. But, there are always skeptics. For one example among several, Fred Singer, retired University of Virginia professor of physics

[116] William J Cromie (April 24, 2003). "Global warming is not so hot: 1003 was worse, researchers find". Harvard University Gazette. Retrieved August 26, 2011. there's increasingly strong evidence that previous research conclusions [...] may have been biased by underestimation of natural climate variations.

[117] Rowland, Christopher (5 November 2013). "Researcher helps sow climate-change doubt". The Boston Globe. Retrieved 30 October 2014. Willie Soon is a hero of the skeptical movement

[118] "Testimony of Roy W. Spencer". *before the Senate Environment and Public Works Committee.* 22 July 2008. Retrieved 31 August 2012. I predict that [scientists will realise] most of the climate change we have observed is natural, and that mankind's role is relatively minor

[119] Bachelard, Michaellast (11 September 2011). "Majority report: why consensus is all the rage". Sydney Morning Herald. Retrieved 30 October 2014. Internationally, sceptics look to climatologist Dr Roy Spencer

[120] Svensmark, Henrik (2007). "Cosmoclimatology: a new theory emerges" (PDF). *Astronomy & Geophysics* **48** (1): 18–24. Retrieved December 19, 2011. The case for anthropogenic climate change during the 20th century rests primarily on the fact that concentrations of carbon dioxide and other greenhouse gases increased and so did global temperatures. Attempts to show that certain details in the climatic record confirm the greenhouse forcing (e.g. Mitchell et al. 2001) have been less than conclusive. By contrast, the hypothesis that changes in cloudiness obedient to cosmic rays help to force climate change predicts a distinctive signal that is in fact very easily observed, as an exception that proves the rule.

[121] Nuccitelli, Dana (12 November 2013). "Cosmic rays fall cosmically behind humans in explaining global warming". The Guardian. Retrieved 30 October 2014. Henrik Svensmark of the Danish National Space Institute is the main proponent of the hypothesis linking [comic rays] to global climate change

[122] Tomlinson, Stuart (21 February 2008). "Update: Controversial "State Climatologist" Steps Aside". OregonLive.com. Retrieved 20 March 2014. Taylor said he believes climate change is a combination of natural factor and human factors. "I don't deny that human activities affect climate change," he said. "But I believe up to now, natural variations have played a more important role than human activities.

[123] Learn, Scott (26 January 2012). "Presentation by global warming skeptics draws big crowd in Portland". The Oregonian. Retrieved 30 October 2014.

[124] Veizer, Ján (2005). "Celestial Climate Driver: A Perspective from Four Billion Years of the Carbon Cycle". *Geoscience Canada.* 1 **32**. Retrieved 26 August 2012. At this stage, two scenarios of potential human impact on climate appear feasible: (1) the standard IPCC model that advocates the leading role of greenhouse gases, particularly of CO_2, and (2) the alternative model that argues for celestial phenomena as the principal climate driver. The two scenarios are likely not even mutually exclusive, but a prioritization may result in different relative impact. Models and empirical observations are both indispensable tools of science, yet when discrepancies arise, observations should carry greater weight than theory. If so, the multitude of empirical observations favours celestial phenomena as the most important driver of terrestrial climate on most time scales, but time will be the final judge.

[125] "Esteemed Ottawa scientist says cosmic rays, not greenhouse gases, cause global warming". Ottawa Citizen. 16 March 2006. Retrieved 30 October 2014.

[126] Syun-Ichi, Akasofu (June 15, 2007). "On the Fundamental Defect in the IPCC's Approach to Global Warming Research by Syun-Ichi Akasofu". *Climate Science: Roger Pielke Sr.* wordpress.com. Retrieved 31 August 2012. [T]he method of study adopted by the International Panel of Climate Change (IPCC) is fundamentally flawed, resulting in a baseless conclusion: *Most of the observed increase in globally averaged temperatures since the mid-20th century is very likely due to the observed increase in anthropogenic greenhouse gas concentrations.* Contrary to this statement ..., there is so far no definitive evidence that 'most' of the present warming is due to the greenhouse effect. ... [The IPCC] should have recognized that the range of observed natural changes should not be ignored, and thus their conclusion should be very tentative. The term 'most' in their conclusion is baseless.

[127] Alford, Peter (14 March 2009). "Japanese scientists cool on theories". The Australian. Retrieved 30 October 2014. Dr Akasofu and Tokyo Institute of Technology geology professor Shigenori Maruyama are highly critical of the UN Intergovernmental Panel on Climate Change's acceptance that hazardous global warming results mainly from man-made gas emissions.

[128] "Climat: la prévention, oui, la peur, non" (in French). L'Express. May 10, 2006. Archived from the original on November 17, 2006. Retrieved August 26, 2011. :The increase in the CO_2 content of the atmosphere is an observed fact and mankind is most certainly responsible. In the long term, this increase will without doubt become harmful, but its exact role in the climate is less clear. Various parameters appear more important than CO_2. Consider the water cycle and formation of various types of clouds, and the complex effects of industrial or agricultural dust. Or fluctuations of the intensity of the solar radiation on annual and century scale, which seem better correlated with heating effects than the variations of CO_2 content.

[129] Lean, Geoffrey (19 June 2009). "Conservatives have always been green". The Daily Telegraph. Retrieved 30 October 2014. France's foremost climate sceptic, Claude Allègre

[130] Balling, Robert (September 2003). "The Increase in Global Temperature: What it Does and Does Not Tell Us" (PDF). George C. Marshall Institute. [I]t is very likely that the recent upward trend [in global surface temperature] is very real and that the upward signal is greater than any noise introduced from uncertainties in the record. However, the general error is most likely to be in the warming direction, with a maximum possible (though unlikely) value of 0.3 °C. ... At this moment in time we know only that: (1) Global surface temperatures have risen in recent decades. (2) Mid-tropospheric temperatures have warmed little over the same period. (3) This difference is not consistent with predictions from numerical climate models.

[131] Carroll, Vincent (20 June 2009). "Carroll: Skeptical of climate alarmists". The Denver Post. Retrieved 30 October 2014.

[132] Jacobsen, Siw Ellen (February 29, 2008). "Pål Brekke: Internationally renowned climate sceptic and solar expert". The Research Council of Norway. The UN Intergovernmental Panel on Climate Change (IPPC) has determined that the earth's temperature has risen by about 0.7° C since 1901. According to Dr Brekke, this time period coincides not only with an increase in human-caused greenhouse gas emissions, but also with a higher level of solar activity, which makes it complicated to separate the effects of these two phenomena. [...] Dr Brekke has published more than 40 scientific articles on the sun and on the interaction between the sun and the earth. "We could be in for a surprise," he cautions. "It's possible that the sun plays an even more central role in global warming than we have suspected. Anyone who claims that the debate is over and the conclusions are firm has a fundamentally unscientific approach to one of the most momentous issues of our time."

[133] Brekke, Paal (November 16, 2000). "Viewpoint: The Sun and climate change". BBC News. Natural processes involving changes in the Sun could have at least as powerful an effect on global temperature as increased emissions of carbon dioxide (CO_2)...

[134] Christy, John R.; Douglass, David H. (2009). "Limits on CO_2 Climate Forcing from Recent Temperature Data of Earth" (PDF). Energy & Environment 20: 177–189. doi:10.1260/095830509787689277. Retrieved June 17, 2011. ...the data show a small underlying positive trend that is consistent with CO_2 climate forcing with no-feedback. [...] There is disagreement in regard to the validity of the global warming hypothesis that states that there are positive feedback processes leading to gains g that are larger than 1, perhaps as large as 3 or 4. However, recent studies suggest that the values of g is much smaller.

[135] Christy, John (November 1, 2007). "My Nobel Moment". Wall Street Journal. Retrieved November 2, 2007. ...I see neither the developing catastrophe nor the smoking gun proving that human activity is to blame for most of the warming we see. Rather, I see a reliance on climate models (useful but never "proof") and the coincidence that changes in carbon dioxide and global temperatures have loose similarity over time.

[136] Sullivansept, Margaret (6 September 2014). "Meant as Portraits, Seen as Hagiography". The New York Times. Retrieved 30 October 2014. John Christy — a prominent climate-change skeptic

[137] Petr Chylek (April 2002). "A Long Term Perspective on Climate Change" (PDF). Heartland.org. Archived from the original (PDF) on September 29, 2007. Retrieved August 26, 2011. Carbon dioxide should not be considered as a dominant force behind the current warming...how much of the [temperature] increase can be ascribed to CO_2, to changes in solar activity, or to the natural variability of climate is uncertain

[138] Borenstein, Seth (31 October 2011). "Noted skeptic finds climate change real". The Journal Gazette. Retrieved 30 October 2014. Petr Chylek of Los Alamos National Lab, a noted skeptic

[139] Dr. David Deming (6 December 2006). "U.S. Senate Committee on Environment & Public Works, Hearing Statements". epw.senate.gov. Retrieved 31 August 2012. The amount of climatic warming that has taken place in the past 150 years is poorly constrained, and its cause – human or natural – is unknown. There is no sound scientific basis for predicting future climate change with any degree of certainty. If the climate does warm, it is likely to be beneficial to humanity rather than harmful. In my opinion, it would be foolish to establish national energy policy on the basis of misinformation and irrational hysteria.

[140] Davis, Tony (6 December 2009). "UA prof involved in Climategate replies to critics". Arizona Daily Star. Retrieved 30 October 2014. longtime warming skeptic David Deming

[141] Ivar Giaever (26 June 2011). "De forunderlige klimamytene". Retrieved 17 June 2013. Therefore, it certainly is not likely that the temperature rise is due to CO2, because the correlation is weak.

[142] Jacoby, Jeff (25 September 2011). "Climate skeptics don't 'deny science'". The Boston Globe. Retrieved 28 October 2014. Giaever is only one of many distinguished scientists who dissent from the alarmist view on climate change

[143] Gray, Vincent R. (April 2008). "The Global Warming Scam" (PDF). Retrieved 13 February 2014.

[144] Barton, Chris (4 Nov 2006). "It's hype, hysteria and hot air says climate change nay-sayers". The New Zealand Herald. Retrieved 28 October 2014.

[145] Idso, Craig D., Idso, Keith E. (1998). "Carbon Dioxide and Global Warming". *CO2science.org*. Archived from the original on February 24, 2007. Retrieved 16 March 2014. ...there is no compelling reason to believe that the rise in temperature was caused by the rise in CO2. Furthermore, it is highly unlikely that future increases in the air's CO2 content will produce any global warming; for there are numerous problems with the popular hypothesis that links the two phenomena.

[146] "U.S. House of Representatives Joint Hearing Before the Subcommittee on National Economic Growth, Natural Resources, and Regulatory Affairs of the Committee on Government Reform and the Subcommittee on Energy and the Environment of the Committee on Science: Is CO2 a Pollutant and does the EPA Have the Power to Regulate It?" (PDF). United States Government Printing Office. 6 October 1999. Retrieved 1 November 2014.

[147] "ZENIT - Global Warming Natural, Says Expert". *zenit.org*. 2007-04-27. Retrieved August 31, 2012. it is not possible to exclude the idea that climate changes can be due to natural causes

[148] Solomon, Lawrence (22 August 2013). "Lawrence Solomon: Model mockery?". National Post. Retrieved 28 October 2014. climate change skeptics such as... Antonino Zichichi

[149] Goklany, Indur M., "A Climate Policy for the Short and Medium Term: Stabilization or Adaptation?", *Energy & Environment*, 16: 667-680 (2005).

[150] Goklany, Indur M., "Is Climate Change the "Defining Challenge of Our Age?"", *Energy & Environment*, 20(3): 279-302 (2009).

[151] Goklany, Indur M., "Discounting the Future", *Regulation*, 32: 36-40 (Spring 2009).

[152] Craig Idso. "A Science- Based Rebuttal to the Testimony of Al Gore before the United States Senate Environment & Public Works Committee" (PDF). Center for the Study of Carbon Dioxide and Global Change. Retrieved 26 August 2012. The rising CO_2 content of the air should boost global plant productivity dramatically, enabling humanity to increase food, fiber and timber production and thereby continue to feed, clothe, and provide shelter for their still-increasing numbers ... this atmospheric CO_2-derived blessing is as sure as death and taxes.

[153] Carpenter, Zoë (10 April 2014). "This Sham Report Is What the Climate Movement Is Up Against?". The Nation. Retrieved 28 October 2014. climate change skeptic Craig Idso

[154] Sherwood B. Idso, Craig D. Idso and Keith E. Idso (November 2003). "Enhanced or Impaired? Human Health in a CO2-Enriched Warmer World" (PDF). Center for the Study of Carbon Dioxide and Global Change. p. 30. Retrieved 26 August 2012. [W]arming has been shown to positively impact human health, while atmospheric CO_2 enrichment has been shown to enhance the health-promoting properties of the food we eat, as well as stimulate the production of more of it. ... [W]e have nothing to fear from increasing concentrations of atmospheric CO_2 and global warming.

[155] Gelbspan, Ross (March 22, 2001). "Bush's Global Warmers". The Nation. Retrieved 28 October 2014. Sherwood Idso, a longtime coal-sponsored global warming skeptic

[156] Michaels, Patrick (October 16, 2003). "Posturing and Reality on Warming". CATO Institute. Retrieved June 10, 2009. Scientists know quite precisely how much the planet will warm in the foreseeable future, a modest three-quarters of a degree (Celsius), plus or minus a mere quarter-degree ... a modest warming is a likely benefit... human warming will be strongest and most obvious in very cold and dry air, such as in Siberia and northwestern North America in the dead of winter.

[157] Gillis, Justin (10 February 2014). "Freezing Out the Bigger Picture". New York Times. Retrieved 28 October 2014. Patrick J. Michaels, a climate skeptic at the Cato Institute

[158] *Auer explains why he backs climate science coalition*, New Zealand Press Association, April 30, 2006, the global warming argument, particularly with all the disastrous consequences that are being promulgated ... this is all a non-sustainable argument. In other words the facts will, in time, prove them to be wrong

[159] "Wisconsin's Energy Cooperative". May 2007. Retrieved September 2012. It's absurd. Of course [temperature's] going up. It has gone up since the early 1800s, before the Industrial Revolution, because we're coming out of the Little Ice Age, not because we're putting more carbon dioxide into the air.

[160] Oreskes, Naomi; Conway, Erik M. (2010). *Merchants of doubt : how a handful of scientists obscured the truth on issues from tobacco smoke to global warming* (1st U.S. ed.). New York: Bloomsbury Press. ISBN 9781596916104.

[161] Seitz, F. and Jastrow, R. (Dec 2001) Retrieved July 16, 2010 Do people cause global warming?

[162] Lewis: My Resignation From The American Physical Society, 08 October 2010, GWPF site (archived copy accessed 12/30/13)

[163] Seitz, Frederick (1 December 2001). "Do people cause global warming?". Heartland Institute Environment News. Retrieved 25 August 2012. "So we see that the scientific facts indicate that all the temperature changes observed in the last 100 years were largely natural changes and were not caused by carbon dioxide produced in human activities.

[164] WG1. "Chap 10, Executive Summary". IPCC.

2.9.9   Further reading

- Boykoff, Maxwell (2009). "Ch. 39: Carbonundrums: The Role of the Media: Contemporary Media Courtesans: Climate Contrarians". In Schneider, Stephen H.; Rosencranz, Armin; Mastrandrea, Michael D.; et al. *Climate change science and policy*. Island Press. p. 401. ISBN 978-1-59726-567-6.

- Fleming, James Rodger (2005). *Historical Perspectives on Climate Change*. Oxford University Press. ISBN 978-0-19-518973-5.

- Oreskes, Naomi; Conway, Erik M. (2010). "The Denial of Global Warming". *Merchants of Doubt*. Bloomsbury. pp. 169–215. ISBN 978-1-59691-610-4.

- Solomon, Lawrence (2010). *The Deniers*. Richard Vigilante Books. ISBN 978-0-9800763-7-0.

- Powell, James Lawrence (2011). "The Scientist Deniers". *The Inquisition of Climate Science*. Columbia University Press. ISBN 978-0-231-15718-6.

- Dunlap, Riley E.; McCright, Aaron M. (2011). "Ch. 10: Organized Climate Change Denial: 2.4 Contrarian Scientists". In Dryzek, John S.; Norgaard, Richard B.; Schlosber, David. *The Oxford Handbook of Climate Change and Society*. Oxford University Press. p. 151. ISBN 978-0-19-956660-0.

2.9.10   External links

- Intergovernmental Panel on Climate Change (IPCC) — Consensus, mainstream assessment of climate change

- Nongovernmental International Panel on Climate Change (NIPCC) — Compilation of some non-consensus views and study references

2.10   Media coverage of climate change

Media coverage of climate change has had effects on public opinion on climate change,[1] as it mediates the scientific opinion on climate change that the global instrumental temperature record shows increase in recent decades and that the trend is caused mainly by human-induced emissions of greenhouse gases. Almost all scientific bodies of national or international standing agree with this view,[2][3] although a few organisations hold non-committal positions.

The way the media report on climate change in the English-speaking media, especially in the United States, has been widely studied, while studies of reporting in other countries have been fewer.[4] A number of studies have shown that particularly in the United States and in the UK tabloid press, the media significantly understated the strength of scientific consensus on climate change established in IPCC Assessment Reports in 1995 and in 2001.

A peak in media coverage occurred in early 2007, driven by the IPCC Fourth Assessment Report and Al Gore's documentary *An Inconvenient Truth*.[5] A subsequent peak in late 2009, which was 50% higher,[6] may have been driven by a combination of the November 2009 Climatic Research Unit email controversy and December 2009 United Nations Climate Change Conference.[5]

Some researchers and journalists believe that media coverage of political issues is adequate and fair, while a few feel that it is biased (see, for example, Bozel & Baker, 1990; Lichter & Rothman, 1984).[7][8] However, most studies on media coverage of the topic are neither recent nor concerned with coverage of environmental issues. Moreover, they are only rarely concerned specifically with the question of bias (cf., Bell, 1994; Trumbo, 1996; Wilkins, 1993).[9][10]

2.10.1 Factual distortions

Bord et al. claim that a substantial portion of the United States public has a flawed understanding of global warming, seeing it as linked to general "pollution" and causally connected in some way to atmospheric ozone depletion.[11] News reporters have been labeled by some scientists as ignorant about the science of climate change. Scientists and media scholars who express frustrations with inadequate science reporting[12][13][14][15][16][17] argue that it can lead to at least three basic distortions. First, journalists distort reality by making scientific errors. Second, they distort by keying on human-interest stories rather than scientific content. And third, journalists distort by rigid adherence to the construct of balanced coverage. Bord, O'Connor, & Fisher (2000)[18] argue that responsible citizenry necessitates a concrete knowledge of causes and that until, for example, the public understands what causes climate change it cannot be expected to take voluntary action to mitigate its effects.

In 2015, Media notice of adjustments to historical temperature raw-data, use-of algorithms where data was unavailable, and omitted data produced headlines calling global warming "sciences biggest scandal".[19]

2.10.2 Narrative distortions

Journalists are attracted to risk controversies, what interests them are not the intellectual arguments so much as the underlying human-interest drama.[20] When a group of parents believes their children are at risk from some "agent" in the environment, those parents become scared and angry. Their predicament, as journalists know, grabs the audience's attention. As readers and viewers are drawn into a conflict, and come to feel as though they know the protagonists, they are motivated to learn more and more about the issues that are important to the main characters.

Human-interest controversies that pit "innocent victim" against "alleged perpetrator" are a popular story type. According to Shoemaker and Reese,[21] controversy is one of the main variables affecting story choice among news editors,along with human interest, prominence, timeliness, celebrity, and proximity. But controversy raises editorial issues, such as, what is the fairest way to report such hotly disputed versions of reality? The culture of political journalism has long used the notion of balanced coverage. In this construct, it is permissible to air a highly partisan opinion, provided this view is accompanied by a competing opinion. But recently scientists and scholars have challenged the legitimacy of this journalistic core value.

2.10.3 Distortions of balance

The notion of balanced coverage may make perfect sense when covering a political convention, but in the culture of science, balancing opposing views may be neither fair nor truthful. To quote climate scientist Stephen Schneider (Schneider, 2005): "In science, it's different." Extreme examples bring this point home. Does a flat-Earth proponent deserve equal time to a modern astrophysicist? Surely not. Following this logic, some experts argue that it is misleading to give scientific mavericks or advocates equal time with established mainstream scientists.

Yet there is evidence that this is exactly what the media is doing. In a survey of 636 articles from four top United States newspapers between 1988 and 2002, two scholars[22] (M.T. Boykoff & J.M. Boykoff, 2004) found that most articles gave as much time to the small group of climate change doubters as to the scientific consensus view. Given the real consensus among climatologists over global warming, many scientists find the media's desire to portray the topic as a scientific controversy to be a gross distortion. As Stephen Schneider put it:[15]

> "a mainstream, well-established consensus may be 'balanced' against the opposing views of a few extremists, and to the uninformed, each position seems equally credible."

The subgenre of science journalism concerns itself with gathering and evaluating various types of relevant evidence and rigorously checking sources and facts. As Boyce Rensberger,[23] the director of the Massachusetts Institute of Technology (MIT) Knight Center for Science Journalism, put it "balanced coverage of science does not mean giving equal weight to both sides of an argument. It means apportioning weight according to the balance of evidence."

The claims of scientists also get distorted by the media by a tendency to seek out extreme views, which can result in portrayal of risks well beyond the claims actually being made by scientists.[24] Journalists tend to overemphasize the most extreme outcomes from a range of possibilities reported in scientific articles. A study that tracked press reports about a climate change article in the journal *Nature* found that "results and conclusions of the study were widely misrepresented, especially in the news media, to make the consequences seem more catastrophic and the timescale shorter."[25]

Powers of perception

Where the risk of global warming lacks traction, a serious global warming catastrophe—such as a succession of years with super storms or a large sea level rise that drowned a United States city—would change perceptions, alter media traction, and influence public opinion. Just as the disaster at Chernobyl offered an enduring example of the realities of a reactor accident, a global warming catastrophe could offer a striking image of the dangers of affecting the climate.

Unlike advocates, journalists are not supposed to persuade but to report. It is inappropriate for them to use these insights to manipulate their audience but, in order to make their stories relevant narratives include not just the facts, but also how people feel about the risks presented and why. In essence, they report two dimensions of the risk story—the physical narrative of global warming, and the psychological subtext that discusses how the public thinks about those risks. Journalists, of course, should avoid distorting the science, but getting to the heart of risk tales involves something more: in sum it requires not only understanding the objective facts of the danger, but also navigating the way their audience feels about the risk issue, while telling an accurate story.

Claims of alarmism

Alarmism is described as the use of a linguistic repertoire which communicates climate change using inflated language, an urgent tone and imagery of doom. In a report produced for the Institute for Public Policy Research Gill Ereaut and Nat Segnit reported that alarmist language is frequently employed by newspapers, popular magazine and in campaign literature put out by government and environment groups.[26]

The term *alarmist* can be used as a pejorative by critics of mainstream climate science to describe those that endorse it. MIT meteorologist Kerry Emanuel wrote that labeling someone as an "alarmist" is "a particularly infantile smear considering what is at stake." He continued that using this "inflammatory terminology has a distinctly Orwellian flavor."[27]

Instead of motivating people to action, using sensational and alarming techniques often evoke "denial, paralysis [or] apathy"[28] and do not motivate people to become engaged with the issue of climate change.[29] In the context of climate refugees — the potential for climate change to displace people—it has been reported that "alarmist hyperbole" is frequently employed by private military contractors and think tanks.[30]

One way media reports attack global warming is to compare it with a purported episode of alarmism related to global cooling. In the 1970s, global cooling, a claim with limited scientific support (even during the height of the media frenzy over global cooling, "the possibility of anthropogenic warming dominated the peer-reviewed literature"[31]) was widely reported in the press. Several media pieces have claimed that, since the even-at-the-time-poorly-supported theory of global cooling was shown to be false, that the well-supported theory of global warming can also be dismissed. For example, an article in The Hindu by Kapista and Bashkirtsev wrote: "Who remembers today, they query, that in the 1970s, when global temperatures began to dip, many warned that we faced a new ice age? An editorial in The Time magazine on June 24, 1974, quoted concerned scientists as voicing alarm over the atmosphere 'growing gradually cooler for the past three decades', 'the unexpected persistence and thickness of pack ice in the waters around Iceland,' and other harbingers of an ice age that could prove 'catastrophic.' Man was blamed for global cooling as he is blamed today for global warming".,[32] and Irish Independent published an article claiming that "The widespread alarm over global warming is only the latest scare about the environment to come our way since the 1960s. Let's go through some of them. Almost exactly 30 years ago the world was in another panic about climate change. However, it wasn't the thought of global warming that concerned us. It was the fear of its opposite, global cooling. The doom-sayers were wrong in the past and it's entirely possible they're wrong this time as well."[33] Numerous other examples exist.[34][35][36]

2.10.4   Coverage by country

Australia

See also: Climate change in Australia

Canada

Further information: Environmental policy of the Harper government § Media coverage of climate change

Sweden

See also: Climate change in Sweden

Japan

See also: Climate change in Japan

In Japan, a study of newspaper coverage of climate change from January 1998 to July 2007 found coverage increased dramatically from January 2007.[37]

India

See also: Climate change in India

A 2010 study of four major, national circulation English-language newspapers in India examined "the frames through which climate change is represented in India", and found that "The results strongly contrast with previous studies from developed countries; by framing climate change along a 'risk-responsibility divide', the Indian national press set up a strongly nationalistic position on climate change that divides the issue along both developmental and postcolonial lines."[38]

On the other hand, a qualitative analysis of some mainstream Indian newspapers (particularly opinion and editorial pieces) during the release of the IPCC 4th Assessment Report and during the Nobel Peace Prize win by Al Gore and the IPCC found that Indian media strongly pursue scientific certainty in their coverage of climate change. This is in contrast to the scepticism displayed by American newspapers at the time. Indian media highlights energy challenges, social progress, public accountability and looming disaster.[39]

New Zealand

See also: Climate change in New Zealand

A six-month study in 1988 on climate change reporting in the media found that 80% of stories were no worse than slightly inaccurate. However, one story in six contained significant misreporting.[40] Al Gore's film *An Inconvenient Truth* in conjunction with the Stern Review generated an increase in media interest in 2006.

The popular media in New Zealand often give equal weight to the those supporting anthropogenic climate change and those who deny it. This stance is out of step with the findings of the scientific community where the vast majority support the climate change scenarios. A survey carried out in 2007 on climate change gave the following responses:[41]

United Kingdom

See also: Climate change in the United Kingdom

A study of the UK tabloid press (*The Sun*, *Daily Express*, *Daily Mail*, *Daily Mirror* and their Sunday equivalents) covering the years 2000 to 2006 found that "UK tabloid coverage significantly diverged from the scientific consensus that humans contribute to climate change. Moreover, there was no consistent increase in the percentage of accurate coverage throughout the period of analysis and across all tabloid newspapers, and these findings are not consistent with recent trends documented in United States and UK 'prestige press' or broadsheet newspaper reporting. Findings from interviews indicate that inaccurate reporting may be linked to the lack of specialist journalists in the tabloid press."[42] Another study of the same dataset found that "news articles on climate change were predominantly framed through weather events, charismatic megafauna and the movements of political actors and rhetoric, while few stories focused on climate justice and risk. In addition, headlines with tones of fear, misery and doom were most prevalent."[43]

A two-year study of media coverage of climate change feedback loops found that "non-US news organizations, especially in the UK, are at the forefront of the discourse on climate feedback loops. Poor US press coverage on such climate thresholds might be understood not only as self-censorship, but as a "false negative" error."[1]

A 2010 study looked at "prominent, disruptive direct action around the climate change issue, in the context of comparable activity across a range of political groupings" and found that "they garner significant but unflattering attention from [the conventional mass media], partly as a consequence of the persistent pressures and imperatives that drive conventional journalism."[44]

United States

See also: Climate change in the United States and Propaganda model § Applications

According to Peter J. Jacques et al., the mainstream news media of the United States is an example of the effectiveness of environmental skepticism as a tactic.[45] A 2005 study reviewed and analyzed the US mass-media coverage of the environmental issue of climate change from 1988 to 2004. The authors confirm that within the journalism industry there is great emphasis on eliminating the presence of media bias. In their study they found that — due to this practice of objectivity — "Over a 15-year period, a majority (52.7%) of prestige-press articles featured balanced accounts that gave 'roughly equal attention' to the views that humans were contributing to global warming and that exclusively natural fluctuations could explain the earth's temperature increase." As a result, they observed that it is easier for people to conclude that the issue of global warming and the accompanying scientific evidence is still hotly debated.[46]

A study of US newspapers and television news from 1995 to 2006 examined "how and why US media have represented conflict and contentions, despite an emergent consensus view regarding anthropogenic climate science." The IPCC Assessment Reports in 1995 and in 2001 established an increasingly strong scientific consensus, yet the media continued to present the science as contentious. The study noted the influence of Michael Crichton's 2004 novel *State of Fear*, which "empowered movements across scale, from individual perceptions to the perspectives of US federal powerbrokers regarding human contribution to climate change."[47]

A 2010 study concluded that "Mass media in the U.S. continue to suggest that scientific consensus estimates of global climate disruption, such as those from the Intergovernmental Panel on Climate Change (IPCC), are 'exaggerated' and overly pessimistic. By contrast, work on the Asymmetry of Scientific Challenge (ASC) suggests that such consensus assessments are likely to understate climate disruptions. ... new scientific findings were more than twenty times as likely to support the ASC perspective than the usual framing of the issue in the U.S. mass media. The findings indicate that supposed challenges to the scientific consensus on global warming need to be subjected to greater scrutiny, as well as showing that, if reporters wish to discuss "both sides" of the climate issue, the scientifically legitimate 'other side' is that, if anything, global climate disruption may prove to be significantly worse than has been suggested in scientific consensus estimates to date."[48]

Gallup's annual update on Americans; attitudes toward the environment shows a public that over the last two years has become less worried about the threat of global warming, less convinced that its effects are already happening, and more likely to believe that scientist themselves are uncertain about it occurrence. In response to one key question, 48% of Americans now believe that the seriousness of global warming is generally exaggerated, up from 41% in 2009 and 31% in 1997, when Gallup first asked the question.[49]

According to Moti Nissani, when the devastating effects of El Niño were reported, the likelihood that El Niño itself is caused by global warming was either whispered in passing, and always attached to an emphatic question mark, or flatly denied.[50]

On August 12, 1997, the New York Times promised its readers: "Between now and December, when representatives of most nations will meet in Japan to discuss limits on greenhouse gases, The Times will examine the science, politics and economics of climate change." [NYT 8/12].[51]

2.10.5 Media, politics and public opinion

As McCombs et al.'s 1972 study of the political function of mass media first showed, media coverage of an issue can "play an important part in shaping political reality".[52] Research into media coverage of climate change has demonstrated the significant role of the media in determining climate policy formation.[53] The media has considerable bearing on public opinion, and the way in which issues are reported, or framed, establishes a particular discourse.[54] Discourse, broadly defined, is a linguistic or communicative regularity, which creates particular norms and determines the way we understand an issue, and "help[s] shape institutional considerations of policy".[55]

Media-policy interface

The relationship between media and politics is reflexive (see:reflexivity (social theory)). As Feindt & Oels neatly state, "[media] discourse has material and power effects as well as being the effect of material practices and power relations".[56] Public support of climate change research ultimately decides whether or not funding for the research is made available to scientists and institutions.

As highlighted above, media coverage in the United States during the Bush Administration often emphasised and exaggerated scientific uncertainty over climate change, reflecting the interests of the political elite.[55] Hall et al. suggest that government and corporate officials enjoy privileged access to the media, so their line quickly becomes the 'primary definer' of an issue.[57] Furthermore, media sources and their institutions very often have political leanings which determine their reporting on climate change, mirroring the views of a particular party.[58] However, media also has the capacity to challenge political norms and expose corrupt behaviour,[59] as demonstrated in 2007 when *The Guardian* revealed that American Enterprise Institute received $10,000 from petrochemical giant Exxon Mobil to publish articles undermining the IPCC's 4th assessment report.

Ever-strengthening scientific consensus on climate change means that scepticism is becoming less prevalent in the media (although the email scandal in the build up to Copenhagen reinvigorated climate scepticism in the media[60]), however in terms of weighting impacts and positing responses, climate change remains a discursive battleground.

Discourses of action - 'Creating a Climate for Change'[61]

Commentators have argued that the climate change discourses constructed in the media have not been conducive to generating the political will for swift action. The polar bear has become a powerful discursive symbol in the fight against climate change. However, such images, it is argued, create a perception of climate change impacts as geographically distant,[62] and MacNagton argues that climate change needs to be framed as an issue 'closer to home'.[63] On the other hand, Beck suggests that a major benefit of global media is that it brings distant issues within our consciousness.[64]

Furthermore, media coverage of climate change (particularly in tabloid journalism but also more generally), is concentrated around extreme weather events and projections of catastrophe, creating "a language of imminent terror"[65] which some commentators argue has instilled policy-paralysis and inhibited response. Moser et al. suggest using solution-orientated frames will help inspire action to solve climate change.[61] The predominance of catastrophe frames over solution frames[66] may help explain the apparent value-action gap with climate change; the current discursive setting has generated concern over climate change but not inspired action.

The Polar Bear has become a powerful symbol for those attempting to generate support for addressing climate change

Compared to what experts know about traditional media's and tabloid journalism's impacts on the formation of public perceptions of climate change and willingness to act, there is comparatively little knowledge of the impacts of social media, including message platforms like Twitter, on public attitudes toward climate change.[67]

2.10.6   References

[1] Antilla, L. (2010). "Self-censorship and science: A geographical review of media coverage of climate tipping points". *Public Understanding of Science* **19** (2): 240–256. doi:10.1177/0963662508094099.

[2] Julie Brigham-Grette; et al. (September 2006). "Petroleum Geologists' Award to Novelist Crichton Is Inappropriate" (PDF). *Eos* **87** (36). The AAPG stands alone among scientific societies in its denial of human-induced effects on global warming.

[3] DiMento, Joseph F. C.; Doughman, Pamela M. (2007). *Climate Change: What It Means for Us, Our Children, and Our Grandchildren*. MIT Press. p. 68. ISBN 978-0-262-54193-0.

[4] Lyytimäki, J., Tapio, P. (2009). "Climate change as reported in the press of Finland: From screaming headlines to penetrating background noise". *International Journal of Environmental Studies* **66** (6): 723–735. doi:10.1080/00207230903448490.

[5] Boykoff, M. (2010). "Indian media representations of climate change in a threatened journalistic ecosystem" (PDF). *Climatic Change* **99** (1): 17–25. doi:10.1007/s10584-010-9807-8.

[6] "2004–2010 World Newspaper Coverage of Climate Change or Global Warming". *Center for Science and Technology Policy Research*. University of Colorado at Boulder.

[7] Lichter, S.R.; Rothman (1984). "The media and national defense". *National Security Policy*: 265–282.

[8] Bozell, L.B.; Baker, B.H. (1990). "Thats the way it is(n't)". *Alexandria, VA.*

[9] Bell, A (1994). "Media (mis)communication on the science of Climate change.". *Public Understanding of Science*: 3, 259–275.

[10] Trumbo, C. (1996). "Constructing climate change: Claims and frames in US news coverage of an environmental issue". *Public Understanding of Science* (5): 269–283.

[11] Bord et al. 1998

[12] Boykoff, M.T.; Boykoff, J.M. (2004). "Balance as bias: Global warming and the US prestige press". *Global Environmental Change* **14**: 125–136. doi:10.1016/j.gloenvcha.2003.10.001.

[13] Moore, B; Singletary, M. (1985). "Scientific sources' perceptions of network news accuracy". *Journalism Quarterly* (Association for Education in Journalism and Mass Communication) **62** (4): 816–823. doi:10.1177/107769908506200415.

[14] Nelkin, D (1995). "Selling science: How the press covers science and technology". *New York: W.H. Freeman.*

[15] Schneider, S. "Mediarology: The role of citizens, journalists, and scientists in debunking climate change myths". Retrieved 2011-04-03.

[16] Singer, E., & Endreny, P. M. (1993). *Reporting on risk: How the mass media portray accidents, diseases, disasters and other hazards.* New York: Russell Sage.

[17] Tankard, J. W.,; Ryan, M. (1974). "News source perceptions of accuracy in science coverage". *Journalism Quarterly* **51**: 219–225. doi:10.1177/107769907405100204.

[18] Bord, R.J.; O'Connor; Fisher (1998). "Public perceptions of global warming: United States and international perspectives". *Climate Research* **11** (1): 75–84. doi:10.3354/cr011075.

[19] "Global warming: Science's biggest scandal". the official temperature records were systematically adjusted to show the earth as having warmed much more than the actual data justified

[20] Mazur, A.; Lee, J. (1993). "Sounding the global alarm: Environmental issues in the U.S. national news". *Social Studies of Science* **23**: 681–720. doi:10.1177/030631293023004003.

[21] Shoemaker, P. J., & Reese, S. D. (1996). *Mediating the message: Theories of influence on mass media content.* New York: Longman. p. 261.

[22] Boykoff, M.T.; Boykoff, J.M. (2004). "Balance as bias: Global warming and the US prestige press". *Global Environmental Change* **14**: 125–136. doi:10.1016/j.gloenvcha.2003.10.001.

[23] Rensberger, B (2002). "Reporting Science Means Looking for Cautionary Signals". *Nieman Reports Special Issue on Science Journalism,*: 12–14.

[24] Boykoff, Maxwell T. (2009). "We Speak for the Trees: Media Reporting on the Environment". *Annual Review of Environment and Resources* **34** (1): 431–457. doi:10.1146/annurev.environ.051308.084254.

[25] Ladle, R. J.; Jepson, P.; Whittaker, R. J. (2005). "Scientists and the media: the struggle for legitimacy in climate change and conservation science". *Interdisciplinary Science Reviews* **30** (3): 231. doi:10.1179/030801805X42036.

[26] Ereaut, Gill; Segrit, Nat (2006). "Warm Words: How are we Telling the Climate Story and can we Tell it Better?" (PDF). London: Institute for Public Policy Research.

[27] "Climategate": A Different Perspective, by Kerry Emanuel, National Association of Scholars, July 19, 2010

[28] Lisa Dilling; Susanne C. Moser (2007). "Introduction". *Creating a climate for change: communicating climate change and facilitating social change.* Cambridge, UK: Cambridge University Press. pp. 1–27. ISBN 0-521-86923-4.

[29] O'Neill, S.; Nicholson-Cole, S. (2009). ""Fear Won't Do It": Promoting Positive Engagement with Climate Change Through Visual and Iconic Representations". *Science Communication* **30** (3): 355. doi:10.1177/1075547008329201.

[30] Hartmann, Betsy (2010). "Rethinking climate refugees and climate conflict: Rhetoric, reality and the politics of policy discourse". *Journal of International Development* **22** (2): 233–246. doi:10.1002/jid.1676. ISSN 0954-1748.

[31] Peterson, Thomas & Connolley, William & Fleck, John (September 2008). "The Myth of the 1970s Global Cooling Scientific Consensus" (PDF). *Bulletin of the American Meteorological Society* (American Meteorological Society) **89** (9): 1325–1337. Bibcode:2008BAMS...89.1325P. doi:10.1175/2008BAMS2370.1.

[32] Kapitsa, Andrei, and Vladimir Bashkirtsev, "Challenging the basis of Kyoto Protocol", *The Hindu*, 10 July 2008,

[33] *Irish Independent*, "Don't believe doomsayers that insist the world's end is nigh", 16 March 2007, p. 1.

[34] Schmidt, David, "It's curtains for global warming", *Jerusalem Post*, 28 June 2002, p. 16B. "If there is one thing more remarkable than the level of alarm inspired by global warming, it is the thin empirical foundations upon which the forecast rests. Throughout the 1970s, the scientific consensus held that the world was entering a period of global cooling, with results equally catastrophic to those now predicted for global warming."

[35] Wilson, Francis, "The rise of the extreme killers", *Sunday Times*, 19 April 2009, p. 32. "Throughout history there have been false alarms: "shadow of the bomb", "nuclear winter", "ice age cometh" and so on. So it's no surprise that today many people are skeptical about climate change. The difference is that we have hard evidence that increasing temperatures will lead to a significant risk of dangerous repercussions."

[36] *National Post*, "The sky was supposed to fall: The '70s saw the rise of environmental Chicken Littles of every shape as a technique for motivating public action", 5 April 2000, p. B1. "One of the strange tendencies of modern life, however, has been the institutionalization of scaremongering, the willingness of the mass media and government to lend plausibility to wild surmises about the future. The crucial decade for this odd development was the 1970s. Schneider's book excited a frenzy of glacier hysteria. The most-quoted ice-age alarmist of the 1970s became, in a neat public-relations pivot, one of the most quoted global-warming alarmists of the 1990s."

[37] Sampei, Y., Aoyagi-Usui, M. (2009). "Mass-media coverage, its influence on public awareness of climate-change issues, and implications for Japan's national campaign to reduce greenhouse gas emissions". *Global Environmental Change* **19** (2): 203–212. doi:10.1016/j.gloenvcha.2008.10.005.

[38] Billett, Simon (2010). "Dividing climate change: global warming in the Indian mass media". *Climatic Change* **99** (1–2): 1–16. doi:10.1007/s10584-009-9605-3.

[39] Mittal, Radhika (2012). "Climate Change Coverage in Indian Print Media: A Discourse Analysis". *The International Journal of Climate Change: Impacts and Responses* **3** (2): 219–230.

[40] Bell, Allan (1994). "Media (mis)communication on the science of climate change". *Public Understanding of Science (journal)* (IOP Publishing) (3): 259–275.

[41] ShapeNZ research report. 13 April 2007, *New Zealanders' views on climate change and related policy options*

[42] Boykoff, M.T., Mansfield, M. (2008). "'Ye Olde Hot Aire': Reporting on human contributions to climate change in the UK tabloid press". *Environmental Research Letters* (PDF) **3** (2): 024002. Bibcode:2008ERL.....3d4002B. doi:10.1088/1748-9326/3/2/024002.

[43] Boykoff, M.T. (2008). "The cultural politics of climate change discourse in UK tabloids". *Political Geography* **27** (5): 549–569. doi:10.1016/j.polgeo.2008.05.002.

[44] Gavin, N.T. (2010). "Pressure group direct action on climate change: The role of the media and the web in Britain — a case study". *British Journal of Politics and International Relations* **12** (3): 459–475. doi:10.1111/j.1467-856X.2010.00411.x.

[45] Environmental skepticism is "a tactic of an elite-driven counter-movement designed to combat environmentalism, and ... the successful use of this tactic has contributed to the weakening of US commitment to environmental protection." — Jacques, P.J.; Dunlap, R.E.; Freeman, M. (June 2008). "The organization of denial: Conservative think tanks and environmental skepticism". *Environmental Politics* **17** (3): 349–385. doi:10.1080/09644010802055576.

[46] Boykoff, M. T.; Boykoff, J. M. (2007). "Climate change and journalistic norms: A case-study of US mass-media coverage" (PDF). *Geoforum* **28** (6): 1190–1204. doi:10.1016/j.geoforum.2007.01.008. Retrieved 2009-10-15.

[47] Boykoff, M.T. (2007). "From convergence to contention: United States mass media representations of anthropogenic climate change science". *Transactions of the Institute of British Geographers* **32** (4): 477–489. doi:10.1111/j.1475-5661.2007.00270.x. CiteSeerX: 10.1.1.132.9906.

[48] Freudenburg, W.R., Muselli, V. (2010). "Global warming estimates, media expectations, and the asymmetry of scientific challenge". *Global Environmental Change* **20** (3): 483–491. doi:10.1016/j.gloenvcha.2010.04.003.

[49] Newport, Frank (11 March 2010). "Americans'Global Warming Concerns Continue to Drop: Multiple indicators show less concern, more feelings that global warming is exaggerated". *Gallup Poll News Service.*

[50] Nissani, Moti (Sep 1999). "Media Coverage of the Green House Effect". *Population and Environment* **21** (1): 27–43, 36. doi:10.1007/BF02436119.

[51] *New York Times.* 1997-12-08. Missing or empty |title= (help)

[52] McCombs, M; Shaw, D. (1972). "The Agenda Setting Function of Mass Media". *Public Opinion Quarterly* **36** (2): 176–187. doi:10.1086/267990.

[53] Boykoff, M (2007). "Flogging a Dead Norm? Newspaper Coverage of Anthropogenic Climate Change in the United States and United Kingdom from 2003-2006". *Area* **39** (2): 000–000, 200. doi:10.1111/j.1475-4762.2007.00769.x.

[54] Hajer, M; Versteeg, W (2005). "A Decade of Discourse Analysis of Environmental Politics: Achievements, Challenges, Perspectives". *Journal of Environmental Policy & Planning* **7** (3): 175–184. doi:10.1080/15239080500339646.

[55] Boykoff, M (2007). "Flogging a Dead Norm? Newspaper Coverage of Anthropogenic Climate Change in the United States and United Kingdom from 2003-2006". *Area* **39** (2): 000–000, 200. doi:10.1111/j.1475-4762.2007.00769.x.

[56] Feindt, P; Oels, A (2005). "Does Discourse Matter? Discourse Analysis in Environmental Policy Making". *Journal of Environmental Policy and Planning* **7** (3): 161–173. doi:10.1080/15239080500339638.

[57] Hall, S; et al. (1978). *Policing the Crisis - Mugging, the State, and Law and Order.* New York: Holmes and Meier. p. 438.

[58] Carvalho, A; Burgess, J (December 2005). "Cultural Circuits of Climate Change in UK Broadsheet Newspapers". *Risk Analysis* **25** (6): 1457–1469. doi:10.1111/j.1539-6924.2005.00692.x.

[59] Anderson, A (2009). "Media, Politics and Climate Change: Towards a New Research Agenda". *Sociology Compass* **3** (2): 166–182. doi:10.1111/j.1751-9020.2008.00188.x.

[60] Monibot, George (29 April 2009). "The media laps up fake controversy over climate change". London: The Guardian. Retrieved 2011-11-05.

[61] Moser & Dilling, M., and L. (2007). *Creating a Climate for Change.* Cambridge University Press. ISBN 978-0-521-86923-2.

[62] Lorenzoni, I; Pidgeon (2006). "Public Views on Climate Change: European and USA Perspectives". *Climatic Change* **77** (1): 73–95. doi:10.1007/s10584-006-9072-z.

[63] MacNaghten, P (2003). "Embodying the Environmental in Everyday Life Practices". *The Sociological Review* **77** (1).

[64] Beck, U (1992). *Risk Society - Towards a New Modernity.* Frankfurt: Sage. ISBN 978-0-8039-8345-8.

[65] Hulme, M (2009). *Why We Disagree About Climate Change.* Cambridge University Press. p. 432. ISBN 978-0-521-72732-7.

[66] Boykoff, M; Boykoff, J (November 2007). "Climate Change and Journalistic Norms: A case study of US mass-media coverage". *Geoforum* **38** (6): 1190–1204. doi:10.1016/j.geoforum.2007.01.008.

[67] Auer M.; et al. (2014). "The Potential of Microblogs for the Study of Public Perceptions of Climate Change". *WIREs Climate Change.* doi:10.1002/wcc.273.

2.10.7 Further reading

- *The Climate War* (2010) Eric Pooley ISBN 978-1-4013-2326-4

- Michael Specter (2009). Denialism: How Irrational Thinking Hinders Scientific Progress, Harms the Planet, and Threatens Our Lives. Penguin Press HC, The. ISBN 978-1-59420-230-8

- Mike Hulme (2009). *Why we disagree about climate change: understanding controversy, inaction and opportunity.* Cambridge, UK: Cambridge University Press. ISBN 0-521-72732-4.

- Tammy Boyce; Lewis, Justin, ed. (2009). *Climate Change and the Media (Global Crises and the Media).* Peter Lang Publishing. ISBN 1-4331-0460-1.

- Uusi-Rauva, C., Tienari, J. (2010). "On the relative nature of adequate measures: Media representations of the EU energy and climate package". *Global Environmental Change* **20** (3): 492–501. doi:10.1016/j.gloenvcha.2010.03.001.

- Anderson, Alison (March 2009). "Media, Politics and Climate Change: Towards a New Research Agenda". *Sociology Compass* **3** (2): 166–182. doi:10.1111/j.1751-9020.2008.00188.x.

- *Who Speaks for the Climate?: Making Sense of Media Reporting on Climate Change* by Maxwell T. Boykoff, Cambridge University Press; 1 edition (September 30, 2011) ISBN 978-0-521-13305-0

2.10.8   See also

- Global warming controversy
- Climate change denial
- Merchants of Doubt
- The Age of Stupid
- Requiem for a Species

2.11   Merchants of Doubt

This article is about the book. For the film based on the book, see Merchants of Doubt (film).

Merchants of Doubt is a 2010 non-fiction book by American historians of science Naomi Oreskes and Erik M. Conway. It identifies parallels between the global warming controversy and earlier controversies over tobacco smoking, acid rain, DDT, and the hole in the ozone layer. Oreskes and Conway write that in each case "keeping the controversy alive" by spreading doubt and confusion after a scientific consensus had been reached, was the basic strategy of those opposing action.[1] In particular, they say that Fred Seitz, Fred Singer, and a few other contrarian scientists joined forces with conservative think tanks and private corporations to challenge the scientific consensus on many contemporary issues.[2]

The George C. Marshall Institute and Fred Singer, two of the subjects, have been critical of the book, but most reviewers received it favorably. One reviewer said that *Merchants of Doubt* is exhaustively researched and documented, and may be one of the most important books of 2010. Another reviewer saw the book as his choice for best science book of the year.[3] It was made into a film, *Merchants of Doubt*, directed by Robert Kenner, released in 2014.[4]

2.11.1   Themes

Oreskes and Conway write that a handful of politically conservative scientists, with strong ties to particular industries, have "played a disproportionate role in debates about controversial questions".[5] The authors write that this has resulted in "deliberate obfuscation" of the issues which has had an influence on public opinion and policy-making.[5]

The book criticizes the so-called Merchants of Doubt, some predominantly American science key players, above all Bill Nierenberg, Fred Seitz, and Fred Singer. All three are physicists: Singer was a rocket scientist, whereas Nierenberg and Seitz worked on the atomic bomb.[6] They have been active on topics like acid rain, tobacco smoking, global warming and pesticides. The book claims that these scientists have challenged and diluted the scientific consensus in the various fields, as of the dangers of smoking, the effects of acid rain, the existence of the ozone hole, and the existence of anthropogenic climate change.[5] Seitz and Singer been involved with institutions such as The Heritage Foundation, Competitive Enterprise Institute and George C. Marshall Institute in the United States. Funded by corporations and conservative foundations, these organizations have opposed many forms of state intervention or regulation of U.S. citizens. The book lists similar tactics in each case: "discredit the science, disseminate false information, spread confusion, and promote doubt".[7]

The book states that Seitz, Singer, Nierenberg and Robert Jastrow were all fiercely anti-communist and they viewed government regulation as a step towards socialism and communism. The authors argue that, with the collapse of the Soviet

Union, they looked for another great threat to free market capitalism and found it in environmentalism. They feared that an over-reaction to environmental problems would lead to heavy-handed government intervention in the marketplace and intrusion into people's lives.[8] Oreskes and Conway state that the longer the delay the worse these problems get, and the more likely it is that governments will need to take the draconian measures that conservatives and market fundamentalists most fear. They say that Seitz, Singer, Nierenberg and Jastrow denied the scientific evidence, contributed to a strategy of delay, and thereby helped to bring about the situation they most dreaded.[8] The authors have a strong doubt about the ability of the media to differentiate between false truth and the actual science in question, however stop short of endorsing censorship in the name of science.[9] The journalistic norm of balanced reporting has helped, according the authors, to amplify the misleading messages of the contrarians. Oreskes and Conway state: "small numbers of people can have large, negative impacts, especially if they are organised, determined and have access to power".[7]

The main conclusion of the book is that there would have been more progress in policymaking, if not for the influence of the contrarian "experts", which tried on ideological reasons to undermine trust in the science base for regulation.[9] Similar conclusions were already drawn, among others on Frederick Seitz and William Nierenberg in the book *Requiem for a Species: Why We Resist the Truth about Climate Change* (2010) by Australian academic Clive Hamilton.

2.11.2   Reception

Philip Kitcher in *Science* says that Naomi Oreskes and Erik Conway are "two outstanding historians".[5] He calls *Merchants of Doubt* a "fascinating and important study". Kitcher says that the apparently harsh claims against Nierenberg, Seitz, and Singer are "justified through a powerful dissection of the ways in which prominent climate scientists, such as Roger Revelle and Ben Santer, were exploited or viciously attacked in the press".[5]

In *The Christian Science Monitor*, Will Buchanan says that *Merchants of Doubt* is exhaustively researched and documented, and may be one of the most important books of 2010. Oreskes and Conway are seen to demonstrate that the doubt merchants are not "objective scientists" as the term is popularly understood. Instead, they are "science-speaking mercenaries" hired by corporations to process numbers to prove that the corporations' products are safe and useful. Buchanan says they are salesmen, not scientists.[10]

Bud Ward published a review of the book in *The Yale Forum on Climate and the Media*. He wrote that Oreskes and Conway use a combination of thorough scholarly research combined with writing reminiscent of the best investigative journalism, to "unravel deep common links to past environmental and public health controversies".[11] In terms of climate science, the authors' leave "little doubt about their disdain for what they regard as the misuse and abuse of science by a small cabal of scientists they see as largely lacking in requisite climate science expertise".[11]

Phil England writes in *The Ecologist* that the strength of the book is the rigour of the research and the detailed focus on key incidents. He said, however, that the climate change chapter is only 50 pages long, and recommends several other books for readers who want to get a broader picture of this aspect: Jim Hoggan's *Climate Cover-Up*, George Monbiot's *Heat: How to Stop the Planet Burning* and Ross Gelbspan's *The Heat is On* and *Boiling Point*. England also said that there is little coverage about the millions of dollars which Exxon Mobil has put into funding groups actively involved in promoting climate change denial and doubt.[12]

A review in *The Economist* calls this a powerful book which articulates the politics involved and the degree to which scientists have sometimes manufactured and exaggerated environmental uncertainties, but opines that the authors fail to fully explain how environmental action has still often proved possible despite countervailing factors.[13]

Robert N. Proctor, who coined the term "agnotology" to describe the study of culturally induced ignorance or doubt, wrote in *American Scientist* that *Merchants of Doubt* is a detailed and artfully written book. He set it in the context of other books which cover the "history of manufactured ignorance":[14] David Michaels's *Doubt is their Product* (2008), Chris Mooney's *The Republican War on Science* (2009), David Rosner and Gerald Markowitz's *Deceit and Denial* (2002), and his own book *Cancer Wars* (1995).[14]

Robin McKie in *The Guardian* states that Oreskes and Conway deserve considerable praise for exposing the influence of a small group of Cold War ideologues. Their tactic of spreading doubt has confused the public about a series of key scientific issues such as global warming, even though scientists have actually become more certain about their research results. McKie says that *Merchants of Doubt* includes detailed notes on all sources used, is carefully paced, and is "my runaway contender for best science book of the year".[3]

Sociologist Reiner Grundmann's review in *BioSocieties* journal, acknowledges that the book is well researched and factually based, but criticizes the book as being written in a black an white manner whereas historians should write a more nuanced description. The book depicts special interests and contrarians misleading the public as being mainly responsible for stopping action on policy. He says this shows a lack of basic understanding of the political process and the mechanisms of knowledge policy, because the authors assume that public policy would follow on from an understanding of the science. While the book provides *all the (formal) hallmarks of science*, Grundmann sees it less as a scholarly work than a passionate attack and overall as a problematic book.[9]

William O'Keefe and Jeff Kueter from the George C. Marshall Institute, which was founded by Seitz,[15] say that although *Merchants of Doubt* has the appearance of a scholarly work, it discredits and undermines the reputations of people who in their lifetime contributed greatly to the American nation. They say that it does this by questioning their integrity, impugning their character, and questioning their judgement.[16]

2.11.3   Authors

Naomi Oreskes is Professor of History and Science Studies at Harvard University. She has degrees in geological science and a Ph.D. in Geological Research and the History of Science. Her work came to public attention in 2004 with the publication of "The Scientific Consensus on Climate Change," in *Science*, in which she wrote that there was no significant disagreement in the scientific community about the reality of global warming from human causes.[17] Erik M. Conway is the historian at NASA's Jet Propulsion Laboratory at the California Institute of Technology in Pasadena.[17]

2.11.4   See also

- Climate change controversy

- Climate change policy of the United States

- Fear, uncertainty and doubt

- Health effects of tobacco

- List of books about the politics of science

- List of scientists opposing the mainstream scientific assessment of global warming in contrast with Scientific opinion on climate change

- Media coverage of climate change

- Tobacco control movement

- Tobacco politics

- *Climate Capitalism*

2.11.5   References

[1] Steketee, Mike (November 20, 2010). "Some sceptics make it a habit to be wrong". *The Australian*.

[2] Oreskes, Naomi; Conway, Erik M. (2010). *Merchants of Doubt: How a Handful of Scientists Obscured the Truth on Issues from Tobacco Smoke to Global Warming*. Bloomsbury Press. p. 6. ISBN 978-1-59691-610-4. merchantsofdoubt.org

[3] McKie, Robin (August 8, 2010). "Merchants of Doubt by Naomi Oreskes and Erik M Conway". *The Guardian*.

[4] "Merchants of Doubt". Sony Pictures Classics. Retrieved 8 March 2015.

[5] Kitcher, Philip (June 4, 2010). "The Climate Change Debates". *Science* **328** (5983): 1231–2. doi:10.1126/science.1189312.

[6] Brown, Seth (May 31, 2010). "'Merchants of Doubt' delves into contrarian scientists". *USA Today*.

[7] McKie, Robin (August 1, 2010). "A dark ideology is driving those who deny climate change". *The Guardian*.

[8] Oreskes & Conway 2010, pp. 248–255

[9] Debunking sceptical propaganda Book review by Reiner Grundmann, BioSocieties (2013) 8, 370–374. doi:10.1057/biosoc. 2013.15

[10] Buchanan, Will (June 22, 2010). Merchants of Doubt: How "scientific" misinformation campaigns sold untruths to consumers *The Christian Science Monitor*.

[11] Ward, Bud (July 8, 2010). Reviews: Leaving No Doubt on Tobacco, Acid Rain, Climate Change, *The Yale Forum on Climate and the Media*.

[12] England, Phil (September 10, 2010). Merchants of Doubt *The Ecologist*.

[13] All guns blazing: A question of dodgy science, (June 17, 2010), *The Economist*.

[14] Proctor, Robert (September–October 2010). Book Review: Manufactured Ignorance, *American Scientist*.

[15] *"...a central cog in the denial machine..."*, August 13, 2007, Newsweek

[16] O'Keefe, William; Kueter, Jeff (June 2010). "Clouding the Truth: A Critique of Merchants of Doubt" (PDF). *Policy Outlook*. George C. Marshall Institute. Although cloaked in the appearance of scholarly work, the book constitutes an effort to discredit and undermine the reputations of three deceased scientists who contributed greatly to our nation... This book questions their integrity, impugns their character, and questions their judgment on the basis of little more than faulty logic and preconceived opinion

[17] Collins Literary Agency Rights Guide/March 2008

2.11.6 External links

- Merchants of Doubt, Public Lecture (2010), University of NSW, *The Science Show*, ABC Radio National, 8 January 2011.

2.12 National Association of Manufacturers

The **National Association of Manufacturers** (**NAM**) is an advocacy group headquartered in Washington, D.C., United States, with 10 additional offices across the country. It is the nation's largest manufacturing industrial trade association, representing 11,000 small and large manufacturing companies in every industrial sector and in all 50 states.[1]

2.12.1 Policy issues

The NAM's policy issue work is focused in the areas of labor, employment, health care, energy, climate, corporate finance, tax, bilateral trade, multilateral trade, export controls, technology, regulatory and infrastructure policy.[2] According to Bloomberg, Duke Energy did not renew its membership with the NAM partly because of differences over climate policy.[3]

NAM recently partnered with the National Council of La Raza to support legalizing 11 million illegal immigrants and increasing annual immigration numbers "because it provides the skilled workers manufacturers need, and it is simply the right thing to do."[4]

According to NAM, manufacturing employs nearly 12 million workers, contributes more than $1.6 trillion to the U.S. economy annually, is the largest driver of economic growth in the nation and accounts for the lion's share of private sector research and development.[1]

Legislation

NAM supported the EPS Service Parts Act of 2014 (H.R. 5057; 113th Congress), a bill that would exempt certain external power supplies from complying with standards set forth in a final rule published by the United States Department of Energy in February 2014.[5][6] The United States House Committee on Energy and Commerce describes the bill as a bill that "provides regulatory relief by making a simple technical correction to the 2007 Energy Independence and Security Act to exempt certain power supply (EPS) service and spare parts from federal efficiency standards."[7]

2.12.2   Board of Directors

The NAM's Board of Directors includes President Jay Timmons, CEO of NAM; Executive Committee Member Mary Vermeer Andringa, President and CEO, Vermeer Corporation; and Chair of the Board Douglas Oberhelman, CEO, Caterpillar Inc., among others.[8]

2.12.3   History

The National Association of Manufacturers (NAM) "was founded in Cincinnati, Ohio in 1895. The U.S. was in the midst of a deep recession and many of the nation's manufacturers saw a strong need to export their products in other countries. One of the NAM's earliest efforts was to call for the creation of the U.S. Department of Commerce".[9] The organization's first president was Thomas Dolan of Philadelphia [10] (not, as erroneously listed in some sources, Samuel P. Bush).

The early history of NAM was marked by frank verbal attacks on labor. In 1903, then-president David MacLean Parry[11] delivered a speech at its annual convention which argued that unions' goals would result in "despotism, tyranny, and slavery." Parry advocated the establishment of a great national anti-union federation under the control of the NAM, and the NAM responded by initiating such an effort.[12]

In an address at its 1911 convention, NAM president John Kirby, Jr. proclaimed, "The American Federation of Labor is engaged in an open warfare against Jesus Christ and his cause." [13]

The NAM also encouraged the creation and propagation of a network of local anti-union organizations, many of which took the name Citizens' Alliance.[14] In October 1903 the local Citizens' Alliance groups were united by a national called the Citizens' Industrial Alliance of America.[15]

NAM, in the late 1930s, used one of the earliest versions of a modern multi-faceted public relations campaign to promote the benefits of capitalism and to combat the policies of President Roosevelt.[16] NAM made efforts to undermine organized labor in the United States before the New Deal.[17] NAM lobbied successfully for the 1947 Taft-Hartley Act to restrict unions' power.[18]

The advent of commercial television led to the NAM's own 15-minute television program, "Industry on Parade",[19] which aired from 1950–1960.[20]

2.12.4   Affiliates

The NAM has one affiliate. According to its website,[21] the Manufacturing Institute is the 501(c) 3 affiliate of the National Association of Manufacturers. The Manufacturing Institute describes its priorities as the development of a world-class manufacturing workforce, the growth of individual U.S. manufacturing companies and the expansion of the manufacturing sector in regional economies. The Manufacturing Institute is the authority on the attraction, qualification, and development of world-class manufacturing talent.[22]

2.12.5   Footnotes

[1] About the NAM - National Association of Manufacturers

[2] Policy Issues - National Association of Manufacturers

[3] Duke Energy to Leave Trade Group Over Climate Policy

[4] http://www.nam.org/Communications/Articles/2013/08/Manufacturers-Launch-Radio-Ads-in-Support-of-Immigration-Reform. aspx#

[5] "CBO - H.R. 5057". Congressional Budget Office. Retrieved 9 September 2014.

[6] Hankin, Christopher (15 July 2014). "House Energy & Commerce Committee passes bipartisan regulatory relief for external power supplies". Information Technology Industry Council. Retrieved 10 September 2014.

[7] "Committee to Build on #RecordOfSuccess with Nine Bills On the House Floor This Week". House Energy and Commerce Committee. 8 September 2014. Retrieved 10 September 2014.

[8] NAM Board of Directors

[9] NAM website's history page

[10] American industries, Volume 13 By National Association of Manufacturers (U.S.), May 1913, page 33

[11] For more on Parry and his views, see *The Scarlet Empire*.

[12] George G. Suggs, Jr., *Colorado's War on Militant Unionism: James H. Peabody and the Western Federation of Miners.* Detroit, MI: Wayne State University Press, 1972, pp. 66-67.

[13] Violations of free speech and assembly and interference with rights of labor: hearings before a subcommittee, Seventy-fourth Congress, second session, on S. Res. 266, a resolution to investigate violations of the right of free speech and assembly and interference with the right of labor to organize and bargain collectively. April 10–11, 14-17, 21, 23, 1936

[14] Colorado's War on Militant Unionism, James H. Peabody and the Western Federation of Miners, George G. Suggs, Jr., 1972, page 67-68.

[15] Stuart B. Kaufman, Peter J. Albert, and Grace Palladino (eds.), *The Samuel Gompers Papers: Volume 6: The American Federation of Labor and the Rise of Progressivism, 1902-6.* Urbana, IL: University of Illinois Press, 1997; pg. 193, fn. 1.

[16] Burton St. John III, "Press Professionalization and Propaganda: The Rise of Journalistic Double-Mindedness, 1917-1941." Amherst, NY: Cambria Press, 2010; p. 12.

[17] LJ Griffin, ME Wallace, and BA Rubin. 1986. "Capitalist Resistance to the Organization of Labor Before the New Deal: Why? How? Success?" American Sociological Review. 51:2:147-67.

[18] Anna McCarthy, *The Citizen Machine: Governing by Television in 1950s America*, New York: The New Press, 2010, p. 54. ISBN 978-1-59558-498-4.

[19] National Archives

[20] Susan B. Strange and Wendy Shay, "Industry on Parade Film Collection, 1950–1960: #507", *National Museum of American History: Archives Center*, 10 September 2001.

[21] Manufacturing Institute

[22] "The Manufacturing Institute".

2.12.6 Further reading

- John N. Stalker, *The National Association of Manufacturers: A Study in Ideology.* Ph.D. dissertation. University of Wisconsin, Madison, 1950.

- Sarah Lyons Watts, *Order Against Chaos: Business Culture and Labor Ideology in America, 1880-1915.* New York: Greenwood Press, 1991.

- Burton St. John III, "Press Professionalization and Propaganda: The Rise of Journalistic Double-Mindedness, 1917-1941." Amherst, NY: Cambria Press, 2010.

2.12.7   External links

- National Association of Manufacturers

- Shopfloor Manufacturing Blog

- Manufacturing Institute

- online description of historical NAM pamphlets, 1908-1969

- National Association of Manufacturers. World War I posters. 5190+. Kheel Center for Labor-Management Documentation and Archives, Martin P. Catherwood Library, Cornell University.

2.13   Oregon Petition

The **Global Warming Petition Project**, also known as the **Oregon Petition**, is a petition urging the United States government to reject the global warming Kyoto Protocol of 1997 and similar policies.[1] It was organized and circulated by Arthur B. Robinson, president of the Oregon Institute of Science and Medicine in 1998, and again in 2007.[2][3] Past National Academy of Sciences president Frederick Seitz wrote a cover letter endorsing the petition.[4]

According to Robinson, the petition has over 31,000 signatories. Over 9,000 report to have a Ph.D.,[1][2][3] mostly in engineering.[5] The NIPCC (2009) Report lists 31,478 degreed signatories, including 9,029 with Ph.D.s.[6] The list has been criticized for its lack of verification, with pranksters successfully submitting Charles Darwin, members of the Spice Girls and characters from Star Wars, and getting them briefly included on the list.[7]

2.13.1   Petition text

The text of the petition reads, in its entirety:[4][8]

The petition included a covering letter from Frederick Seitz, chairman of the George C. Marshall Institute, and made reference to his former position as president of the US National Academy of Sciences; together with a manuscript plus a reprint of a December 1997 *Wall Street Journal* op-ed, "Science Has Spoken: Global Warming Is a Myth", by Arthur and Zachary Robinson. The current version of Seitz's letter describes the summary as "a twelve page review of information on the subject of 'global warming'."[9] The article is titled "Environmental Effects of Increased Atmospheric Carbon Dioxide" by Arthur B. Robinson, Noah E. Robinson, Sallie Baliunas and Willie Soon.[10][11][12]

As of October 2007, the petition project website includes an article by Arthur Robinson, Noah E. Robinson and Willie Soon, published in 2007 in the *Journal of American Physicians and Surgeons*.[13]

2.13.2   Signatories

The Oregon Petition Project clarified their verification process as follows:

- The petitioners could submit responses only by physical mail, not electronic mail, to limit fraud. Older signatures submitted via the web were not removed. The verification of the scientists was listed at 95%,[14] but the means by which this verification was done was not specified.

- Signatories to the petition were requested to list an academic degree.[15] The petition sponsors stated that approximately two thirds held higher degrees.[14] As of 2013, the petition's website states, "The current list of 31,487 petition signers includes 9,029 PhD; 7,157 MS; 2,586 MD and DVM; and 12,715 BS or equivalent academic degrees. Most of the MD and DVM signers also have underlying degrees in basic science."[16]

- Petitioners were also requested to list their academic discipline. As of 2007, about 2,400 people in addition to the original 17,100 signatories were "trained in fields other than science or whose field of specialization was not specified on their returned petition."[14] The petition sponsors state the following numbers of individuals from each discipline:[16]

- Atmospheric, Environmental and Earth sciences: 3,805 (Climatology: 39)

- Computer and Mathematical sciences: 935

- Physics & Aerospace sciences: 5,812

- Biochemistry, Biology, and Agriculture: 2,965

- Medicine: 3,046

- Engineering and General Science: 10,102

Credentials and authenticity

The credentials, verification process, and authenticity of the signatories have been questioned.

Jeff Jacoby promoted the Oregon Institute petition as delegates convened for the United Nations Framework Convention on Climate Change in 1998. Jacoby, a columnist for the *Boston Globe*, said event organizers "take it for granted" that global warming is real when scientists do not agree "that greater concentrations of CO2 would be harmful" or "that human activity leads to global warming in the first place."[17] George Woodwell and John Holdren, two members of the National Academy of Sciences, responded to Jacoby in the *International Herald Tribune*, describing the petition as a "farce" in part because "the signatories are listed without titles or affiliations that would permit an assessment of their credentials."[18] Myanna Lahsen said, "Assuming that all the signatories reported their credentials accurately, credentialed climate experts on the list are very few." The problem is made worse, Lahsen notes, because critics "added bogus names to illustrate the lack of accountability the petition involved".[19] Approved names on the list included fictional characters from the television show *M*A*S*H*,[20] the movie *Star Wars*,[19] Spice Girls group member Geri Halliwell, English naturalist Charles Darwin (d. 1882) and prank names such as "I. C. Ewe".[21] When questioned about the pop singer during a telephone interview with Joseph Hubert of the *Associated Press*, Robinson acknowledged that her endorsement and degree in microbiology was inauthentic, remarking "When we're getting thousands of signatures there's no way of filtering out a fake".[20] A cursory examination by Todd Shelly of the *Hawaii Reporter* revealed duplicate entries, single names lacking any initial, and even corporate names. "These examples underscore a major weakness of the list: there is no way to check the authenticity of the names. Names are given, but no identifying information (e.g., institutional affiliation) is provided."[22] According to the Petition Project website, the issue of duplication has been resolved.[23] Kevin Grandia offered similar criticism, saying although the Petition Project website provides a breakdown of "areas of expertise", it fails to assort the 0.5% of signatories who claim to have a background in Climatology and Atmospheric Science by name, making independent verification difficult. "This makes an already questionable list seem completely insignificant".[24]

In 2001, *Scientific American* took a random sample "of 30 of the 1,400 signatories claiming to hold a Ph.D. in a climate-related science."

> Of the 26 we were able to identify in various databases, 11 said they still agreed with the petition — one was an active climate researcher, two others had relevant expertise, and eight signed based on an informal evaluation. Six said they would not sign the petition today, three did not remember any such petition, one had died, and five did not answer repeated messages. Crudely extrapolating, the petition supporters include a core of about 200 climate researchers – a respectable number, though rather a small fraction of the climatological community.[25]

Former *New Scientist* correspondent Peter Hadfield says scientists are not experts on every topic, as depicted by the character Brains in *Thunderbirds*. Rather, they must specialize:

> "In between Aaagard and Zylkowski, the first and last names on the petition, are an assortment of metallurgists, botanists, agronomists, organic chemists and so on. ... The vast majority of scientists who signed the petition have never studied climatology and don't do any research into it. It doesn't matter if you're a Ph.D. A Ph.D in metallurgy just makes you better at metallurgy. It does not transform you into some kind of expert in paleoclimatology. ... So the petition's suggestion that everyone with a degree in metallurgy or geophysics knows a lot about climate change, or is familiar with all the research that's been done, is patent crap."[26][27]

2.13.3 NAS incident

A manuscript accompanying the petition was presented in a near identical style and format to contributions that appear in *Proceedings of the National Academy of Sciences*, a scientific journal,[28] but upon careful examination was distinct from a publication by the U.S. National Academy of Sciences. Raymond Pierrehumbert, an atmospheric scientist at the University of Chicago, said the presentation was "designed to be deceptive by giving people the impression that the article … is a reprint and has passed peer review." Pierrehumbert also said the publication was full of "half-truths".[29] F. Sherwood Rowland, who was at the time foreign secretary of the National Academy of Sciences, said that the Academy received numerous inquiries from researchers who "are wondering if someone is trying to hoodwink them."[29]

After the petition appeared, the National Academy of Sciences said in a 1998 news release that "The NAS Council would like to make it clear that this petition has nothing to do with the National Academy of Sciences and that the manuscript was not published in the *Proceedings of the National Academy of Sciences* or in any other peer-reviewed journal."[30] It also said "The petition does not reflect the conclusions of expert reports of the Academy." The NAS further noted that its own prior published study had shown that "even given the considerable uncertainties in our knowledge of the relevant phenomena, greenhouse warming poses a potential threat sufficient to merit prompt responses. Investment in mitigation measures acts as insurance protection against the great uncertainties and the possibility of dramatic surprises."[30]

Robinson responded in a 1998 article in *Science*, "I used the Proceedings as a model, but only to put the information in a format that scientists like to read, not to fool people into thinking it is from a journal."[29] A 2006 article in the magazine *Vanity Fair* stated: "Today, Seitz admits that 'it was stupid' for the Oregon activists to copy the academy's format. Still, he doesn't understand why the academy felt compelled to disavow the petition, which he continues to cite as proof that it is "not true" there is a scientific consensus on global warming".[31]

2.13.4 See also

- Leipzig Declaration
- An Evangelical Declaration on Global Warming
- Climate change denial
- Global warming controversy
- Global Warming Policy Foundation
- Heartland Institute
- Public opinion on climate change
- Scientific opinion on climate change

2.13.5 References

[1] Brennan, Phil (May 19, 2008). "31,000 Scientists Debunk Al Gore and Global Warming". *Newsmax*. Retrieved 2012-08-25.

[2] Avery, Dennis (May 24, 2008). "31000 scientists sign Oregon GW Skeptic Petition". *Canada Free Press*. Retrieved 2012-08-25.

[3] Henry, Devin (May 28, 2008). "Climate change petition pits scientists against each other". *Minnesota Daily*. Retrieved 2012-08-25.

[4] "What warming consensus?". *The Washington Times*. November 16, 1998. Retrieved 2012-08-25.

[5] Morrison, David (September–October 2011). "Reports of the National Center for Science Education". National Center for Science Education. ISSN 2159-9270.

[6] Idso, Craig and S. Fred Singer (2009). "Climate Change Reconsidered: 2009 Report of the Nongovernmental Panel on Climate Change (NIPCC), Appendix 4, The Petition Project" (PDF). The Heartland Institute. ISBN 978-1-934791-28-8. Retrieved 2012-08-25.

[7] Mann, Michael E. (2012). *The Hockey Stick and the Climate Wars*. Columbia University Press. p. 66.

[8] "Global Warming Petition Project". Oregon Institute of Science and Medicine. Retrieved 2012-08-25.

[9] Frederick Seitz. "Letter from Frederick Seitz". OISM. Retrieved 2010-01-11.

[10] A. B. Robinson, S. L. Baliunas, W. Soon, & Z. W. Robinson (1998). "Environmental Effects of Increased Atmospheric Carbon Dioxide". *J. Am. Physicians and Surgeons 3, 171-178*.

[11] A. B. Robinson, N. E. Robinson, W. Soon (2007). "Environmental Effects of Increased Atmospheric Carbon Dioxide". *J. Am. Physicians and Surgeons 12, 79-90*.

[12] W. Soon, S. L. Baliunas, A. B. Robinson, and Z. W. Robinson (1999). "Environmental Effects of Increased Atmospheric Carbon Dioxide". *Climate Research 13, 149-164*.

[13] Environmental Effects of Increased Atmospheric Carbon Dioxide by Arthur B. Robinson, Noah E. Robinson, and Willie Soon. Published in *The Journal of American Physicians and Surgeons*, 2007; 12(3), 79.

[14] "Explanation". OISM. Archived from the original on 2007-08-20. Retrieved 2008-07-14.

[15] OISM Mail-in Petition

[16] "Qualification of Signers". OISM. Archived from the original on 2013-10-03. Retrieved 2013-10-20.

[17] Jeff Jacoby. Scientists don't agree on global warming, *The Boston Globe*. 5 November 1998.

[18] George Woodwell and John Holdren. Climate-Change Skeptics Are Wrong *New York Times*. November 14, 1998.

[19] Myanna Lahsen (Winter 2005). "The Example of the 1998 Petition Campaign". *Technocracy, Democracy, and U.S. Climate Politics: The Need for Demarcations* (PDF). *Science, Technology, & Human Values* **30** (1). p. 137. doi:10.1177/0162243904270710.

[20] Joseph H. Hubert Odd Names Added to Greenhouse Plea *Associated Press*. (abridged version) 1 May 1998.

[21] David McNeely. It's easy for pseudoscientists to mislead people, *Edmond Sun*. February 22, 2006.

[22] Todd Shelly. Bashing the Scientific Consensus on Global Warming, *Hawaii Reporter*. 14 July 2005.

[23] "Frequently Asked Questions". Global Warming Petition Project. Retrieved 2010-09-10.

[24] Kevin Grandia. The 30,000 Global Warming Petition Is Easily-Debunked *The Huffington Post*. October 27, 2012.

[25] "Skepticism About Skeptics (sidebar of Climate of Uncertainty)". Scientific American. Archived from the original on 2006-08-23., October 2001

[26] Peter Hadfield. How my YouTube channel is converting climate change sceptics *The Guardian*. 29 March 2010.

[27] Peter Hadfield. Meet the Scientists. 25 May 2010.

[28] Arthur B. Robinson; Sallie L. Baliunas; Willie Soon; Zachary W. Robinson (January 1998). "Environmental Effects of Increased Atmospheric Carbon Dioxide". OISM and the George C. Marshall Institute. Archived from the original on 2007-01-14. Retrieved 2008-07-14.

[29] David Malakoff (10 April 1998). "Climate Change: Advocacy Mailing Draws Fire". *Science* **195** (5361): 195. doi:10.1126/science.280.5361.1

[30] "Statement by the Council of the National Academy of Sciences regarding Global Change Petition" (Press release). National Academy of Sciences. April 20, 1998. Retrieved 2010-03-04.

[31] Mark Hertsgaard (May 2006). "While Washington Slept".

2.13.6 Further reading

- Environmental Effects of Increased Carbon Dioxide Arthur B. Robinson, Noah E. Robinson, and Willie Soon, OISM

- Environmental Effects of Increased Atmospheric Carbon Dioxide 2007, AB Robinson, NE Robinson, & W. Soon, J. American Physicians & Surgeons

- Environmental effects of increased atmospheric carbon dioxide 1999, W. Soon, SL Beliunas, AB Robinson, ZW Robinson, Climate Research V. 13, 149-164, Oct. 26, 1999, Harvard Unv.

- Thomas R. Karl; Kevin Trenberth; James E. Hansen (December 18, 1997). "Op-Ed Science a Myth: Global Warming is happening". Natural Science Open Forum. Retrieved 2007-03-31.

- Prof. Joseph E. Armstrong (1998). "*The Wall Street Journal* Blurs the Lines Between Science, Opinion, & Politics on Global Warming". Reall (newsletter) . Retrieved 2007-03-31.

2.13.7 External links

- Global Warming Petition Project at the Oregon Institute of Science and Medicine

- Global Warming Petition Project

- Collected debunkings of the Oregon Petition

2.14 Pattern Recognition in Physics

Pattern Recognition in Physics was an open-access journal originally published by Copernicus Publications which was established in March 2013 and terminated in January 2014. The editors-in-chief were Sid-Ali Ouadfeul (Algerian Petroleum Institute) and Nils-Axel Mörner, the latter of whom is a well-known global warming skeptic.[1] Copernicus ceased its publication due to concerns over the publications views towards the scientific consensus of global climate change and the method of peer review. In March 2014 Ouadfeul reopened the journal, "run on private founding" [sic] [2]

2.14.1 History

Copernicus agreed to publish the journal because the editors claimed that its aim would be "to publish articles about patterns recognized in the full spectrum of physical disciplines rather than to focus on climate-research-related topics."[3] Concerns regarding the journal's peer-review process were first raised in July 2013 by Jeffrey Beall, an American librarian and critic of predatory open access publishing. Beall wrote on his blog that Ouadfeul's research has "only been cited a couple times," and went on to accuse him of self-plagiarizing from a book he had written in 2012, entitled "Wavelet Transforms and Their Recent Applications in Biology and Geoscience." Beall concluded that "This is not a good start for a journal, and the publisher ought to be concerned and take action."[4]

After a special issue of the journal was published in December 2013, which contained a paper in which the authors said they "doubt the continued, even accelerated, warming as claimed by the IPCC project,"[5] managing director Martin Rasmussen expressed concern regarding this journal; he also said that "the editors selected the referees on a nepotistic basis, which we regard as malpractice in scientific publishing."[3] On January 17, 2014, Copernicus Publications announced that they were terminating the journal, citing both the statement that questioned the IPCC's prediction of continued global warming and the "nepotistic" appointing of similarly-minded scientists to the journal's editorial board.[6]

2.14.2 References

[1] Castelvecchi, Davide (20 January 2014). "Climate comments push open-access publisher to terminate journal". *Nature News Blog*. Retrieved 11 February 2014.

[2] "Pattern Recognition in Physics". March 2014.

[3] Stokstad, Erik (17 January 2014). "Alleging 'Malpractice' With Climate Skeptic Papers, Publisher Kills Journal". *Science Insider*. Retrieved 11 February 2014.

[4] Beall, Jeffrey (16 July 2013). "Recognizing a Pattern of Problems in "Pattern Recognition in Physics"". *Scholarlyoa.com*. Retrieved 11 February 2014.

[5] Mörner, Nils-Axel (December 2013). "General conclusions regarding the planetary–solar–terrestrial interaction" (PDF). *Pattern Recognition in Physics* **1** (1): 205–206.

[6] Adler, Jonathan H. (20 January 2014). "Was a scientific journal canned for disagreeing with the IPCC?". *The Washington Post*. Retrieved 11 February 2014.

2.15 Politics of global warming

The **politics of global warming** are complex due to numerous factors that arise from the global economy's interdependence on carbon dioxide emitting hydrocarbon energy sources and because carbon dioxide is directly implicated in global warming[1] - making global warming a non-traditional environmental challenge:

1. **Implications to all aspects of a nation-state's economy** - The vast majority of the world economy relies on energy sources or manufacturing techniques that release greenhouse gases at almost every stage of production, transportation, storage, delivery & disposal while a consensus of the world's scientists attribute global warming to the release of carbon dioxide and other greenhouse gases. This intimate linkage between global warming and economic vitality implicates almost every aspect of a nation-state's economy;[2]

2. **Perceived lack of adequate advanced energy technologies** - Fossil fuel abundance and low prices continue to put pressure on the development of adequate advanced energy technologies that can realistically replace the role of fossil fuels - as of 2010, over 91% of the worlds energy is derived from fossil fuels and non carbon-neutral technologies.[3] Developing countries do not have cost effective access to the advanced energy technologies that they need for development (most advanced technologies has been developed by and exist in the developed world). Without adequate and cost effective post-hydrocarbon energy sources, it is unlikely the countries of the developed or developing world would accept policies that would materially affect their economic vitality or economic development prospects;

3. **Industrialization of the developing world** - As developing nations industrialize their energy needs increase and since conventional energy sources produce carbon dioxide, the carbon dioxide emissions of developing countries are beginning to rise at a time when the scientific community, global governance institutions and advocacy groups are telling the world that carbon dioxide emissions should be decreasing. Without access to cost effective and abundant energy sources many developing countries see climate change as a hindrance to their unfettered economic development;

4. **Metric selection (transparency) and perceived responsibility / ability to respond** - Among the countries of the world, disagreements exist over which greenhouse gas emission metrics should be used like total emissions per year, per capita emissions per year, CO2 emissions only, deforestation emissions, livestock emissions or even total historical emissions. Historically, the release of carbon dioxide has not been historically even among all nation-states and nation-states have challenges with determining who should restrict emissions and at what point of their industrial development they should be subject to such commitments;

5. **Vulnerable developing countries and developed country legacy emissions** - Some developing nations blame the developed world for having created the global warming crisis because it was the developed countries that emitted most of the carbon dioxide over the twentieth century and vulnerable countries perceive that it should be the developed countries that should pay to address the challenge;

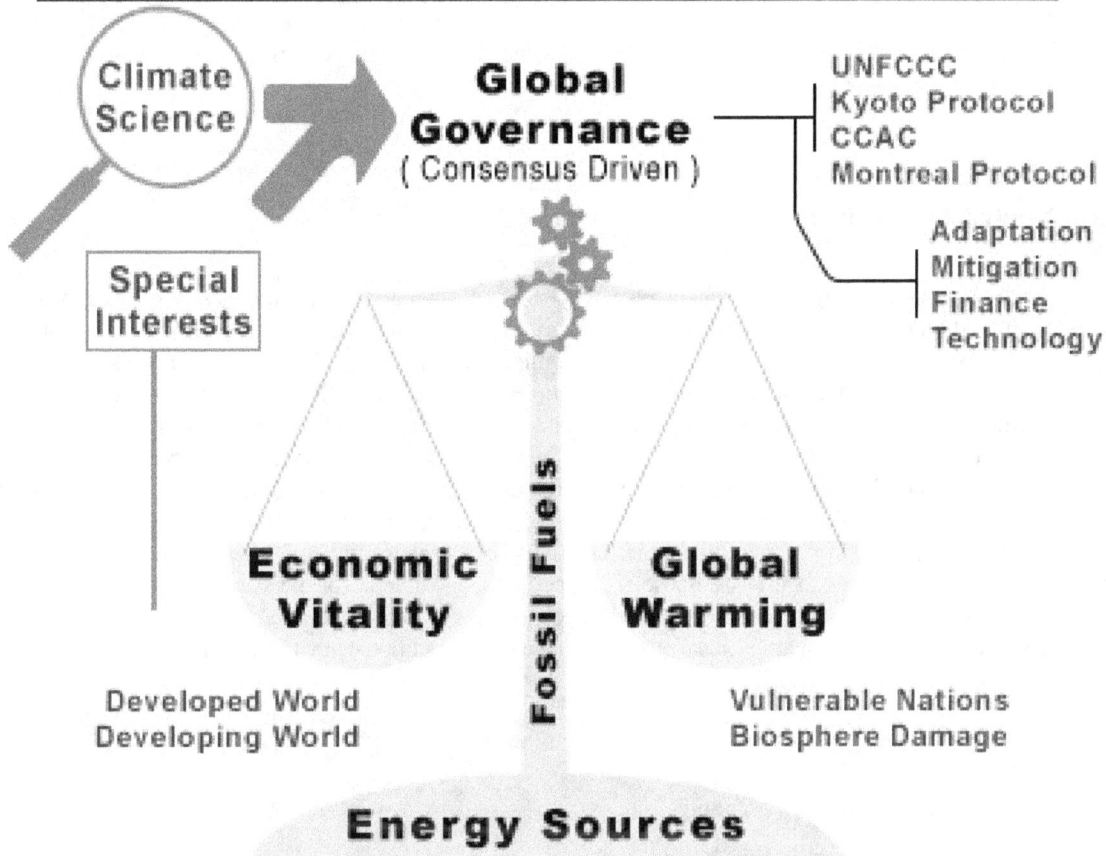

Politics of Global Warming Pictogram.

6. **Consensus-driven global governance models** - The global governance institutions that evolved during the 20th century are all consensus driven deliberative forums where agreement is difficult to achieve and even when agreement is achieved it is almost impossible to enforce;

7. **Well organized and funded special-interest lobbying bodies** - Special interest lobbying by well organized groups distort and amplify aspects of the challenge (environmental lobbying, energy industry lobbying, other special interest lobbying);

8. **Politicization of climate science** - Although there is a consensus on the science of global warming and its likely effects - some special interests groups work to suppress the consensus while others work to amplify the alarm of global warming. All parties that engage in such acts add to the politicization of the science of global warming. The result is a clouding of the reality of the global warming problem.

The focus areas for global warming politics are Adaptation, Mitigation, Finance, Technology and Losses which are well quantified and studied but the urgency of the global warming challenge combined with the implication to almost every facet of a nation-state's economic interests places significant burdens on the established largely-voluntary global institutions that have developed over the last century; institutions that have been unable to effectively reshape themselves and move fast enough to deal with this unique challenge. Rapidly developing countries who see traditional energy sources as a means to fuel their development, well funded aggressive environmental lobbying groups and an established fossil fuel energy paradigm boasting a mature and sophisticated political lobbying infrastructure all combine to make global warming politics extremely polarized. Distrust between developed and developing countries at most international conferences that seek to address the topic add to the challenges. Further adding to the complexity is the advent of the Internet and the development of media technologies like blogs and other mechanisms for disseminating information that enable the exponential growth in production and dissemination of competing points of view which make it nearly impossible for the development and dissemination of an objective view into the enormity of the subject matter and its politics.

2.15.1   Nontraditional Environmental Challenge

Traditional environmental challenges generally involve behavior by a small group of industries who create products or services for a limited set of consumers in a manner that causes some form of damage to the environment which is clear. As an example, a gold mine might release a dangerous chemical byproduct into a waterway that kills the fish in the waterway: a clear environmental damage.[4] By contrast, carbon dioxide is a naturally occurring colorless odorless trace gas that is essential to the biosphere. Carbon dioxide is produced by all animals and utilized by plants and algae to build their body structures. Plant structures buried for tens of millions of years sequester carbon to form coal, oil and gas which modern industrial societies find essential to economic vitality. Over 80% of the worlds energy is derived from carbon dioxide emitting fossil fuels and over 91% of the world's energy is derived from non carbon-neutral energy sources. Scientists attribute the increases of carbon dioxide in the atmosphere to industrial emissions and scientists have linked carbon dioxide to global warming. However, the scientific consensus is difficult for the average individual layperson to readily see and grasp. This essential nature to the world's economies combined with the complexity of the science and the interests of countless interested parties make climate change a non-traditional environmental challenge.

Carbon dioxide and a nation-state's economy

Energy Consumption Linked to CO2 Emissions

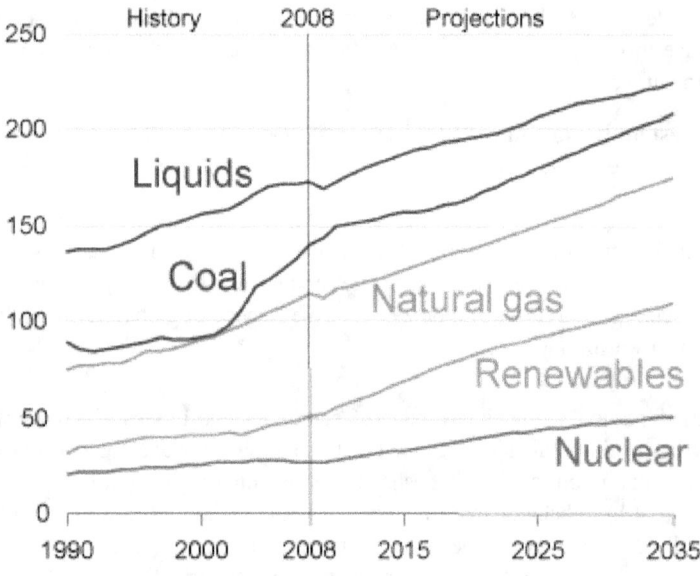

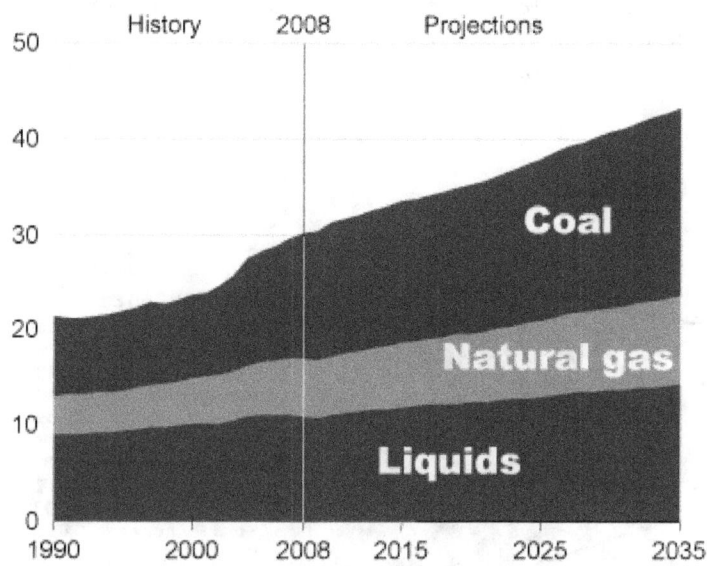

source: USDOE EIA IEO 2011
Main article: World energy consumption

The vast majority of developed countries rely on carbon dioxide emitting energy sources for large components of their economic activity.[5] Fossil fuel energy generally dominates the following areas of an OECD economy:

- agriculture (fertilizers, irrigation, plowing, planting, harvesting, pesticides)

- transportation & distribution (automobiles, shipping, trains, airplanes)

- storage (refrigeration, warehousing)

- national defense (armies, tanks, military aircraft, manufacture of munitions)

In addition, carbon dioxide emitting fossil fuels many times dominate the utilities aspect of an economy that provide electricity for:

- lighting

- heating & cooling

- refrigeration

- production of products

- computing and telecommunications

Also, activities like cement production, deforestation, brick production, livestock raising, refrigeration and other industrial activity contributes greenhouse gases that together are believed to account for 1/3 of global warming.

Because carbon dioxide emitting fossil fuels are intrinsically connected to a developed nation-state's economy, the taxation of fossil fuels or policies that decrease the availability of cost-effective fossil fuels is a significant political matter for fear that those taxes might precipitate a decrease in economic vitality. The replacement of cost-effective fossil fuels with more expensive renewable energy sources are seen by many as a hidden tax that would achieve the same result of depressing economic vitality and lead to impoverishment. Beyond the economic vitality of a single nation, some are concerned that taxation would depress economic activity in a manner that could affect the geopolitical order by providing incentives to one set of countries over another.

In developing countries the challenges are slightly different. Developing countries see carbon dioxide emitting fossil fuels as a cost effective and proven energy source to fuel their growing economies. Sometimes renewable energy technologies are not readily available to developing countries because of cost or due to export restrictions from developed countries who own those technologies.

Perceived lack of adequate advanced energy technologies

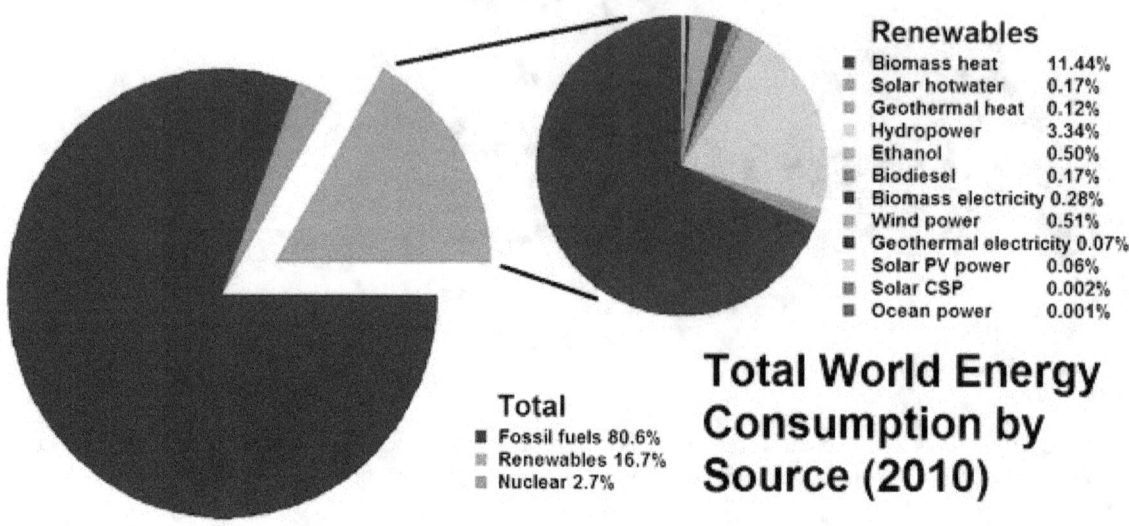

In 2010 renewable energy accounted for 16.7% of total energy consumption but Biomass energy which releases carbon dioxide and is not carbon-neutral accounted for 11.4%, leaving only 4.9% for carbon-neutral renewables (hydropower 3.3% and 1.56% for all others).

See also: Peak Oil, Fossil Fuels, Solar Energy, Wind Energy and Renewable Energy

Carbon dioxide emitting fossil fuels continue to be abundant and their prices are consequently low accounting in 2010 for over 80% of the world's energy needs.[3] Advanced recovery technologies like horizontal drilling, offshore oil production, and oil sand recovery technologies continue to push back the threshold of Peak Oil and with it the high prices seen necessary to foster the development of viable alternative energy technologies that can replace fossil fuels in a post-hydrocarbon economy. Renewables in 2010 accounted for 16.7% of the world's energy, however Biomass energy accounted for 11.4% meaning that non-carbon dioxide producing renewables accounted for only 4.9% of the world's energy use with the vast majority of that renewable energy coming from hydroelectric production at 3.34% further leaving 1.56% of renewable energy derived from newer advanced technologies like ethanol, biodiesel, wind, solar, ocean power and geothermal (see graph to right).

The biomass-is-carbon-neutral proposal put forward in the early 1990s has been superseded by more recent science that recognizes that mature, intact forests sequester carbon more effectively than cut-over areas. When a tree's carbon is released into the atmosphere in a single pulse, it contributes to climate change much more than woodland timber rotting slowly over decades.[3] Current studies indicate that "even after 50 years the forest has not recovered to its initial carbon storage" and "the optimal strategy is likely to be protection of the standing forest".[6]

After adjustment, carbon neutral renewables account for 4.9% of the world's energy needs in 2010 with solar accounting for .23% and wind for 0.51% of total world energy need.[7] Recent optimistic projections from the EIA and the IEA show that renewables will account for one sixth of global energy production in the next few decades (which includes Biomass energy), substantially below what is needed to significantly curtail carbon dioxide emissions.

Without help developing countries usually do not have access to the advanced energy technologies like wind and solar that they require for development forcing them to rely on hydrocarbon energy sources like fossil fuels and biomass. Without adequate and cost effective post-hydrocarbon energy sources, it is very unlikely the countries in developed or developing world would accept policies that would materially affect their economic vitality or economic development prospects. To date, developing countries have resisted adopting verifiable carbon dioxide targets for fear of impacts to their economies and the The United States, Russia, Canada, Japan, New Zealand, Belarus and Ukraine have either not ratified the Kyoto Protocol, withdrawn from the Kyoto Protocol or have chosen to not accept a second commitment period leaving the Kyoto Protocol extension covering only 15% of global carbon dioxide emissions. A strong contributor to these decisions is that the existing technologies are not yet adequate to replace the role of fossil hydrocarbon fuels.

Arguments have been made that fostering renewable energy through subsidies and other adoption-mechanisms are the path towards increasing the percentage of carbon-neutral renewable technologies that are used. According to IEA (2011) energy subsidies artificially lower the price of energy paid by consumers, raise the price received by producers or lower the cost of production. "Fossil fuels subsidies costs generally outweigh the benefits. Subsidies to renewables and low-carbon energy technologies can bring long-term economic and environmental benefits".[8] In November 2011, an IEA report entitled *Deploying Renewables 2011* said "subsidies in green energy technologies that were not yet competitive are justified in order to give an incentive to investing into technologies with clear environmental and energy security benefits". The IEA's report disagreed with claims that renewable energy technologies are only viable through costly subsidies and not able to produce energy reliably to meet demand. "A portfolio of renewable energy technologies is becoming cost-competitive in an increasingly broad range of circumstances, in some cases providing investment opportunities without the need for specific economic support," the IEA said, and added that "cost reductions in critical technologies, such as wind and solar, are set to continue."[9]

By contrast, Fossil-fuel consumption subsidies were $409 billion in 2010, oil products ca half of it. Renewable-energy subsidies were $66 billion in 2010 and will reach according to IEA $250 billion by 2035. Renewable energy is subsidized in order to compete in the market, increase their volume and develop the technology so that the subsidies become unnecessary with the development. Eliminating fossil-fuel subsidies could bring economic and environmental benefits. Phasing out fossil-fuel subsidies by 2020 would cut primary energy demand 5%. Since the start of 2010, *at least 15 countries have taken steps to phase out fossil-fuel subsidies.* According to IEA onshore wind may become competitive around 2020 in the European Union.[8]

Industrialization of the developing world

Economic Development Linked to Energy Consumption

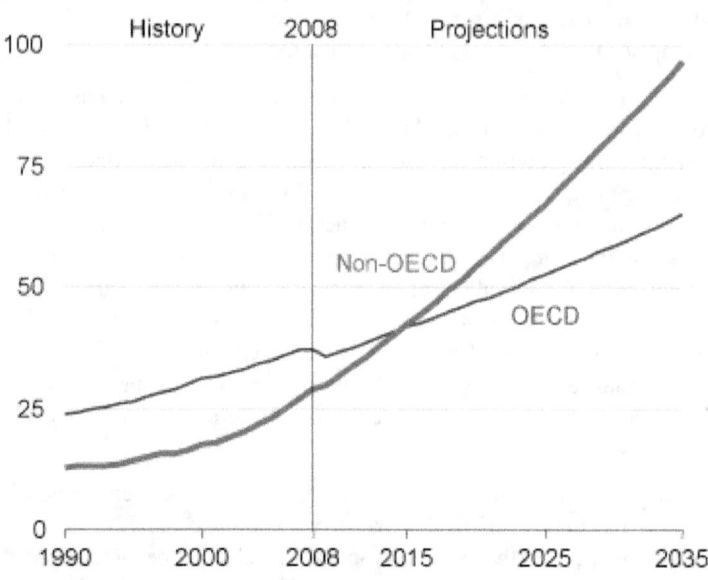

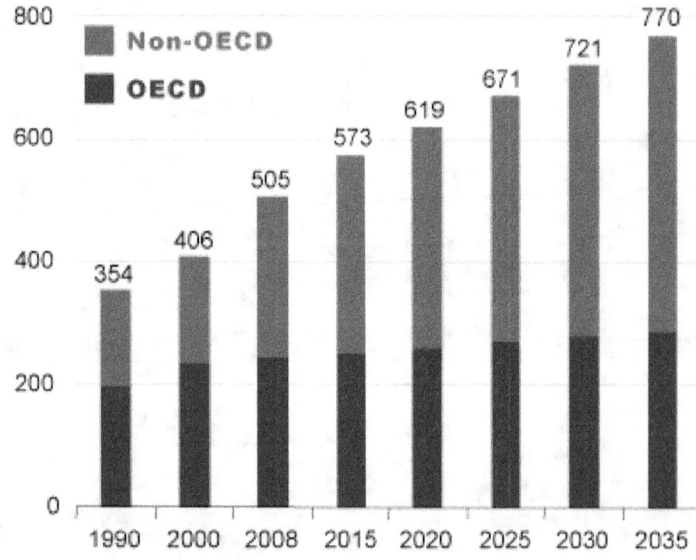

source: USDOE EIA IEO 2011

Main articles: developing world and Newly industrialized country

The developing world sees economic and industrial development as a natural right and the evidence shows that the developing world is industrializing. The developing world is leveraging the use of carbon dioxide emitting fossil fuels as one of the primary energy sources to fuel their development. At the same time the scientific consensus on climate change and the existing global governance bodies like the United Nations are urging all countries to decrease their carbon dioxide emissions. Developing countries logically resist this lobbying to decrease their use of fossil fuels without significant concessions like:

- advanced energy technologies

- advanced adaptation technologies

- monies to adapt to climate change.

Metric selection and perceived responsibility / ability to respond

Main article: Political economy of climate change

There are significant disagreements over which metrics to use when tracking global warming and there are also disagreements over which countries should be subject to emissions restrictions.

While the biosphere is indifferent to whether the greenhouse gases are produced by one country or by a multitude, the countries of the world do express an interest in such matters. As such disagreements arise on whether per capita emissions should be used or whether total emissions should be used as a metric for each individual country. Countries also disagree over whether a developing country should share the same commitment as a developed country that has been emitting carbon dioxide and other greenhouse gases for close to a century.

Some developing countries expressly state that they require assistance if they are to develop, which is seen as a right, in a fashion that does not contribute carbon dioxide or other greenhouse gases to the atmosphere. Many times, these needs materialize as profound differences in global conferences by countries on the subject and the debates quickly turn to pecuniary matters.

Most developing countries are unwilling to accept limits on their carbon dioxide and other greenhouse gas emissions while most developed countries place very modest limits on their willingness to assist developing countries. In addition, most developed countries would rather not participate in greenhouse gas reduction treaties if those would lead to decreased economic activity, transfers of wealth to developing countries, or significant shifts in the geopolitical balance of power of the world.

Vulnerable developing countries and developed country legacy emissions

See also: Political economy of climate change and Climate Vulnerable Forum

Some developing countries fall under the category of vulnerable to climate change. These countries involve small, sometimes isolated, island nations, low lying nations, nations who rely on drinking water from shrinking glaciers etc. These vulnerable countries see themselves as the victims of climate change and some have organized themselves under groups like the Climate Vulnerable Forum. These countries seek mitigation monies from the developed and the industrializing countries to help them adapt to the impending catastrophes that they see climate change will bring upon them.[10] For these countries climate change is seen as an existential threat and the politics of these countries is to seek reparation and adaptation monies from the developed world and some see it as their right.

2.15.2   Governance

Global warming politics focus areas

See also: Adaptation to global warming and Mitigation of global warming

Government politics regarding climate change and many official reports on the subject usually revolve around addressing one of the following topic areas:

- **Adaptation** – social and other changes that must be undertaken to successfully adapt to climate change. Adaptation might encompass, but is not limited to, changes in agriculture and urban planning.

- **Finance** – how countries will finance adaptation to and mitigation of climate change, whether from public or private sources or from wealth/technology transfers from developed countries to developing countries and the management mechanisms for those monies.

- **Mitigation** – steps and actions that the countries of the world can take to mitigate the effects of climate change.

- **Technology** – the technologies that are needed lower carbon emissions through increasing energy efficiency or replacement or carbon dioxide emitting technologies and technologies needed to adapt or mitigate climate change. Also encompasses ways that developed countries can support developing countries in adopting new technologies or increasing efficiency.

- **Loss and damage** – first articulated at the 2012 conference and in part based on the agreement that was signed at the 2010 United Nations Climate Change Conference in Cancun. It introduces the principle that countries vulnerable to the effects of climate change may be financially compensated in future by countries that fail to curb their carbon emissions.

Consensus-driven global political institutions

See also: United Nations Framework Convention on Climate Change

The primary mechanism for the world to tackle global warming is through a process established under the United Nations Framework Convention on Climate Change (UNFCCC) treaty. The current state of global warming politics is that there is frustration over a perceived lack of progress with the establish UNFCCC overall process which has progressed over eighteen years but which has been unable to curb global greenhouse gas emissions. Todd Stern - the US Climate Change envoy - has expressed the challenges with the UNFCCC process as follows, "Climate change is not a conventional environmental issue...It implicates virtually every aspect of a state's economy, so it makes countries nervous about growth and development. This is an economic issue every bit as it is an environmental one." He went on to explain that, the United Nations Framework Convention on Climate Change is a multilateral body concerned with climate change and can be an inefficient system for enacting international policy. Because the framework system includes over 190 countries and because negotiations are governed by consensus, small groups of countries can often block progress.[2]

The eighteenth conference of the parties held in Doha, Qatar, 2012 United Nations Climate Change Conference, yielded minor to modest restults. At the 2012 Doha climate change talks, Parties to the Kyoto Protocol agreed to an extension of the Kyoto Protocol to 2020.[11][12][13] Participants in the extension to the Kyoto Protocol have taken on targets for the period 2013-2020, and include Australia, the European Union, and a number of other developed countries.[14] Canada, who withdrew from the Kyoto Protocol in 2011, and The United States - who never ratified the Kyoto Protocol - have been joined by New Zealand, Japan, Russia, Belarus, and Ukraine who have stated that they would not sign up to a second Kyoto Protocol commitment period or extension due lack of commitments from the developing world which today include the world's largest carbon dioxide emitters. Japan and New Zealand also added that their country's CO2 emissions are minor when compared to the emissions of China, The United States, and the European Union.[15] These defections place significant pressures on the UNFCCC process which to date has not been able to curtail carbon dioxide emissions, whose latest Kyoto Protocol extension only accounts for 15% of greenhouse gas emissions,[11][12] and whose process is seen by some as slow, cumbersome, expensive and an inefficient use of taxpayer money: in the UK alone the climate change department has taken over 3,000 flights over the course of two years at a cost to the taxpayer of over £1,500,000 (British Pounds).[14][16] The outcome of the Doha talks has received a mixed response,[11][12][13] with small island states critical of the overall package.[12] Other results of the conference include a timetable for a global agreement to be adopted by 2015 which includes all countries.[17]

As a result, some have argued that perhaps the consensus driven model could be replaced with a majority vote model. However, that model would likely drive disagreement at the country-level-ratification by countries who disagreed with any global treaties that might passed through a majority vote at such restructured institutions.

Voluntary emissions reductions

See also: Climate and Clean Air Coalition to Reduce Short-Lived Climate Pollutants

The perceived slow process of efforts for countries to agree to a comprehensive global level binding agreements has led some countries to seek independent/voluntary steps and focus on alternative high-value voluntary activities like the creation of the Climate and Clean Air Coalition to Reduce Short-Lived Climate Pollutants by the United States, Canada, Mexico, Bangladesh, and Sweden which seeks to regulate short-lived pollutants such as methane, black carbon and hydrofluorocarbons (HFCs) which together are believed to account for up to 1/3 of current global warming but whose regulation is not as fraught with wide economic impacts and opposition.[18] The Climate and Clean Air Coalition to Reduce Short-Lived Climate Pollutants (CCAC) was launched on February 16, 2012 to regulate short-lived climate pollutants (SLCPs) that together contribute up to 1/3 of global warming. The coalition's creation is seen as a necessary and pragmatic step given the slow pace of global climate change agreements under the UNFCCC.[1]

As part of the 2010 Cancún agreements, 76 developed and developing countries have made voluntary pledges to control their emissions of greenhouse gases.[19] These voluntary steps are seen by some as a new model where countries pledge to voluntarily take action against global warming outside of international treaties or obligations to other parties. This voluntary mechanism, while promising, does not address many of the challenges seen by the developing world in their efforts to mitigate global warming, adapt to global warming, and increasingly to deal with losses and damages that they directly attribute to global warming that they blame on the developed world's historical emissions.

2.15.3   Special interests and lobbying by non-country interested parties

There are numerous special interest groups, organizations, corporations who have public and private positions on the multifaceted topic of global warming. The following is a partial list of the types of special interest parties that have demonstrated an interest in the politics of global warming:

- **Financial Institutions** - Financial institutions generally support restrictive policies regarding global warming, particularly the implementation of carbon trading schemes and the creation of market mechanisms that associate a price with carbon. These new markets would require trading infrastructures which banking institutions are well positioned to provide. Financial institutions would also be positioned well to invest, trade and develop various financial instruments that they could profit from through speculative positions on carbon prices and the use of brokerage and other financial functions like insurance and derivative instruments.[20]

- **Environmental groups** - Environmental groups generally take ideological positions on global warming and favor strict restrictions on carbon dioxide emissions. Environmental groups, as activists, engage in raising awareness and attracting investment into the advocacy movement to further raise awareness.[21]

- **Fossil fuel companies** - Traditional fossil fuel corporations could benefit or lose from stricter global warming regulations. A reduction in the use of fossil fuels could negatively impact fossil fuel corporations.[22][23] However, the fact that fossil fuel companies are a large source of energy, are also the primary source of carbon dioxide, and are engaged in energy trading might mean that their participation in trading schemes and other such mechanisms might give them a unique advantage and makes it unclear whether traditional fossil fuel companies would all and always be against stricter global warming policies. As an example, Enron, a traditional gas pipeline company with a large trading desk heavily lobbied the government for the EPA to regulate CO2: they thought that they would dominate the energy industry if they could be at the center of energy trading.[24]

- **Alternative energy companies** - alternative energy companies like wind and solar generally support stricter global warming policies. They would expect their share of the energy market to expand as fossil fuels are made more expensive through trading schemes or taxes.[25]

- **Nuclear energy companies** - nuclear energy companies could see a renaissance in a world where fossil fuels are taxed directly or through a carbon trading mechanism. For this reason, it is likely that nuclear energy companies would likely support stricter global warming policies.[26]

- **Traditional retailers and marketers** - traditional retailers, marketers, and the general corporations respond by adopting policies that resonate with their customers. If "being green" helps a general corporation, then they could undertake modest programs to please and better align with their customers. However, since the general corporation does not make a profit from their particular position, it is unlikely that they would strongly lobby either for or against a stricter global warming policy position.[27]

- **Governments** - On the Australian Sunday morning political discussion show *The Bolt Report*, Richard Lindzen said in a 2011 interview that governments might use global warming as a rationale for additional taxes.[28]

The various interested parties sometimes align with one another to reinforce their message. Sometimes industries will fund specialty nonprofit organizations to raise awareness and lobby on their behest.[29][30] The combinations and tactics that the various interested parties use are nuanced and sometimes unlimited in the variety of their approaches to promote their positions onto the general public.

2.15.4   Interaction of climate science and actual policy

Main articles: Global warming controversy, Politicization of science and Knowledge policy

In the scientific literature, there is a strong consensus that global surface temperatures have increased in recent decades and that the trend is caused primarily by human-induced emissions of greenhouse gases.[31][32][33] With regard to the **global warming controversy**, the scientific mainstream puts neither doubt on the existence of global warming nor on its causes and effects.

The **politicization of science** in the sense of a manipulation of science for political gains is a part of the political process. It is part of the controversies about intelligent design[34][35] (compare the Wedge strategy) or Merchants of Doubt, scientists that are under suspicion to willingly obscure findings. e.g. about issues like tobacco smoke, ozone depletion, global warming or acid rain.[36][37] However, e.g. in case of the Ozone depletion, global regulation based on the Montreal Protocol has been successful, in a climate of high uncertainty and against strong resistance[38] while in case of Climate Change, the Kyoto Protocol failed.[39]

While the IPCC process tries to find and orchestrate the findings of global (climate) change research to shape a worldwide consensus on the matter[40] it has been itself been object of a strong politicization.[41] Anthropogenic climate change evolved from a mere science issue to a top global policy topic.[41]

The IPCC process faces currently a paradox lockstep[41] where having built a broad science consensus does not hinder governments to follow different, if not opposing goals.[42] In case of the ozone depletion challenge, there was global regulation already being installed before a scientific consensus was established.[38]

A linear model of policy-making, based on a *more knowledge we have, the better the political response will be* does therefore not apply.**Knowledge policy**,[41] successfully managing knowledge and uncertainties as base of political decision making requires a better understanding of the relation between science, public (lack of) understanding and policy instead.[39][42][43] Michael Oppenheimer confirms limitations of the IPCC consensus approach and asks for concurring, smaller assessments of special problems instead of large scale attempts as in the previous IPCC assessment reports.[44] He claims that governments require a broader exploration of uncertainties in the future.[44]

2.15.5   Historical Reference

History of global warming politics

Historically, the politics of climate change dates back to several conferences in the late 1960s and the early 1970s under NATO and President Richard Nixon. 1979 saw the world's first World Climate Conference. 1985 was the year that the Vienna Convention for the Protection of the Ozone Layer was created and two years later in 1987 saw the signing of the Montreal Protocol under the Vienna convention. This model of using a Framework conference followed by Protocols under the Framework was seen as a promising governing structure that could be used as a path towards a functional governance approach that could be used to tackle broad global multi-nation/state challenges like global warming.

One year later in 1988 the Intergovernmental Panel on Climate Change was created by the World Meteorological Organization and the United Nations Environment Programme to assess the risk of human-induced climate change. Margaret Thatcher 1988 strongly supported IPCC and 1990 was instrumental to found the Hadley Centre for Climate Prediction and Research in Exeter.[45][46]

1991 saw the publishing of the book The First Global Revolution by the Club of Rome report which sought to connect environment, water availability, food production, energy production, materials, population growth and other elements into a blueprint for the twenty-first century: political thinking was evolving to look at the world in terms of an integrated global system not just in terms of weather and climate but in terms of energy needs, food, population, etc.

1992 was the year that the United Nations Framework Convention on Climate Change (UNFCCC) was agreed at the Earth Summit in Rio de Janeiro and the framework entered into force 21 March 1994. The conference established a yearly meeting, a conference of the parties or COP meeting to be held to continue work on Protocols which would be enforceable treaties.

1995 saw the creation of the phrase preventing dangerous anthropogenic interference with the climate system (also called *avoiding dangerous climate change*) first appeared in a policy document of a governmental organization, the IPCC's Second Assessment Report: Climate Change 1995.[47] and in 1996 the European Union adopt a goal of limiting temperature rises to a maximum 2 °C rise in average global temperature.

1997 saw the creation of the Kyoto Protocol under the United Nations Framework Convention on Climate Change (UNFCCC) in a very similar structure as the Montreal Protocol was under the Vienna Convention for the Protection of the Ozone Layer which would have yearly meetings of the members or CMP meetings. However, in the same year, the U.S. Senate passed Byrd–Hagel Resolution rejecting Kyoto without more commitments from developing countries [48]

Since the 1992 UNFCCC treaty, eighteen COP sessions and eight CMP sessions have been held under the existing structure. In that time, global CO2 emissions have risen significantly and developing countries have grown significantly with China replacing the United States as the largest emitter of greenhouse gases. To some, the UNFCCC has made significant progress in helping the world become aware of the perils of global warming and has moved the world forward in the addressing of the challenge. To others, the UNFCCC process has been a failure due to its inability to control the rise of greenhouse gas emissions.

A number of proposals for a Global Climate Regime are currently discussed, as the Durban Platform for Enhanced Action calls for a comprehensive new agreement in 2015 that includes both Annex-I and Non-Annex-I parties.

Selective historical timeline of significant climate change political events

- 1969, on Initiative of US President Richard Nixon, NATO tried to establish a third civil column and planned to establish itself as a hub of research and initiatives in the civil region, especially on environmental topics.[49] Daniel Patrick Moynihan, Nixons NATO delegate for the topic[49] named Acid Rain and the Greenhouse effect as suitable international challenges to be dealt by NATO. NATO had suitable expertise in the field, experience with international research coordination and a direct access to governments.[49] After an enthusiastic start on authority level, the German government reacted sceptical.[49] The initiative was seen as an American attempt[49] to regain international terrain after the lost Vietnam War. The topics and the internal coordination and preparation effort however gained momentum in civil conferences and institutions in Germany and beyond during the Brandt government.[49]

- 1972 United Nations Conference on the Human Environment,[49] leading role of Nobel Prize winner Willy Brandt and Olof Palme,[50] Germany saw enhanced international research cooperation on the Greenhouse topic as necessary[49]

- 1978 Brandt Report, the greenhouse effect dealt with in the energy section[51]

- 1979: First World Climate Conference [52]

- 1987: Brundtland Report[51]

- 1987: Montreal Protocol on restricting ozone layer-damaging CFCs demonstrates the possibility of coordinated international action on global environmental issues.

Daniel Patrick Moynihan, Pioneer of the political treatment of the Greenhouse Effect

- 1988: Intergovernmental Panel on Climate Change set up to coordinate scientific research, by two United Nations organizations, the World Meteorological Organization and the United Nations Environment Programme (UNEP) to assess the "risk of human-induced climate change".

- 1992: United Nations Framework Convention on Climate Change was formed to "prevent dangerous anthropogenic interference with the climate system"

- 1996: European Union adopts target of a maximum 2 °C rise in average global temperature

- 25 June 1997: U.S. Senate passes Byrd–Hagel Resolution rejecting Kyoto without more commitments from developing countries [48]

- 1997: Kyoto Protocol agreed

- 2001: George W. Bush withdraws from the Kyoto negotiations

- 16 February 2005: Kyoto Protocol comes into force (not including the US or Australia)

- 2005: first carbon emissions trading scheme (EU) implemented

- July 2005: 31st G8 summit has climate change on the agenda, but makes relatively little concrete progress

- November/December 2005: United Nations Climate Change Conference; the first meeting of the Parties of the Kyoto Protocol, alongside the 11th Conference of the Parties (COP11), to plan further measures for 2008–2012 and beyond.

- October 30, 2006: The Stern Review is published. It is the first comprehensive contribution to the global warming debate by an economist and its conclusions lead to the promise of urgent action by the UK government to further curb Europe's CO_2 emissions and engage other countries to do so. It discusses the consequences of climate change, mitigation measures to prevent it, possible adaptation measures to deal with its consequences, and prospects for international cooperation.

- June 26, 2009: U.S. House of Representatives passes the American Clean Energy and Security Act, the "first time either house of Congress had approved a bill meant to curb the heat-trapping gases scientists have linked to climate change."[53]

2.15.6   See also

- 2012 United Nations Climate Change Conference

- 2013 United Nations Climate Change Conference* Climate change denial

- Avoiding Dangerous Climate Change

- Climate and Clean Air Coalition to Reduce Short-Lived Climate Pollutants

- Economics of global warming

- Environmental impact of aviation

- Green Climate Fund

- Post–Kyoto Protocol negotiations on greenhouse gas emissions

- Scorcher: The Dirty Politics of Climate Change

- United Nations Framework Convention on Climate Change and accompanying Kyoto Protocol (CO2 Regulations)

- Vienna Convention for the Protection of the Ozone Layer and accompanying Montreal Protocol (Ozone Regulation - Context)

2.15.7 References

[1] Rudd, Kevin (25 May 2015). "Paris Can't Be Another Copenhagen". *New York Times*. Retrieved 26 May 2015.

[2] ""Voices" speaker talks climate change". The Dartmouth. Retrieved 29 November 2012.

[3] Mary S. Booth. "Biomass Briefing, October 2009" (PDF). *massenvironmentalenergy.org*. Massachusetts Environmental Energy Alliance. Retrieved 12 December 2010.

[4] "Arsenic poisoning stalks India's gold mines". *SciDev*.

[5] Global Energy Review in 2011, Enerdata Publication

[6] Edmunds, Joe; Richard Richets; Marshall Wise, "Future Fossil Fuel Carbon Emissions without Policy Intervention: A Review". In T. M. L. Wigley, David Steven Schimel, *The carbon cycle*. Cambridge University Press, 2000, pp.171–189

[7] "Biomass Briefing, October 2009" (PDF). *REN21*. Renewables Energy Policy Network for the 21st Century. Retrieved 12 December 2012.

[8] World Energy Outlook 2011 Factsheet How will global energy markets evolve to 2035? IEA November 2011 6 pages

[9] Henning Gloystein (Nov 23, 2011). "Renewable energy becoming cost competitive, IEA says". *Reuters*.

[10] Vidal, John (3 December 2012). "Climate change compensation emerges as major issue at Doha talks". London: The Guardian. Retrieved 3 December 2012.

[11] Ritter, K. and M. Casey. "UN conference adopts extension of Kyoto accord". CTPost. Retrieved 8 December 2012.

[12] Harrabin, R. (8 December 2012), "UN climate talks extend Kyoto Protocol, promise compensation", *BBC News*

[13] Harvey, F. (8 December 2012), "Doha climate change deal clears way for 'damage aid' to poor nations", *The Observer* (London, UK)

[14] UNFCCC. Conference of the Parties serving as the meeting of the Parties to the Kyoto Protocol (CMP) (8 December 2012), *Outcome of the work of the Ad Hoc Working Group on Further Commitments for Annex I Parties under the Kyoto Protocol. Draft decision proposed by the President (EN). Notes: Agenda item 4: Report of the Ad Hoc Working Group on Further Commitments for Annex I Parties under the Kyoto Protocol. Meeting: Conference of the Parties serving as the meeting of the Parties to the Kyoto Protocol (CMP), Eighth session, 26 November - 7 December 2012, Doha, Qatar. FCCC/KP/CMP/2012/L.9* (PDF), Geneva, Switzerland: United Nations Office, pp.6-7. Other languages available.

[15] McCarthy, Michael (2 December 2010). "Japan derails climate talks by refusing to renew Kyoto treaty". London: The Independent. Retrieved 29 November 2012.

[16] "UK climate change department takes over 3000 flights at a cost of over £1.3m". The Commenator. Retrieved 29 November 2012.

[17] UN Climate Change Secretariat (8 December 2012), *Doha climate conference opens gateway to greater ambition and action on climate change (press release)* (PDF), Bonn, Germany: UN Climate Change Secretariat, p.2.

[18] "Secretary Clinton To Announce a Climate and Clean Air Initiative To Reduce Short-Lived Climate Pollutants". US Dept of State. Retrieved 29 November 2012.

[19] King, D.; et al. (July 2011), "Copenhagen and Cancun", *International climate change negotiations: Key lessons and next steps*, Oxford, UK: Smith School of Enterprise and the Environment, University of Oxford, p. 12, doi:10.4210/ssee.pbs.2011.0003 PDF version is also available

[20] "Banking on carbon trading: Can banks stop climate change?". CNN. 20 July 2008. Retrieved 22 February 2013.

[21] "The climate lobby from soup to nuts". Center for Public Integrity. Retrieved 23 February 2013.

[22] David Michaels (2008) *Doubt is Their Product: How Industry's Assault on Science Threatens Your Health*.

[23] Hoggan, James; Littlemore, Richard (2009). *Climate Cover-Up: The Crusade to Deny Global Warming*. Vancouver: Greystone Books. ISBN 978-1-55365-485-8. Retrieved 2010-03-19. See, e.g., p31 *ff*, describing industry-based advocacy strategies in the context of climate change denial, and p73 *ff*, describing involvement of free-market think tanks in climate-change denial.

[24] "Enron Sought Global Warming Regulation, Not Free Markets". Competitive Enterprise Institute. Retrieved 4 December 2012.

[25] "Under Obama, Spain's Solar, Wind Energy Companies Invest Big In U.S.". Huffington Post. 18 January 2013. Retrieved 22 February 2013.

[26] "The Pro-Nukes Environmental Movement". Slate Institute. Retrieved 22 February 2013.

[27] "25 Big Companies That Are Going Green". Business Pundit. Retrieved 22 February 2013.

[28] "26 Interview with Richard Lindzen". Bolt. Retrieved 3 May 2013.

[29] "Climate change lobbying dominated by 10 firms". Politico. Retrieved 23 February 2013.

[30] "Greenpeace informal alliance with Wind and Solar". Retrieved 23 February 2013.

[31] Oreskes, Naomi (December 2004). "BEYOND THE IVORY TOWER: The Scientific Consensus on Climate Change". *Science* **306** (5702): 1686. doi:10.1126/science.1103618. PMID 15576594. Such statements suggest that there might be substantive disagreement in the scientific community about the reality of anthropogenic climate change. This is not the case. [...] Politicians, economists, journalists, and others may have the impression of confusion, disagreement, or discord among climate scientists, but that impression is incorrect.

[32] America's Climate Choices: Panel on Advancing the Science of Climate Change; National Research Council (2010). *Advancing the Science of Climate Change*. Washington, D.C.: The National Academies Press. ISBN 0-309-14588-0. (p1) ... there is a strong, credible body of evidence, based on multiple lines of research, documenting that climate is changing and that these changes are in large part caused by human activities. While much remains to be learned, the core phenomenon, scientific questions, and hypotheses have been examined thoroughly and have stood firm in the face of serious scientific debate and careful evaluation of alternative explanations. * * * (p21-22) Some scientific conclusions or theories have been so thoroughly examined and tested, and supported by so many independent observations and results, that their likelihood of subsequently being found to be wrong is vanishingly small. Such conclusions and theories are then regarded as settled facts. This is the case for the conclusions that the Earth system is warming and that much of this warming is very likely due to human activities.

[33] "Understanding and Responding to Climate Change" (PDF). United States National Academy of Sciences. 2008. Retrieved 30 May 2010. Most scientists agree that the warming in recent decades has been caused primarily by human activities that have increased the amount of greenhouse gases in the atmosphere.

[34] American Association for the Advancement of Science Statement on the Teaching of Evolution

[35] Intelligent Judging — Evolution in the Classroom and the Courtroom George J. Annas, New England Journal of Medicine, Volume 354:2277-2281 May 25, 2006

[36] Oreskes, Naomi; Conway, Erik. *Merchants of Doubt: How a Handful of Scientists Obscured the Truth on Issues from Tobacco Smoke to Global Warming* (first ed.). Bloomsbury Press. ISBN 978-1-59691-610-4.

[37] Boykoff, M.T.; Boykoff, J.M. (2004). "Balance as bias: Global warming and the US prestige press", *Global Environmental Change* (14): 125–136.

[38] Technische Problemlösung, Verhandeln und umfassende Problemlösung, (eng. technical trouble shooting, negotiating and generic problem solving capability) in Gesellschaftliche Komplexität und kollektive Handlungsfähigkeit (Societys complexity and collective ability to act), ed. Schimank, U. (2000). Frankfurt/Main: Campus, p.154-182 book summary at the Max Planck Gesellschaft

[39] Of Montreal and Kyoto: A Tale of Two Protocols by Cass R. Sunstein 38 ELR 10566 8/2008

[40] Aant Elzinga, "Shaping Worldwide Consensus: the Orchestration of Global Change Research", in Elzinga & Landström eds. (1996): 223-255. ISBN 0-947568-67-0.

[41] Climate Change: What Role for Sociology? A Response to Constance Lever-Tracy, Reiner Grundmann and Nico Stehr, doi: 10.1177/0011392110376031 Current Sociology November 2010 vol. 58 no. 6 897-910, see Lever Tracys paper in the same journal

[42] Environmental Politics Climate Change and Knowledge Politics REINER GRUNDMANN Vol. 16, No. 3, 414–432, June 2007

[43] [Knowledge, ignorance and the popular culture: climate change versus the ozone hole, by Sheldon Ungar, doi: 10.1088/0963-6625/9/3/306 Public Understanding of Science July 2000 vol. 9 no. 3 297-312 Abstract

[44] Michael Oppenheimer et al., The limits of consensus, in Science Magazine's State of the Planet 2008-2009: with a Special Section on Energy and Sustainability, Donald Kennedy, Island Press, 01.12.2008, separate as CLIMATE CHANGE, The Limits of Consensus Michael Oppenheimer, Brian C. O'Neill, Mort Webster, Shardul Agrawal, in Science 14 September 2007: Vol. 317 no. 5844 pp. 1505-1506 DOI: 10.1126/science.1144831

[45] How Margaret Thatcher Made the Conservative Case for Climate Action, James West, Mother Jones, Mon Apr. 8, 2013

[46] An Inconvenient Truth About Margaret Thatcher: She Was a Climate Hawk, Will Oremus, Slate (magazine) April 8, 2013

[47] IPCC 1995. Second Assessment Report: Climate change 1995

[48] "Byrd-Hagel Resolution (S. Res. 98) Expressing the Sense of the Senate Regarding Conditions for the U.S. Signing the Global Climate Change Treaty". Nationalcenter.org. Retrieved 2010-08-29.

[49] Die Frühgeschichte der globalen Umweltkrise und die Formierung der deutschen Umweltpolitik(1950-1973) (Early history of the environmental crisis and the setup of German environmental policy 1950-1973), Kai F. Hünemörder, Franz Steiner Verlag, 2004 ISBN 3-515-08188-7

[50] A "scandinavian connection" was alleged by Nils-Axel Mörner who saw an early friendship of Palme and Bert Bolin as reasons for Bolin then being promoted as environmental steward in the Swedish government and later as first head of the IPCC

[51] The Brandt Proposals: A Report Card, Energy and the Environment

[52] http://www.cs.ntu.edu.au/homepages/jmitroy/sid101/uncc/fs213.html

[53] Broder, John (2009-06-26). "House Passes Bill to Address Threat of Climate Change". New York Times. Retrieved 2009-06-27.

2.15.8 Further reading

- Aaron M. McCright and Riley E. Dunlap (2003), "Defeating Kyoto: The Conservative Movement's Impact on U.S. Climate Change Policy", *Social Problems* 50(3)

- *New York Times*, 10 March 2005, "Evangelicals Put Climate Change High on Their Agenda: Evangelical Leaders Swing Influence Behind Effort to Combat Global Warming"

- James Hansen (2009). *Storms of My Grandchildren: The Truth About the Coming Climate Catastrophe and Our Last Chance to Save Humanity*, Bloomsbury Press, New York, ISBN 978-1-60819-200-7.

- Naomi Oreskes and Erik M. Conway (2010). *Merchants of Doubt: How a Handful of Scientists Obscured the Truth on Issues from Tobacco Smoke to Global Warming*, Bloomsbury Press, New York.

- Eric Pooley (2010). *The Climate War: True Believers, Power Brokers, and the Fight to Save the Planet*, Hyperion, New York, ISBN 978-1-4013-2326-4.

2.15.9 External links

- Carbon Emissions World Map 2009 Mark McCormick and Paul Scruton, The Guardian February 2011

- Timeline of events related to the politics of global warming

- Univ. Colorado: Politics and Science Prometheus: Science + Politics Archives

- UNFCCC

- History of global warming

- Global warming and media

- Frontline: Hot Politics

- Spencer Weart, The Discovery of Global Warming

- George Monbiot, *The Guardian*, July 12, 2005, "Faced with this crisis: Instead of denying climate change is happening, the US now denies that we need proper regulation to stop it"

- George Monbiot, *The Guardian*, 20 September 2005, "It would seem that I was wrong about big business: Corporations are ready to act on global warming but are thwarted by ministers who resist regulation in the name of the market"

- John D. Sterman and Linda Booth Sweeney (undated) "Understanding Public Complacency About Climate Change: Adults' mental models of climate change violate conservation of matter"

- Amanda Griscom Little, *Grist Magazine*, July 20, 2005, "The Revolution Will Be Localized"

- Dr Oliver Marc Hartwich, *Free Market Foundation*, April 4, 2006, Weatherproofing African economies against climate change

- Senators sound alarm on climate - Christina Bellantoni, Washington Times — January 31, 2007

Environmental groups

- Panda — the World Wide Fund for Nature (WWF)

- World watch — Worldwatch Institute

- Green peace — Greenpeace

- Stop Climate Chaos — Coalition of UK charities

- Fight global warming — Environmental Defense

- http://www.stopglobalwarming.org/ — Stop Global Warming (Organization)

Business

- Carbon Disclosure Project , supported by over 150 institutional investors, aims for transparency on companies' greenhouse gas emissions

2.16   Public opinion on climate change

Public opinion on climate change is the aggregate of attitudes or beliefs held by the adult population concerning the science, economics, and politics of global warming. It is affected by media coverage of climate change.

2.16.1   Influences on individual opinion

Geographic Region

For a list of countries and their opinion, see Climate change opinion by country.

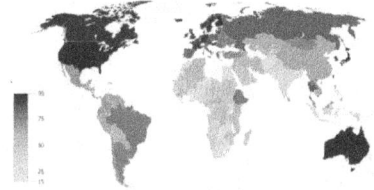

Proportion reporting knowing "something" or "a great deal" about global warming in 2007–08. Darker areas indicate a greater proportion of individuals aware, yellow indicates no data.

Proportion responding yes when asked, "Temperature rise is part of global warming or climate change. Do you think rising temperatures are [...] a result of human activities?"

Proportion responding that global warming is a serious personal threat

A 2007–2008 Gallup Poll surveyed individuals in 128 countries. This poll queried whether the respondent knew of global warming and, for those who were aware of the issue, whether or not they thought it was human-induced. Over a third of the world's population were unaware of global warming, with developing countries less aware than developed, and Africa the least aware. Of those aware, residents of Latin America and developed countries in Asia led the belief that climate change is a result of human activities while Africa, parts of Asia and the Middle East, and a few countries from the former Soviet Union led in the opposite. Opinion within the United Kingdom was divided.[1] Opinions in the United States vary intensely enough to be considered a culture war.[2][3] A Gallup poll in 2014 concluded that 51 percent of Americans were a little or not at all worried about climate change, 24 percent a great deal and 25 percent a fair amount.[4]

Adults in Asia, with the exception of those in developed countries, are the least likely to perceive global warming as a threat. In the western world, individuals are the most likely to be aware and perceive it as a very or somewhat serious threat to themselves and their families;[5] although Europeans are more concerned about climate change than those in the United States.[6] However, the public in Africa, where individuals are the most vulnerable to global warming while producing the least carbon dioxide, is the least aware – which translates into a low perception that it is a threat.[5]

These variations pose a challenge to policymakers, as different countries travel down different paths, making an agreement over an appropriate response difficult. While Africa may be the most vulnerable and produce the least amount of greenhouse gases, they are the most ambivalent. The top five emitters (China, the United States, India, Russia, and Japan), who together emit half the world's greenhouse gases, vary in both awareness and concern. The United States, Russia, and Japan are the most aware at over 85% of the population. Conversely, only two-thirds of people in China and one-third in India are aware. Japan expresses the greatest concern, which translates into support for environmental policies. People in China, Russia, and the United States, while varying in awareness, have expressed a similar proportion of aware individuals concerned. Similarly, those aware in India are likely to be concerned, but India faces challenges spreading this concern to the remaining population as its energy needs increase over the next decade.[7]

An online survey on environmental questions conducted in 20 countries by Ipsos MORI, "Global Trends 2014", shows broad agreement - especially on climate change and if it is caused by humans, though the U.S. ranked with 54% lowest.[8] It has been suggested that the low U.S. ranking is tied to denial campaigns.[9]

Education

In countries varying in awareness, an educational gap translates into a gap in awareness. However an increase in awareness does not always result in an increase in perceived threat. In China, 98% of those who have completed four years or more of college education reported knowing something or a great deal of climate change while only 63% of those who have completed nine years of education reported the same. Despite the differences in awareness in China, all groups perceive a low level of threat from global warming. In India those who are educated are more likely to be aware, and those who are educated there are far more likely to report perceiving global warming as a threat than those who are not educated.[7] In Europe, individuals who have attained a higher level of education perceive climate change as a serious threat. There is also a strong association between education and Internet use. Europeans who use the Internet more are more likely to perceive climate change as a serious threat.[10] However, a survey of American adults found that "...there is in fact little disagreement among culturally diverse citizens on what science knows about climate change. The source of the climate-change controversy and like disputes over societal risks is the contamination of the science-communication environment with forms of cultural status competition that make it impossible for diverse citizens to express their reason as both collective-knowledge acquirers and cultural-identity protectors at the same time.[11][12][13]

Demographics

Residential demographics affect perceptions of global warming. In China, 77% of those who live in urban areas are aware of global warming compared to 52% in rural areas. This trends is mirrored in India with 49% to 29% awareness, respectively.[7]

Of those countries where at least half the population are aware of global warming, those with the greatest proportion believing that global warming is due to human activities spend more on energy.[14]

In Europe, individuals under fifty-five are more likely to perceive both "poverty, lack of food and drinking water" and climate change as a serious threat than individuals over fifty-five. Male individuals are more likely to perceive climate change as a threat than female individuals. Managers, white collar workers, and students are more likely to perceive climate change as a greater threat than house persons and retired individuals.[10]

Political identification

In the United States, support for environmental protection was relatively non-partisan in the past. Republican Theodore Roosevelt established national parks whereas Democrat Franklin Delano Roosevelt established the Soil Conservation Service. US President Richard Nixon was instrumental in founding the United States Environmental Protection Agency and tried to install a third pillar of NATO dealing with environmental challenges as acid rain and greenhouse effect.[15] Daniel Patrick Moynihan, Nixons NATO delegate for the topic.[15] After an enthusiastic start on authority level, the then German government reacted sceptical.[15] The topics and the internal coordination and preparation effort however gained momentum in civil conferences and institutions in Germany and beyond during the Brandt government.[15] This non-partisanship began to erode during the 1980s when the Reagan administration described environmental protection as an economic burden. Views over global warming began to seriously diverge between Democrats and Republicans during the negotiations that led up to the creation of the Kyoto Protocol in 1998. In a 2008 Gallup poll of the American public, 76% of Democrats and only 41% of Republicans said that they believed global warming was already happening. The gap between the opinions of the political elites, such as members of Congress, tends to be even more polarized.[16]

In Europe, opinion is not strongly divided among left and right parties. Although European political parties on the left, and Green parties, strongly support measures to address climate change, conservative European political parties maintain similar sentiments, most notably in Western and Northern Europe. For example, Margaret Thatcher, never a friend of the coal mining industry, had been a strong supporter of an active climate protection policy and was instrumental in founding either the Intergovernmental Panel on Climate Change and the British Hadley Centre for Climate Prediction and Research in Exeter.[17] Some speeches, as to the Royal Society on 27 September 1988[18] and to the UN general assembly, in November 1989 helped to put climate change, acid rain and general pollution in the British mainstream in the early eighties. After her career Thatcher however was less of a climate activists as she doubted climate action a "marvelous excuse for supranational socialism," and called Al Gore an "apocalyptic hyperbole".[19] France's center-right President Chirac pushed key environmental and climate change policies in France in 2005–2007. Conservative German administrations (under the Christian Democratic Union and Christian Social Union) in the past two decades have supported European Union climate change initiatives, concern about Forest dieback and acid rain regulation were initiated under Kohls archconservative minister of the interior Friedrich Zimmermann. In the period after former President Bush announced that the United States was leaving the Kyoto Treaty, European media and newspapers on both the left and right criticized the move. The conservative Spanish *La Razón*, the *Irish Times*, *Irish Independent*, the Danish *Berlingske Tidende*, and the Greek *Kathimerini* all condemned the Bush administration's decision along with left-leaning newspapers.[20]

In Norway, a 2013 poll conducted by TNS Gallup found that 92% of those who vote for the Socialist Left Party and 89% of those who vote for the Liberal Party believe that global warming is caused by humans, while the percentage who held this belief is 60% among voters for the Conservative Party and 41% among voters for the Progress Party.[21]

The shared sentiments between the political left and right on climate change further illustrate the divide in perception between the United States and Europe on climate change. As an example, conservative German Prime Ministers Helmut Kohl and Angela Merkel have differed with other parties in Germany only on "how to meet emissions reduction targets, not whether or not to establish or fulfill them."[20]

Individual risk assessment and assignment

Main article: Ozone depletion and global warming

The IPCC attempts to orchestrate global (climate) change research to shape a worldwide consensus.[22] However the consensus approach has been dubbed more a liability than an asset in comparison to other environmental challenges.[23][23] The linear model of policy-making, based on a *more knowledge we have, the better the political response will be* has not been working and is in the meantime rejected by sociology.[24]

Sheldon Ungar, a Canadian sociologist, compares the different public reactions towards Ozone depletion and global warming.[25] The public opinion failed to tie Climate change to concrete events which could be used as a threashold or beacon to signify immediate danger.[25] Scientific predictions of an temperature rise of two to three degrees Celsius over several decades do not respond with people, e.g. in North America, that experience similar swings during a single day.[25] As Scientists define global warming a problem of the future, a liability in "attention economy", pessimistic outlooks in general and assigning extreme weather to climate change have often been discredited or ridiculed (compare Gore effect) in the public arena.[26] While the greenhouse effect, per se, is essential for life on earth, the case was quite different with the Ozone shield and other metaphors about the ozone depletion. The scientific assessment of the Ozone problem had as well large uncertainties. But the metaphors used in the discussion (ozone shield, ozone hole) reflected better with lay people and their concerns.

> The idea of rays penetrating a damaged "shield" meshes nicely with abiding and resonant cultural motifs, including "Hollywood affinities." These range from the shields on the Starship Enterprise to Star Wars, ... It is these pre-scientific bridging metaphors built around the penetration of a deteriorating shield that render the ozone problem relatively simple. That the ozone threat can be linked with Darth Vader means that it is encompassed in common sense understandings that are deeply ingrained and widely shared. (Sheldon Ungar 2000)[25]

The Chlorofluorocarbon (CFC) regulation attempts end of the 1980s profited from those easy to grasp metaphors and the personal risk assumptions taken from them. As well the fate of celebrities like President Ronald Reagan, which had skin cancer removal in 1985 and 1987 was of high importance. In case of the public opinion on climate change no imminent danger is perceived. [25]

Ideology

In the United States, ideology is an effective predictor of party identification, where conservatives are more prevalent among Republicans, and moderates and liberals among independents and Democrats. A shift in ideology is often associated with in a shift in political views.[27] For example, when the number of conservatives rose from 2008 to 2009, the number of individuals who felt that global warming was being exaggerated in the media also rose.[28]

The pursuit of green energy is an ideology that defines hydroelectric dams, natural gas power plants and nuclear power as unacceptable alternative energies for the eight billion tons of coal burnt each year. While there is popular support for wind, solar, biomass and geothermal even though all these sources combined only supplied 1.3% of global energy in 2013.[29][30]

2.16.2 Issues

Science

See also: Scientific opinion on climate change

A scientific consensus on climate change exists, as recognized by national academies of science and other authoritative bodies.[31] The opinion gap between scientists and the public in 2009 stands at 84% to 49% that global temperatures are increasing because of human-activity.[32]

September 2011 Angus Reid Public Opinion poll found that Britons (43%) are less likely than Americans (49%) or Canadians (52%) to say that "global warming is a fact and is mostly caused by emissions from vehicles and industrial facilities." The same poll found that 20% of Americans, 20% of Britons and 14% of Canadians think "global warming is a theory that has not yet been proven."[33]

A March 2013 Public Policy Polling poll about widespread and infamous conspiracy theories found that 37% of American voters believe that global warming is a hoax, while 51% do not.[34]

A 2013 poll in Norway conducted by TNS Gallup found that 66% of the population believe that climate change is caused by humans, while 17% do not believe this.[35]

Economics

See also: Economics of global warming

Economic debates weigh the benefits of limiting industrial emissions of mitigating global warming against the costs that such changes would entail. While there is a greater amount of agreement over whether global warming exists, there is less agreement over the appropriate response. Electric or petroleum distribution may be government owned or utilities may be regulated by government. The government owned or regulated utilities may, or may not choose to to make lower emissions a priority over economics, in unregulated counties industry follows economic priorities. An example of the economic priority is Royal Dutch Shell PLC reporting CO_2 emissions of 81 million metric tonnes in 2013.[36]

Media

See also: Media coverage of climate change

The popular media in the U.S. gives greater attention to skeptics relative to the scientific community as a whole, and the level of agreement within the scientific community has not been accurately communicated.[37][38] US popular media coverage differs from that presented in other countries, where reporting is more consistent with the scientific literature.[39] Some journalists attribute the difference to climate change denial being propagated, mainly in the US, by business-centered organizations employing tactics worked out previously by the US tobacco lobby.[40][41][42]

The efforts of Al Gore and other environmental campaigns have focused on the effects of global warming and have managed to increase awareness and concern, but despite these efforts as of 2007, the number of Americans believing humans are the cause of global warming was holding steady at 61%, and those believing the popular media was understating the issue remained about 35%.[43] Between 2010 and 2013, the number of Americans who believe the media under-reports the seriousness of global warming has been increasing, and the number who think media over-states it has been falling. According to a 2013 Gallup US opinion poll, 57% believe global warming is at least as bad as portrayed in the media (with 33% thinking media has downplayed global warming and 24% saying coverage is accurate). Less than half of Americans (41%) think the problem is not as bad as media portrays it.[44]

Politics

See also: Politics of global warming

Public opinion impacts on the issue of climate change because governments need willing electorates and citizens in order to implement policies that address climate change. Further, when climate change perceptions differ between the populace and governments, the communication of risk to the public becomes problematic. Finally, a public that is not aware of the issues surrounding climate change may resist or oppose climate change policies, which is of considerable importance to politicians and state leaders.[45]

Public support for action to forestall global warming is as strong as public support has been historically for many other government actions; however, it is not "intense" in the sense that it overrides other priorities.[45][46]

A 2009 Eurobarometer survey found that, on the average, Europeans rate climate change as the second most serious problem facing the world today, between "poverty, the lack of food and drinking water" and "a major global economic downturn." 87% of Europeans consider climate change to be a "serious" or "very serious" problem, while 10% "do not consider it a serious problem." However, the proportion who believe it to be a problem has dropped in the period 2008/9 when the surveys were conducted.[47] While the small majority believe climate change is a serious threat, 55% percent believe the EU is doing too little and 30% believe the EU is going the right amount.[48] As a result of European Union climate change perceptions, "climate change is an issue that has reached such a level of social and political acceptability across the EU that it enables (indeed, forces) the EU Commission and national leaders to produce all sorts of measures, including taxes."[20] Despite the persistent high level of personal involvement of European citizens, found in another Eurobarometer survey in 2011,[49] EU leaders have begun to downscale climate policy issues on the political agenda since the beginning of the Eurozone crisis.[50]

The proportion of Americans who believe that the effects of global warming have begun or will begin in a few years rose to a peak in 2008 where it then declined, and a similar trend was found regarding the belief that global warming is a threat to their lifestyle within their lifetime.[51] Concern over global warming often corresponds with economic downturns and national crisis such as 9/11 as Americans prioritize the economy and national security over environmental concerns. However the drop in concern in 2008 is unique compared to other environmental issues.[28] Considered in the context of environmental issues, Americans consider global warming as a less critical concern than the pollution of rivers, lakes, and drinking water; toxic waste; fresh water needs; air pollution; damage to the ozone layer; and the loss of tropical rain forests. However, Americans prioritize global warming over species extinction and acid rain issues.[52] Since 2000 the partisan gap has grown as Republican and Democratic views diverge.[53]

2.16.3 See also

- Global warming controversy

- How Global Warming Works

- Scientific opinion on climate change

2.16.4 References

[1] Pelham, Brett (22 April 2009). "Awareness, Opinions About Global Warming Vary Worldwide". The Gallup Organization. Retrieved 22 December 2009.

[2] Gillis, Justin (17 April 2012). "Americans Link Global Warming to Extreme Weather, Poll Says". *The New York Times*.

[3] *Climate Science as Culture War: The public debate around climate change is no longer about science—it's about values, culture, and ideology* Fall 2012 Stanford Social Innovation Review

[4] Riffkin, Rebecca (12 March 2014). "Climate Change Not a Top Worry in U.S.". Gallup. Retrieved 21 July 2014.

[5] Pugliese, Anita; Ray, Julie (11 December 2009). "Awareness of Climate Change and Threat Vary by Region". Gallup. Retrieved 22 December 2009.

[6] Crampton, Thomas (1 January 2007). "More in Europe worry about climate than in U.S., poll shows - Health & Science - International Herald Tribune". The New York Times. Retrieved 26 December 2009.

[7] Pugliese, Anita; Ray, Julie (7 December 2009). "Top-Emitting Countries Differ on Climate Change Threat". Gallup. Retrieved 22 December 2009.

[8] Ipsos MORI. "Global Trends 2014".

[9] MotherJones (July 22, 2014). "The Strange Relationship Between Global Warming Denial and…Speaking English".

[10] TNS Opinion and Social 2009, p. 13

[11] Kahan, Dan (20 February 2015). "Climate-Science Communication and the Measurement Problem". Advances in Political Psychology 36 (s1): 1–43. doi:10.1111/pops.12244. Retrieved 26 March 2015.

[12] "Climate Science Literacy Unrelated to Public Acceptance of Human-Caused Global Warming". Yale Law School. 23 February 2015.

[13] Lott, Maxim (12 February 2015). "Study: Global warming skeptics know more about climate science". Fox News.

[14] Pelham, Brett W. (24 April 2009). "Views on Global Warming Relate to Energy Efficiency". Gallup. Retrieved 22 December 2009.

[15] Die Frühgeschichte der globalen Umweltkrise und die Formierung der deutschen Umweltpolitik(1950-1973) (Early history of the environmental crisis and the setup of German environmental policy 1950-1973), Kai F. Hünemörder, Franz Steiner Verlag, 2004 ISBN

[16] Dunlap, Riley E. (29 May 2009). "Climate-Change Views: Republican-Democratic Gaps Expand". Gallup. Retrieved 22 December 2009.

[17] How Margaret Thatcher Made the Conservative Case for Climate Action, James West, Mother Jones, Mon Apr. 8, 2013

[18] 1988 Sep 27 Tu Margaret Thatcher Speech to the Royal Society

[19] An Inconvenient Truth About Margaret Thatcher: She Was a Climate Hawk, Will Oremus, Slate (magazine) April 8 2013

[20] Schreurs, M. A.; Tiberghien, Y. (Nov 2007). "Multi-Level Reinforcement: Explaining European Union Leadership in Climate Change Mitigation" (Full free text). Global Environmental Politics 7 (4): 19–46. doi:10.1162/glep.2007.7.4.19. ISSN 1526-3800.

[21] "Among those who vote for the Liberal Party or the Socialist Party, the great majority think that humans are behind climate changes (89 and 92%). Only 41% of those who vote for the Progress Party agree, while the number for Conservative Party voters is 60%." (Translated from Norwegian to English) Liv Jorun Andenes and Amalie Kvame Holm: Typisk norsk å være klimaskeptisk (Norwegian) Vårt Land, retrieved 8 July, 2013

[22] [[Aant Elzinga]], "Shaping Worldwide Consensus: the Orchestration of Global Change Research", in Elzinga & Landström eds. (1996): 223-255. ISBN 0-947568-67-0.

[23] Environmental Politics Climate Change and Knowledge Politics REINER GRUNDMANN Vol. 16, No. 3, 414–432, June 2007

[24] Climate Change: What Role for Sociology? A Response to Constance Lever-Tracy, Reiner Grundmann and Nico Stehr, doi: 10.1177/0011392110376031 Current Sociology November 2010 vol. 58 no. 6 897-910, see Lever Tracys paper in the same journal

[25] Knowledge, ignorance and the popular culture: climate change versus the ozone hole, by Sheldon Ungar, doi: 10.1088/0963-6625/9/3/306 Public Understanding of Science July 2000 vol. 9 no. 3 297-312 Abstract

[26] Sheldon Ungar Climatic Change February 1999, Volume 41, Issue 2, pp 133-150 Is Strange Weather in the Air? A Study of U.S. National Network News Coverage of Extreme Weather Events

[27] Saad, Lydia (26 Jun 2009). "Conservatives Maintain Edge as Top Ideological Group". Gallup. Retrieved 22 December 2009.

[28] Saad, Lydia (11 April 2009). "Increased Number Think Global Warming Is "Exaggerated"". Gallup. Retrieved 22 December 2009.

[29] http://www.theguardian.com/environment/2012/apr/23/people-want-more-renewable-energy

[30] http://www.ren21.net/wp-content/uploads/2015/06/REN12-GSR2015_Onlinebook_low1.pdf page 27

[31] Joint Science Academies (2005). "Joint science academies' statement: Global response to climate change" (Full free text). United States National Academies of Sciences. Retrieved 22 December 2009.

[32] "Public Praises Science; Scientists Fault Public, Media" (PDF). Pew Research Center. 9 Jun 2009. pp. 5, 55. Retrieved 13 March 2010.

[33] Angus Reid Public Opinion poll conducted 25 August through 2 September 2011

[34] Williams, Jim (2 April 2013). "Conspiracy Theory Poll Results". Public Policy Polling. Retrieved 28 April 2013.

[35] (Translated from Norwegian to English) "Two of three believe climate change is caused by humans. *I believe that climate change is caused by humans* (n=1001) Percentage that fully agree or disagree: (graph that shows numbers from 2009 to 2013, with 66/17 in 2013.")Presentasjon av resultater fra TNS Gallups Klimabarometer 2013 (7 June, 2013): Klimasak avgjør for hver fjerde velger (Norwegian) (link to pdf, p.29), TNS Gallup, retrieved 8 July, 2013

[36] "Royal Dutch Shell PLC - AMEE".

[37] Boykoff, M.; Boykoff, J. (July 2004). "Balance as bias: global warming and the US prestige press" (PDF). *Global Environmental Change Part A* **14** (2): 125–136. doi:10.1016/j.gloenvcha.2003.10.001.

[38] Antilla, L. (2005). "Climate of scepticism: US newspaper coverage of the science of climate change". *Global Environmental Change Part A* **15** (4): 338–352. doi:10.1016/j.gloenvcha.2005.08.003.

[39] Dispensa, J. M.; Brulle, R. J. (2003). "Media's social construction of environmental issues: focus on global warming – a comparative study" (Full free text). *International Journal of Sociology and Social Policy* **23** (10): 74. doi:10.1108/01443330310790327.

[40] Begley, Sharon (13 August 2007). "The Truth About Denial". Newsweek. Retrieved 11 January 2009.

[41] David, Adam (20 Sep 2006). "Royal Society tells Exxon: stop funding climate change denial". London: The Guardian. Retrieved 12 January 2009.

[42] Sandell, Clayton (3 January 2007). "Report: Big Money Confusing Public on Global Warming". ABC News. Retrieved 12 January 2009.

[43] Saad, Lydia (21 March 2007). "Did Hollywood's Glare Heat Up Public Concern About Global Warming?". Gallup. Retrieved 12 January 2010.

[44] Saad, Lydia. "Americans' Concerns About Global Warming on the Rise".

[45] Lorenzoni, I.; Pidgeon, N. F. (2006). "Public Views on Climate Change: European and USA Perspectives" (Full free text). *Climatic Change* **77** (1–2): 73–95. doi:10.1007/s10584-006-9072-z. ISSN 1573-1480."Despite the relatively high concern levels detected in these surveys, the importance of climate change is secondary in relation to other environmental, personal and social issues." 15 November 2005, accessed April 27, 2015

[46] Roger Pielke, Jr. (September 28, 2010). *The Climate Fix: What Scientists and Politicians Won't Tell You About Global Warming* (hardcover). Basic Books. p. 36-46. ISBN 978-0465020522. ...climate change does not rank high as a public priority in the context of the full spectrum of policy issues.

[47] TNS Opinion and Social 2009, p. 15

[48] TNS Opinion and Social 2009, p. 21

[49] European Commission, Special Eurobarometer 372 - Climate Change Brussels, June 2011

[50] Oliver Geden (2012), The End of Climate Policy as We Knew it, SWP Research Paper 2012/RP01

[51] Newport, Frank (11 Mar 2010). "Americans' Global Warming Concerns Continue to Drop". Gallup. Retrieved 13 Mar 2010.

[52] Saad, Lydia (7 April 2006). "Americans Still Not Highly Concerned About Global Warming". Gallup. Retrieved 7 January 2009.

[53] Dunlap, Riley E. (29 May 2008). "Partisan Gap on Global Warming Grows". Gallup Organization. Retrieved 17 December 2009.

2.16.5   Bibliography

- TNS Opinion and Social (Dec 2009). "Europeans' Attitudes Towards Climate Change" (Full free text). European Commission. Retrieved 24 December 2009.

2.16.6   External links

- Why Are Americans So Ill-Informed about Climate Change? from Scientific American

2.17   Scientific opinion on climate change

This article is about scientific consensus on the current climate change, or global warming. For public perception and controversy, see Public opinion on climate change and Global warming controversy.

The **scientific opinion on climate change** is the overall judgment amongst scientists regarding whether global warming

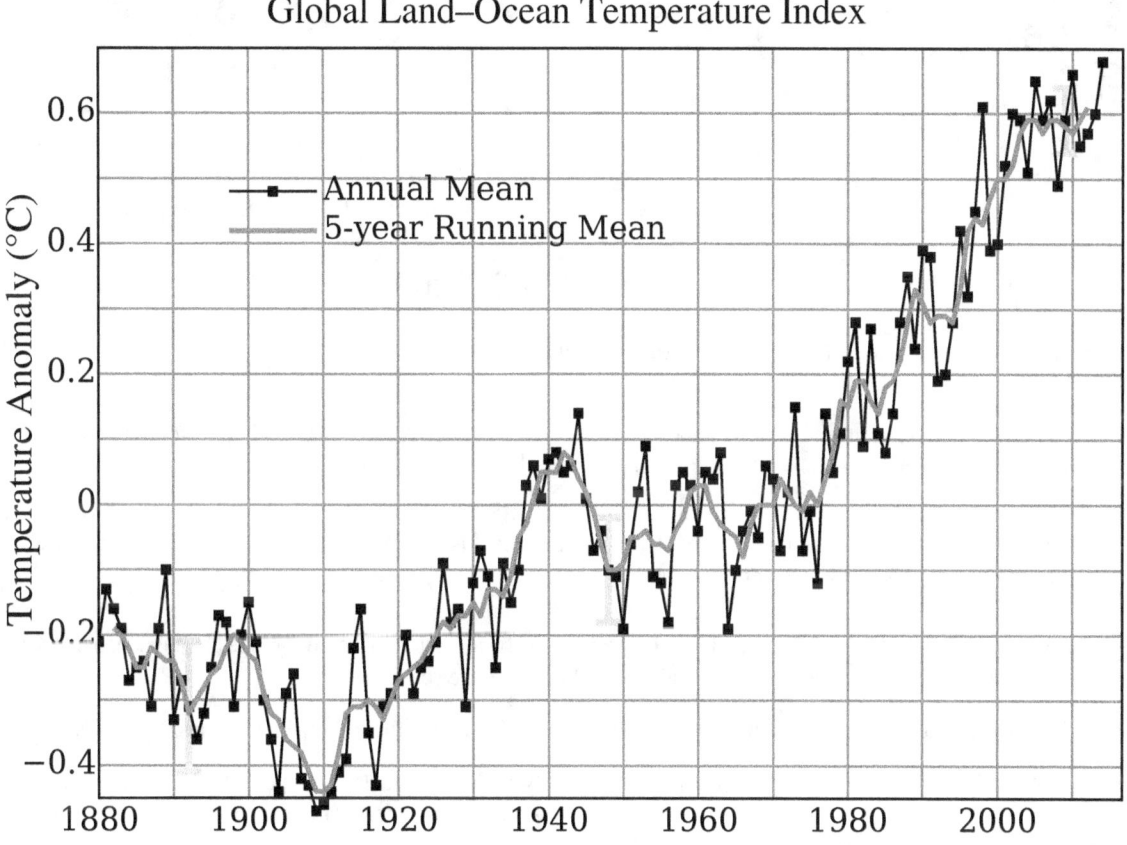

Global Land–Ocean Temperature Index

Global mean land-ocean temperature change since 1880, relative to the 1951–1980 mean. The black line is the annual mean and the red line is the 5-year running mean. The green bars show uncertainty estimates. Source: NASA GISS

is happening, and if so, its causes and probable consequences. This scientific opinion is expressed in synthesis reports, by scientific bodies of national or international standing, and by surveys of opinion among climate scientists. Individual scientists, universities, and laboratories contribute to the overall scientific opinion via their peer-reviewed publications, and the areas of collective agreement and relative certainty are summarised in these high level reports and surveys.[1]

The scientific consensus is that the Earth's climate system is unequivocally warming, and that it is *extremely likely* (at least 95% probability) that humans are causing most of it through activities that increase concentrations of greenhouse

Reconstructed Temperature

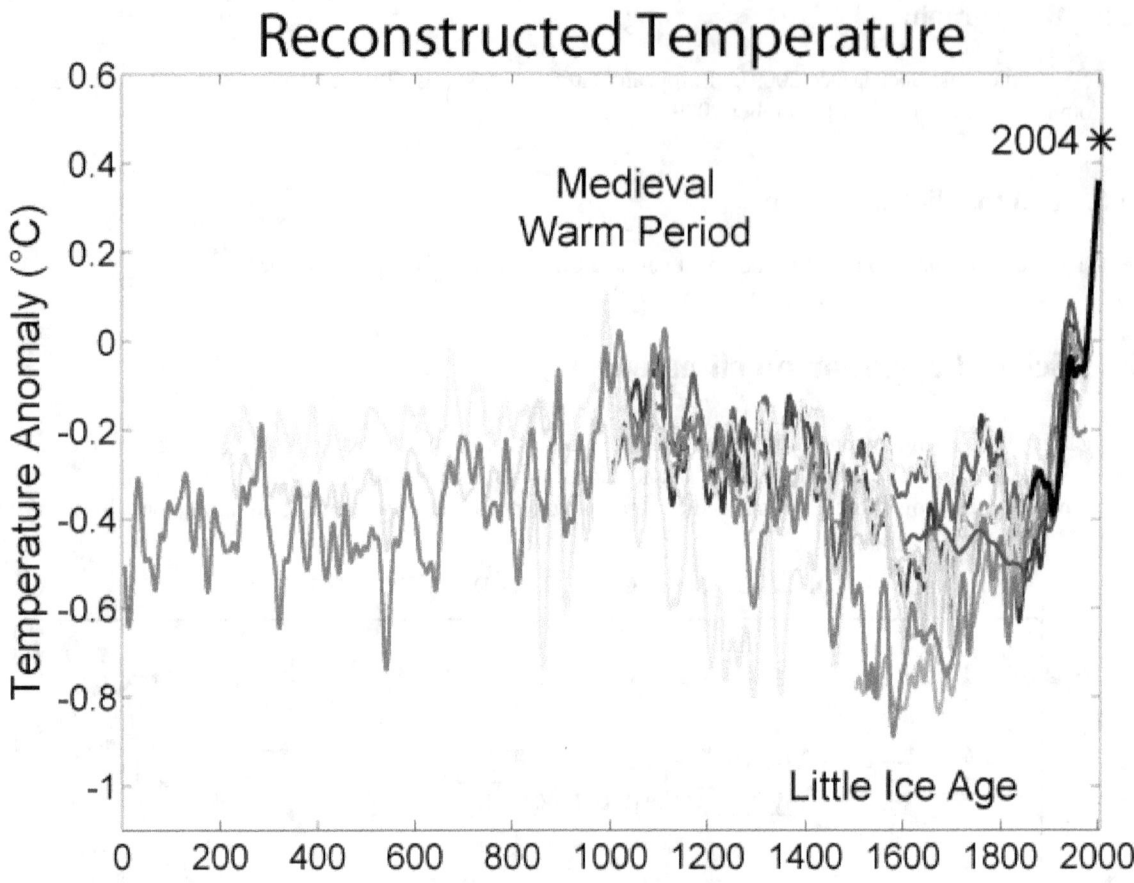

The temperature record of the past 2000 years from several different proxy methods.

gases in the atmosphere, such as deforestation and burning fossil fuels. In addition, it is likely that some potential further greenhouse gas warming has been offset by increased aerosols.[2][3][4][5]

National and international science academies and scientific societies have assessed current scientific opinion on global warming. These assessments are generally consistent with the conclusions of the Intergovernmental Panel on Climate Change, the IPCC Fourth Assessment Report summarized:

- Warming of the climate system is unequivocal, as evidenced by increases in global average air and ocean temperatures, the widespread melting of snow and ice, and rising global average sea level.[6]

- Most of the global warming since the mid-20th century is very likely due to human activities.[7]

- Benefits and costs of climate change for [human] society will vary widely by location and scale.[8] Some of the effects in temperate and polar regions will be positive and others elsewhere will be negative.[8] Overall, net effects are more likely to be strongly negative with larger or more rapid warming.[8]

- The range of published evidence indicates that the net damage costs of climate change are likely to be significant and to increase over time.[9]

- The resilience of many ecosystems is likely to be exceeded this century by an unprecedented combination of climate change, associated disturbances (e.g. flooding, drought, wildfire, insects, ocean acidification) and other global change drivers (e.g. land-use change, pollution, fragmentation of natural systems, over-exploitation of resources).[10]

Some scientific bodies have recommended specific policies to governments and science can play a role in informing an

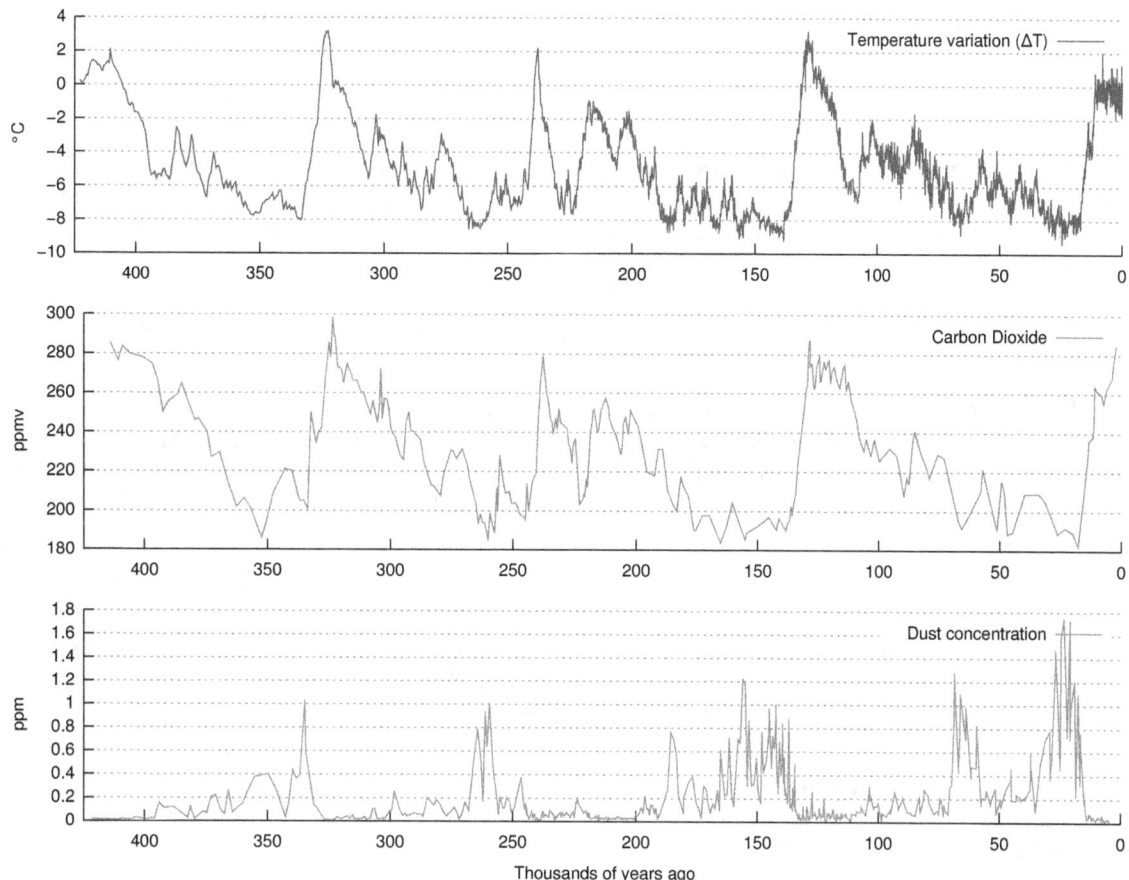

Variations in temperature, CO2 and dust from the Vostok ice core over the last 450,000 years

effective response to climate change. Policy decisions, however, may require value judgements and so are not included in the scientific opinion.[11][12]

No scientific body of national or international standing maintains a formal opinion dissenting from any of these main points. The last national or international scientific body to drop dissent was the American Association of Petroleum Geologists,[13] which in 2007[14] updated its statement to its current non-committal position.[15] Some other organizations, primarily those focusing on geology, also hold non-committal positions.

2.17.1   Synthesis reports

Synthesis reports are assessments of scientific literature that compile the results of a range of stand-alone studies in order to achieve a broad level of understanding, or to describe the state of knowledge of a given subject.[16]

Intergovernmental Panel on Climate Change (IPCC) 2014

Main articles: Intergovernmental Panel on Climate Change and IPCC Fifth Assessment Report

The IPCC Fifth Assessment Report *Summary for Policymakers* stated that warming of the climate system is 'unequivocal' with changes unprecedented over decades to millennia, including warming of the atmosphere and oceans, loss of snow and ice, and sea level rise. Greenhouse gas emissions, driven largely by economic and population growth, have led to greenhouse gas concentrations that are unprecedented in at least the last 800,000 years. These, together with other anthropogenic drivers, are "extremely likely" to have been the dominant cause of the observed global warming since the mid-20th century.[17]

It said that

> Continued emission of greenhouse gases will cause further warming and long-lasting changes in all components of the climate system, increasing the likelihood of severe, pervasive and irreversible impacts for people and ecosystems. Limiting climate change would require substantial and sustained reductions in greenhouse gas emissions which, together with adaptation, can limit climate change risks.[17]

Reporting on the publication of the report, *The Guardian* said that

> In the end it all boils down to risk management. The stronger our efforts to reduce greenhouse gas emissions, the lower the risk of extreme climate impacts. The higher our emissions, the larger climate changes we'll face, which also means more expensive adaptation, more species extinctions, more food and water insecurities, more income losses, more conflicts, and so forth.[18]

The New York Times reported that

> In Washington, President Obama's science adviser, John P. Holdren, cited increased scientific confidence "that the kinds of harm already being experienced from climate change will continue to worsen unless and until comprehensive and vigorous action to reduce emissions is undertaken worldwide."[19]

It went on to say that Ban Ki-moon, the United Nations secretary general, had declared his intention to call a meeting of heads of state in 2014 to develop such a treaty. The last such meeting, in Copenhagen in 2009, the NY Times reported, had ended in disarray.[19]

Intergovernmental Panel on Climate Change (IPCC) 2007

In February 2007, the IPCC released a summary of the forthcoming Fourth Assessment Report. According to this summary, the Fourth Assessment Report found that human actions are "very likely" the cause of global warming, meaning a 90% or greater probability. Global warming in this case was indicated by an increase of 0.75 degrees in average global temperatures over the last 100 years.[20]

The New York Times reported that "the leading international network of climate scientists has concluded for the first time that global warming is 'unequivocal' and that human activity is the main driver, 'very likely' causing most of the rise in temperatures since 1950".[21]

A retired journalist for *The New York Times*, William K. Stevens wrote: "The Intergovernmental Panel on Climate Change said the likelihood was 90 percent to 99 percent that emissions of heat-trapping greenhouse gases like carbon dioxide, spewed from tailpipes and smokestacks, were the dominant cause of the observed warming of the last 50 years. In the panel's parlance, this level of certainty is labeled 'very likely'. Only rarely does scientific odds-making provide a more definite answer than that, at least in this branch of science, and it describes the endpoint, so far, of a progression.".[22]

The Associated Press summarized the position on sea level rise:

> On sea levels, the report projects rises of 7 to 23 inches by the end of the century. An additional 3.9 to 7.8 inches are possible if recent, surprising melting of polar ice sheets continues.[23]

U.S. Global Change Research Program

formerly the Climate Change Science Program

The U.S. Global Change Research Program reported in June 2009[24] that:

Observations show that warming of the climate is unequivocal. The global warming observed over the past 50 years is due primarily to human-induced emissions of heat-trapping gases. These emissions come mainly from the burning of fossil fuels (coal, oil, and gas), with important contributions from the clearing of forests, agricultural practices, and other activities.

The report, which is about the effects that climate change is having in the United States, also says:

Climate-related changes have already been observed globally and in the United States. These include increases in air and water temperatures, reduced frost days, increased frequency and intensity of heavy downpours, a rise in sea level, and reduced snow cover, glaciers, permafrost, and sea ice. A longer ice-free period on lakes and rivers, lengthening of the growing season, and increased water vapor in the atmosphere have also been observed. Over the past 30 years, temperatures have risen faster in winter than in any other season, with average winter temperatures in the Midwest and northern Great Plains increasing more than 7°F. Some of the changes have been faster than previous assessments had suggested.

Arctic Climate Impact Assessment

In 2004, the intergovernmental Arctic Council and the non-governmental International Arctic Science Committee released the synthesis report of the Arctic Climate Impact Assessment:[25]

Climate conditions in the past provide evidence that rising atmospheric carbon dioxide levels are associated with rising global temperatures. Human activities, primarily the burning of fossil fuels (coal, oil, and natural gas), and secondarily the clearing of land, have increased the concentration of carbon dioxide, methane, and other heat-trapping ("greenhouse") gases in the atmosphere...There is international scientific consensus that most of the warming observed over the last 50 years is attributable to human activities.[26]

2.17.2   Policy

See also: Avoiding dangerous climate change

There is an extensive discussion in the scientific literature on what policies might be effective in responding to climate change.[27] Some scientific bodies have recommended specific policies to governments (refer to the later sections of the article).[28] The natural and social sciences can play a role in informing an effective response to climate change.[11] However, policy decisions may require value judgements.[11] For example, the US National Research Council has commented:[12]

The question of whether there exists a "safe" level of concentration of greenhouse gases cannot be answered directly because it would require a value judgment of what constitutes an acceptable risk to human welfare and ecosystems in various parts of the world, as well as a more quantitative assessment of the risks and costs associated with the various impacts of global warming. In general, however, risk increases with increases in both the rate and the magnitude of climate change.

This article mostly focuses on the views of natural scientists. However, social scientists,[27] medical experts,[29] engineers[27] and philosophers[30] have also commented on climate change science and policies. Climate change policy is discussed in several articles: climate change mitigation, climate change adaptation, climate engineering, politics of global warming, climate ethics, and economics of global warming.

2.17.3   Statements by scientific organizations of national or international standing

See also: Global warming controversy § Mainstream scientific position, and challenges to it

This is a list of scientific bodies of national or international standing, that have issued formal statements of opinion, classifies those organizations according to whether they concur with the IPCC view, are non-committal, or dissent from it.

Concurring

Academies of science (general science)   Since 2001, 34 national science academies, three regional academies, and both the international InterAcademy Council and International Council of Academies of Engineering and Technological Sciences have made formal declarations confirming human induced global warming and urging nations to reduce emissions of greenhouse gases. The 34 national science academy statements include 33 who have signed joint science academy statements and one individual declaration by the Polish Academy of Sciences in 2007.

Joint national science academy statements

- 2001 Following the publication of the IPCC Third Assessment Report, seventeen national science academies issued a joint statement, entitled "The Science of Climate Change", explicitly acknowledging the IPCC position as representing the scientific consensus on climate change science. The statement, printed in an editorial in the journal *Science* on May 18, 2001,[31] was signed by the science academies of Australia, Belgium, Brazil, Canada, the Caribbean, China, France, Germany, India, Indonesia, Ireland, Italy, Malaysia, New Zealand, Sweden, Turkey, and the United Kingdom.[32]

- 2005 The national science academies of the G8 nations, plus Brazil, China and India, three of the largest emitters of greenhouse gases in the developing world, signed a statement on the global response to climate change. The statement stresses that the scientific understanding of climate change is now sufficiently clear to justify nations taking prompt action, and explicitly endorsed the IPCC consensus. The eleven signatories were the science academies of Brazil, Canada, China, France, Germany, India, Italy, Japan, Russia, the United Kingdom, and the United States.[33]

- 2007 In preparation for the 33rd G8 summit, the national science academies of the G8+5 nations issued a declaration referencing the position of the 2005 joint science academies' statement, and acknowledging the confirmation of their previous conclusion by recent research. Following the IPCC Fourth Assessment Report, the declaration states, "It is unequivocal that the climate is changing, and it is very likely that this is predominantly caused by the increasing human interference with the atmosphere. These changes will transform the environmental conditions on Earth unless counter-measures are taken." The thirteen signatories were the national science academies of Brazil, Canada, China, France, Germany, Italy, India, Japan, Mexico, Russia, South Africa, the United Kingdom, and the United States.[34]

- 2007 In preparation for the 33rd G8 summit, the Network of African Science Academies submitted a joint "statement on sustainability, energy efficiency, and climate change" :

  > A consensus, based on current evidence, now exists within the global scientific community that human activities are the main source of climate change and that the burning of fossil fuels is largely responsible for driving this change. The IPCC should be congratulated for the contribution it has made to public understanding of the nexus that exists between energy, climate and sustainability.
  > — The thirteen signatories were the science academies of Cameroon, Ghana, Kenya, Madagascar, Nigeria, Senegal, South Africa, Sudan, Tanzania, Uganda, Zambia, Zimbabwe, as well as the African Academy of Sciences, [35]

- 2008 In preparation for the 34th G8 summit, the national science academies of the G8+5 nations issued a declaration reiterating the position of the 2005 joint science academies' statement, and reaffirming "that climate change is happening and that anthropogenic warming is influencing many physical and biological systems." Among other actions, the declaration urges all nations to "(t)ake appropriate economic and policy measures to accelerate transition to a low carbon society and to encourage and effect changes in individual and national behaviour." The thirteen signatories were the same national science academies that issued the 2007 joint statement.[36]

- 2009 In advance of the UNFCCC negotiations to be held in Copenhagen in December 2009, the national science academies of the G8+5 nations issued a joint statement declaring, "Climate change and sustainable energy supply are crucial challenges for the future of humanity. It is essential that world leaders agree on the emission reductions needed to combat negative consequences of anthropogenic climate change". The statement references the IPCC's Fourth Assessment of 2007, and asserts that "climate change is happening even faster than previously estimated; global CO_2 emissions since 2000 have been higher than even the highest predictions, Arctic sea ice has been melting at rates much faster than predicted, and the rise in the sea level has become more rapid." The thirteen signatories were the same national science academies that issued the 2007 and 2008 joint statements.[28]

Polish Academy of Sciences   In December 2007, the General Assembly of the Polish Academy of Sciences (Polska Akademia Nauk), which has not been a signatory to joint national science academy statements issued a declaration endorsing the IPCC conclusions, and stating:

> it is the duty of Polish science and the national government to, in a thoughtful, organized and active manner, become involved in realisation of these ideas.
>
> Problems of global warming, climate change, and their various negative impacts on human life and on the functioning of entire societies are one of the most dramatic challenges of modern times.
>
> PAS General Assembly calls on the national scientific communities and the national government to actively support Polish participation in this important endeavor.[37]

Additional national science academy and society statements

- American Association for the Advancement of Science as the world's largest general scientific society, adopted an official statement on climate change in 2006:

  > The scientific evidence is clear: global climate change caused by human activities is occurring now, and it is a growing threat to society....The pace of change and the evidence of harm have increased markedly over the last five years. The time to control greenhouse gas emissions is now.[38]

- Federation of Australian Scientific and Technological Societies in 2008 published *FASTS Statement on Climate Change*[39] which states:

  > Global climate change is real and measurable...To reduce the global net economic, environmental and social losses in the face of these impacts, the policy objective must remain squarely focused on returning greenhouse gas concentrations to near pre-industrial levels through the reduction of emissions. The spatial and temporal fingerprint of warming can be traced to increasing greenhouse gas concentrations in the atmosphere, which are a direct result of burning fossil fuels, broad-scale deforestation and other human activity.

- United States National Research Council through its Committee on the Science of Climate Change in 2001, published *Climate Change Science: An Analysis of Some Key Questions*.[40] This report explicitly endorses the IPCC view of attribution of recent climate change as representing the view of the scientific community:

  > The changes observed over the last several decades are likely mostly due to human activities, but we cannot rule out that some significant part of these changes is also a reflection of natural variability. Human-induced warming and associated sea level rises are expected to continue through the 21st century... The IPCC's conclusion that most of the observed warming of the last 50 years is likely to have been due to the increase in greenhouse gas concentrations accurately reflects the current thinking of the scientific community on this issue.[40]

- Royal Society of New Zealand having signed onto the first joint science academy statement in 2001, released a separate statement in 2008 in order to clear up "the controversy over climate change and its causes, and possible confusion among the public":

> The globe is warming because of increasing greenhouse gas emissions. Measurements show that greenhouse gas concentrations in the atmosphere are well above levels seen for many thousands of years. Further global climate changes are predicted, with impacts expected to become more costly as time progresses. Reducing future impacts of climate change will require substantial reductions of greenhouse gas emissions.[41]

- The Royal Society of the United Kingdom has not changed its concurring stance reflected in its participation in joint national science academies' statements on anthropogenic global warming. According to the Telegraph, "The most prestigious group of scientists in the country was forced to act after fellows complained that doubts over man made global warming were not being communicated to the public".[42] In May 2010, it announced that it "is presently drafting a new public facing document on climate change, to provide an updated status report on the science in an easily accessible form, also addressing the levels of certainty of key components."[43] The society says that it is three years since the last such document was published and that, after an extensive process of debate and review,[44][45] the new document was printed in September 2010. It summarises the current scientific evidence and highlights the areas where the science is well established, where there is still some debate, and where substantial uncertainties remain. The society has stated that "this is not the same as saying that the climate science itself is in error – no Fellows have expressed such a view to the RS".[43] The introduction includes this statement:

> There is strong evidence that the warming of the Earth over the last half-century has been caused largely by human activity, such as the burning of fossil fuels and changes in land use, including agriculture and deforestation.

International science academies

- African Academy of Sciences in 2007 was a signatory to the "statement on sustainability, energy efficiency, and climate change". This joint statement of African science academies, was organized through the Network of African Science Academies. It's stated goal was "to convey information and spur action on the occasion of the G8 Summit in Heiligendamm, Germany, in June 2007".

> A consensus, based on current evidence, now exists within the global scientific community that human activities are the main source of climate change and that the burning of fossil fuels is largely responsible for driving this change.[46]

- European Academy of Sciences and Arts in 2007 issued a formal declaration on climate change titled *Let's Be Honest*:

> Human activity is most *likely* responsible for climate warming. Most of the climatic warming over the last 50 years is *likely* to have been caused by increased concentrations of greenhouse gases in the atmosphere. Documented long-term climate changes include changes in Arctic temperatures and ice, widespread changes in precipitation amounts, ocean salinity, wind patterns and extreme weather including droughts, heavy precipitation, heat waves and the intensity of tropical cyclones. The above development potentially has dramatic consequences for mankind's future.[47]

- European Science Foundation in a 2007 position paper [48] states:

> There is now convincing evidence that since the industrial revolution, human activities, resulting in increasing concentrations of greenhouse gases have become a major agent of climate change... On-going and increased efforts to mitigate climate change through reduction in greenhouse gases are therefore crucial.

- InterAcademy Council As the representative of the world's scientific and engineering academies,[49][50] the InterAcademy Council issued a report in 2007 titled *Lighting the Way: Toward a Sustainable Energy Future*.

Current patterns of energy resources and energy usage are proving detrimental to the long-term welfare of humanity. The integrity of essential natural systems is already at risk from climate change caused by the atmospheric emissions of greenhouse gases.[51] Concerted efforts should be mounted for improving energy efficiency and reducing the carbon intensity of the world economy.[52]

- International Council of Academies of Engineering and Technological Sciences (CAETS) in 2007, issued a *Statement on Environment and Sustainable Growth*:[53]

As reported by the Intergovernmental Panel on Climate Change (IPCC), most of the observed global warming since the mid-20th century is very likely due to human-produced emission of greenhouse gases and this warming will continue unabated if present anthropogenic emissions continue or, worse, expand without control. CAETS, therefore, endorses the many recent calls to decrease and control greenhouse gas emissions to an acceptable level as quickly as possible.

Physical and chemical sciences

- American Chemical Society[54]

- American Institute of Physics[55]

- American Physical Society[56]

- Australian Institute of Physics[57]

- European Physical Society[58]

Earth sciences

American Geophysical Union   The American Geophysical Union (AGU) statement, adopted by the society in 2003, revised in 2007,[59] and revised and expanded in 2013,[60] affirms that rising levels of greenhouse gases have caused and will continue to cause the global surface temperature to be warmer:

"Human activities are changing Earth's climate. At the global level, atmospheric concentrations of carbon dioxide and other heat-trapping greenhouse gases have increased sharply since the Industrial Revolution. Fossil fuel burning dominates this increase. Human-caused increases in greenhouse gases are responsible for most of the observed global average surface warming of roughly 0.8°C (1.5°F) over the past 140 years. Because natural processes cannot quickly remove some of these gases (notably carbon dioxide) from the atmosphere, our past, present, and future emissions will influence the climate system for millennia.

While important scientific uncertainties remain as to which particular impacts will be experienced where, no uncertainties are known that could make the impacts of climate change inconsequential. Furthermore, surprise outcomes, such as the unexpectedly rapid loss of Arctic summer sea ice, may entail even more dramatic changes than anticipated."

American Society of Agronomy, Crop Science Society of America, and Soil Science Society of America   In May, 2011, the American Society of Agronomy (ASA), Crop Science Society of America (CSSA), and Soil Science Society of America (SSSA) issued a joint position statement on climate change as it relates to agriculture:

A comprehensive body of scientific evidence indicates beyond reasonable doubt that global climate change is now occurring and that its manifestations threaten the stability of societies as well as natural and managed ecosystems. Increases in ambient temperatures and changes in related processes are directly linked to rising anthropogenic greenhouse gas (GHG) concentrations in the atmosphere.

Unless the emissions of GHGs are curbed significantly, their concentrations will continue to rise, leading to changes in temperature, precipitation, and other climate variables that will undoubtedly affect agriculture around the world.

Climate change has the potential to increase weather variability as well as gradually increase global temperatures. Both of these impacts have the potential to negatively impact the adaptability and resilience of the world's food production capacity; current research indicates climate change is already reducing the productivity of vulnerable cropping systems.[61]

European Federation of Geologists In 2008, the European Federation of Geologists[62] (EFG) issued the position paper *Carbon Capture and geological Storage* :

The EFG recognizes the work of the IPCC and other organizations, and subscribes to the major findings that climate change is happening, is predominantly caused by anthropogenic emissions of CO_2, and poses a significant threat to human civilization.

It is clear that major efforts are necessary to quickly and strongly reduce CO_2 emissions. The EFG strongly advocates renewable and sustainable energy production, including geothermal energy, as well as the need for increasing energy efficiency.

CCS [Carbon Capture and geological Storage] should also be regarded as a bridging technology, facilitating the move towards a carbon free economy.[63]

European Geosciences Union In 2005, the Divisions of Atmospheric and Climate Sciences of the European Geosciences Union (EGU) issued a position statement in support of the joint science academies' statement on global response to climate change. The statement refers to the Intergovernmental Panel on Climate Change (IPCC), as "the main representative of the global scientific community", and asserts that the IPCC

represents the state-of-the-art of climate science supported by the major science academies around the world and by the vast majority of science researchers and investigators as documented by the peer-reviewed scientific literature.[64]

Additionally, in 2008, the EGU issued a position statement on ocean acidification which states, "Ocean acidification is already occurring today and will continue to intensify, closely tracking atmospheric CO_2 increase. Given the potential threat to marine ecosystems and its ensuing impact on human society and economy, especially as it acts in conjunction with anthropogenic global warming, there is an urgent need for immediate action." The statement then advocates for strategies "to limit future release of CO_2 to the atmosphere and/or enhance removal of excess CO_2 from the atmosphere."[65]

Geological Society of America In 2006, the Geological Society of America adopted a position statement on global climate change. It amended this position on April 20, 2010 with more explicit comments on need for CO_2 reduction.

Decades of scientific research have shown that climate can change from both natural and anthropogenic causes. The Geological Society of America (GSA) concurs with assessments by the National Academies of Science (2005), the National Research Council (2006), and the Intergovernmental Panel on Climate Change (IPCC, 2007) that global climate has warmed and that human activities (mainly greenhouse-gas emissions) account for most of the warming since the middle 1900s. If current trends continue, the projected increase in global temperature by the end of the twentyfirst century will result in large impacts on humans and other species. Addressing the challenges posed by climate change will require a combination of adaptation to the changes that are likely to occur and global reductions of CO_2 emissions from anthropogenic sources.[66]

Geological Society of London In November 2010, the Geological Society of London issued the position statement *Climate change: evidence from the geological record*:

The last century has seen a rapidly growing global population and much more intensive use of resources, leading to greatly increased emissions of gases, such as carbon dioxide and methane, from the burning of fossil fuels (oil, gas and coal), and from agriculture, cement production and deforestation. Evidence from the geological record is consistent with the physics that shows that adding large amounts of carbon dioxide to the atmosphere warms the world and may lead to: higher sea levels and flooding of low-lying coasts; greatly changed patterns of rainfall; increased acidity of the oceans; and decreased oxygen levels in seawater.

There is now widespread concern that the Earth's climate will warm further, not only because of the lingering effects of the added carbon already in the system, but also because of further additions as human population continues to grow. Life on Earth has survived large climate changes in the past, but extinctions and major redistribution of species have been associated with many of them. When the human population was small and nomadic, a rise in sea level of a few metres would have had very little effect on Homo sapiens. With the current and growing global population, much of which is concentrated in coastal cities, such a rise in sea level would have a drastic effect on our complex society, especially if the climate were to change as suddenly as it has at times in the past. Equally, it seems likely that as warming continues some areas may experience less precipitation leading to drought. With both rising seas and increasing drought, pressure for human migration could result on a large scale.[67]

International Union of Geodesy and Geophysics   In July 2007, the International Union of Geodesy and Geophysics (IUGG) adopted a resolution titled "The Urgency of Addressing Climate Change". In it, the IUGG concurs with the "comprehensive and widely accepted and endorsed scientific assessments carried out by the Intergovernmental Panel on Climate Change and regional and national bodies, which have firmly established, on the basis of scientific evidence, that human activities are the primary cause of recent climate change." They state further that the "continuing reliance on combustion of fossil fuels as the world's primary source of energy will lead to much higher atmospheric concentrations of greenhouse gases, which will, in turn, cause significant increases in surface temperature, sea level, ocean acidification, and their related consequences to the environment and society."[68]

National Association of Geoscience Teachers   In July 2009, the National Association of Geoscience Teachers[69] (NAGT) adopted a position statement on climate change in which they assert that "Earth's climate is changing [and] "that present warming trends are largely the result of human activities":

NAGT strongly supports and will work to promote education in the science of climate change, the causes and effects of current global warming, and the immediate need for policies and actions that reduce the emission of greenhouse gases.[70]

Meteorology and oceanography

American Meteorological Society   The American Meteorological Society (AMS) statement adopted by their council in 2012 concluded:

There is unequivocal evidence that Earth's lower atmosphere, ocean, and land surface are warming; sea level is rising; and snow cover, mountain glaciers, and Arctic sea ice are shrinking. The dominant cause of the warming since the 1950s is human activities. This scientific finding is based on a large and persuasive body of research. The observed warming will be irreversible for many years into the future, and even larger temperature increases will occur as greenhouse gases continue to accumulate in the atmosphere. Avoiding this future warming will require a large and rapid reduction in global greenhouse gas emissions. The ongoing warming will increase risks and stresses to human societies, economies, ecosystems, and wildlife through the 21st century and beyond, making it imperative that society respond to a changing climate. To inform decisions on adaptation and mitigation, it is critical that we improve our understanding of the global climate system and our ability to project future climate through continued and improved monitoring and research. This is especially true for smaller (seasonal and regional) scales and weather and climate extremes, and for important hydroclimatic variables such as precipitation and water availability.

Technological, economic, and policy choices in the near future will determine the extent of future impacts of climate change. Science-based decisions are seldom made in a context of absolute certainty. National and international policy discussions should include consideration of the best ways to both adapt to and mitigate climate change. Mitigation will reduce the amount of future climate change and the risk of impacts that are potentially large and dangerous. At the same time, some continued climate change is inevitable, and policy responses should include adaptation to climate change. Prudence dictates extreme care in accounting for our relationship with the only planet known to be capable of sustaining human life.[71]

Australian Meteorological and Oceanographic Society   The Australian Meteorological and Oceanographic Society has issued a *Statement on Climate Change*, wherein they conclude:

Global climate change and global warming are real and observable ... It is highly likely that those human activities that have increased the concentration of greenhouse gases in the atmosphere have been largely responsible for the observed warming since 1950. The warming associated with increases in greenhouse gases originating from human activity is called the enhanced greenhouse effect. The atmospheric concentration of carbon dioxide has increased by more than 30% since the start of the industrial age and is higher now than at any time in at least the past 650,000 years. This increase is a direct result of burning fossil fuels, broad-scale deforestation and other human activity."[72]

Canadian Foundation for Climate and Atmospheric Sciences   In November 2005, the Canadian Foundation for Climate and Atmospheric Sciences (CFCAS) issued a letter to the Prime Minister of Canada stating that

We concur with the climate science assessment of the Intergovernmental Panel on Climate Change (IPCC) in 2001 ... We endorse the conclusions of the IPCC assessment that 'There is new and stronger evidence that most of the warming observed over the last 50 years is attributable to human activities'. ... There is increasingly unambiguous evidence of changing climate in Canada and around the world. There will be increasing impacts of climate change on Canada's natural ecosystems and on our socio-economic activities. Advances in climate science since the 2001 IPCC Assessment have provided more evidence supporting the need for action and development of a strategy for adaptation to projected changes.[73]

Canadian Meteorological and Oceanographic Society   In November 2009, a letter to the Canadian Parliament by The Canadian Meteorological and Oceanographic Society states:

Rigorous international research, including work carried out and supported by the Government of Canada, reveals that greenhouse gases resulting from human activities contribute to the warming of the atmosphere and the oceans and constitute a serious risk to the health and safety of our society, as well as having an impact on all life.[74]

Royal Meteorological Society (UK)   In February 2007, after the release of the IPCC's Fourth Assessment Report, the Royal Meteorological Society issued an endorsement of the report. In addition to referring to the IPCC as "[the] world's best climate scientists", they stated that climate change is happening as "the result of emissions since industrialization and we have already set in motion the next 50 years of global warming – what we do from now on will determine how worse it will get."[75]

World Meteorological Organization   In its *Statement at the Twelfth Session of the Conference of the Parties to the U.N. Framework Convention on Climate Change* presented on November 15, 2006, the World Meteorological Organization (WMO) confirms the need to "prevent dangerous anthropogenic interference with the climate system." The WMO concurs that "scientific assessments have increasingly reaffirmed that human activities are indeed changing the composition of the atmosphere, in particular through the burning of fossil fuels for energy production and transportation." The WMO concurs that "the present atmospheric concentration of CO_2 was never exceeded over the past 420,000 years;" and that the IPCC "assessments provide the most authoritative, up-to-date scientific advice." [76]

American Quaternary Association   The American Quaternary Association (AMQUA) has stated

> Few credible Scientists now doubt that humans have influenced the documented rise of global tempera-
> tures since the Industrial Revolution," citing "the growing body of evidence that warming of the atmosphere,
> especially over the past 50 years, is directly impacted by human activity.[77]

International Union for Quaternary Research   The statement on climate change issued by the International Union
for Quaternary Research (INQUA) reiterates the conclusions of the IPCC, and urges all nations to take prompt action in
line with the UNFCCC principles.

> Human activities are now causing atmospheric concentrations of greenhouse gases — including carbon
> dioxide, methane, tropospheric ozone, and nitrous oxide — to rise well above pre-industrial levels....Increases
> in greenhouse gases are causing temperatures to rise...The scientific understanding of climate change is now
> sufficiently clear to justify nations taking prompt action....Minimizing the amount of this carbon dioxide
> reaching the atmosphere presents a huge challenge but must be a global priority.[78]

Biology and life sciences   Life science organizations have outlined the dangers climate change pose to wildlife.

- American Association of Wildlife Veterinarians[79]

- American Institute of Biological Sciences. In October 2009, the leaders of 18 US scientific societies and organizations sent an open letter to the United States Senate reaffirming the scientific consensus that climate change
  is occurring and is primarily caused by human activities. The American Institute of Biological Sciences (AIBS)
  adopted this letter as their official position statement.[80][81] The letter goes on to warn of predicted impacts on the
  United States such as sea level rise and increases in extreme weather events, water scarcity, heat waves, wildfires,
  and the disturbance of biological systems. It then advocates for a dramatic reduction in emissions of greenhouse
  gases.[82]

- American Society for Microbiology[83]

- Australian Coral Reef Society[84]

- Institute of Biology (UK)[85]

- Society of American Foresters issued two position statements pertaining to climate change in which they cite the
  IPCC[86] and the UNFCCC.[87]

- The Wildlife Society (international)[88]

Human health   A number of health organizations have warned about the numerous negative health effects of global
warming

- American Academy of Pediatrics[89]

- American College of Preventive Medicine[90]

- American Medical Association[91]

- American Public Health Association[92]

- Australian Medical Association in 2004[93] and in 2008[94]

- World Federation of Public Health Associations[95]

- World Health Organization[96]

There is now widespread agreement that the Earth is warming, due to emissions of greenhouse gases caused by human activity. It is also clear that current trends in energy use, development, and population growth will lead to continuing – and more severe – climate change.

The changing climate will inevitably affect the basic requirements for maintaining health: clean air and water, sufficient food and adequate shelter. Each year, about 800,000 people die from causes attributable to urban air pollution, 1.8 million from diarrhoea resulting from lack of access to clean water supply, sanitation, and poor hygiene, 3.5 million from malnutrition and approximately 60,000 in natural disasters. A warmer and more variable climate threatens to lead to higher levels of some air pollutants, increase transmission of diseases through unclean water and through contaminated food, to compromise agricultural production in some of the least developed countries, and increase the hazards of extreme weather.

Miscellaneous   A number of other national scientific societies have also endorsed the opinion of the IPCC:

- American Astronomical Society[97]

- American Statistical Association[98]

- Canadian Council of Professional Engineers [99]

- The Institution of Engineers Australia[100]

- International Association for Great Lakes Research[101]

- Institute of Professional Engineers New Zealand[102]

- The World Federation of Engineering Organizations (WFEO)

Non-committal

American Association of Petroleum Geologists   As of June 2007, the American Association of Petroleum Geologists (AAPG) Position Statement on climate change stated:

> the AAPG membership is divided on the degree of influence that anthropogenic CO_2 has on recent and potential global temperature increases ... Certain climate simulation models predict that the warming trend will continue, as reported through NAS, AGU, AAAS and AMS. AAPG respects these scientific opinions but wants to add that the current climate warming projections could fall within well-documented natural variations in past climate and observed temperature data. These data do not necessarily support the maximum case scenarios forecast in some models.[103]

Prior to the adoption of this statement, the AAPG was the only major scientific organization that rejected the finding of significant human influence on recent climate, according to a statement by the Council of the American Quaternary Association.[13] Explaining the plan for a revision, AAPG president Lee Billingsly wrote in March 2007:

> Members have threatened to not renew their memberships... if AAPG does not alter its position on global climate change... And I have been told of members who already have resigned in previous years because of our current global climate change position... The current policy statement is not supported by a significant number of our members and prospective members.[104]

AAPG President John Lorenz announced the "sunsetting" of AAPG's Global Climate Change Committee in January 2010. The AAPG Executive Committee determined:

> Climate change is peripheral at best to our science [...] AAPG does not have credibility in that field [...] and as a group we have no particular knowledge of global atmospheric geophysics.[105]

American Institute of Professional Geologists   In 2009, the American Institute of Professional Geologists[106] (AIPG) sent a statement to President Barack Obama and other US government officials:

> The geological professionals in AIPG recognize that climate change is occurring and has the potential to yield catastrophic impacts if humanity is not prepared to address those impacts. It is also recognized that climate change will occur regardless of the cause. The sooner a defensible scientific understanding can be developed, the better equipped humanity will be to develop economically viable and technically effective methods to support the needs of society.[107]

Concerned that the original statement issued in March 2009 was too ambiguous, AIPG's National Executive Committee approved a revised position statement issued in January 2010:

> The geological professionals in AIPG recognize that climate change is occurring regardless of cause. AIPG supports continued research into all forces driving climate change.[108]

In March 2010, AIPG's Executive Director issued a statement regarding polarization of opinions on climate change within the membership and announced that the AIPG Executive had made a decision to cease publication of articles and opinion pieces concerning climate change in AIPG's news journal, *The Professional Geologist*.[109] The Executive Director said that "the question of anthropogenicity of climate change is contentious."[110]

Canadian Federation of Earth Sciences

> The science of global climate change is still evolving and our understanding of this vital Earth system is not as developed as is the case for other Earth systems such as plate tectonics. What is known with certainty is that regardless of the causes, our global climate will continue to change for the foreseeable future... The level of CO_2 in our atmosphere is now greater than at any time in the past 500,000 years; there will be consequences for our global climate and natural systems as a result.[111]

Geological Society of Australia

> After a long and extensive and extended consultation with society members, the GSA executive committee has decided not to proceed with a climate change position statement.[112]

Dissenting

See also: List of scientists opposing the mainstream scientific assessment of global warming

As of 2007, when the American Association of Petroleum Geologists released a revised statement,[14] no scientific body of national or international scientists rejects the findings of human-induced effects on climate change.[13][15]

2.17.4   Surveys of scientists and scientific literature

Main article: Surveys of scientists' views on climate change

Various surveys have been conducted to evaluate scientific opinion on global warming. They have concluded that the majority of scientists support the idea of anthropogenic climate change.

In 2004, the geologist and historian of science Naomi Oreskes summarized a study of the scientific literature on climate change.[116] She analyzed 928 abstracts of papers from refereed scientific journals between 1993 and 2003 and concluded that there is a scientific consensus on the reality of anthropogenic climate change.

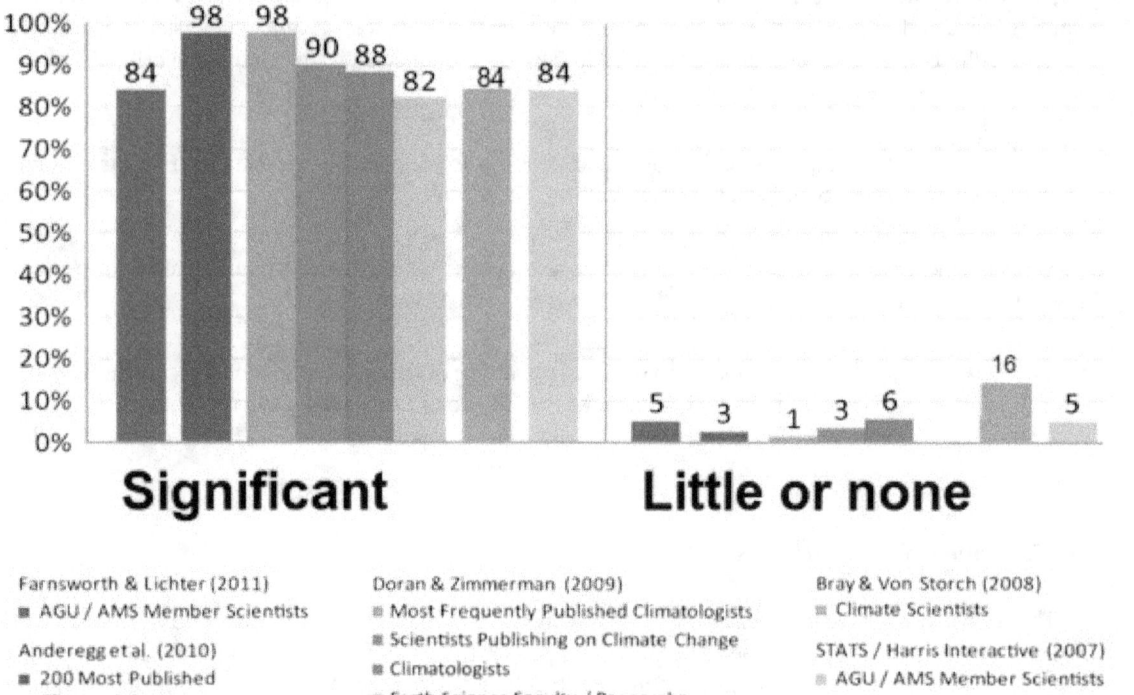

Opinions of Climate and Earth Scientists on Human Role in Global Warming

Farnsworth & Lichter (2011)
■ AGU / AMS Member Scientists

Anderegg et al. (2010)
■ 200 Most Published
Climate Scientists

Doran & Zimmerman (2009)
■ Most Frequently Published Climatologists
■ Scientists Publishing on Climate Change
■ Climatologists
■ Earth Science Faculty / Researchers

Bray & Von Storch (2008)
■ Climate Scientists

STATS / Harris Interactive (2007)
■ AGU / AMS Member Scientists

Summary of opinions from climate and earth scientists regarding climate change. Click to see a more detailed summary of the sources.

Oreskes divided the abstracts into six categories: explicit endorsement of the consensus position, evaluation of impacts, mitigation proposals, methods, paleoclimate analysis, and rejection of the consensus position. Seventy-five per cent of the abstracts were placed in the first three categories (either explicitly or implicitly accepting the consensus view); 25% dealt with methods or paleoclimate, thus taking no position on current anthropogenic climate change. None of the abstracts disagreed with the consensus position, which the author found to be "remarkable". According to the report, "authors evaluating impacts, developing methods, or studying paleoclimatic change might believe that current climate change is natural. However, none of these papers argued that point."

In 2007, Harris Interactive surveyed 489 randomly selected members of either the American Meteorological Society or the American Geophysical Union for the Statistical Assessment Service (STATS) at George Mason University. 97% of the scientists surveyed agreed that global temperatures had increased during the past 100 years; 84% said they personally believed human-induced warming was occurring, and 74% agreed that "currently available scientific evidence" substantiated its occurrence. Catastrophic effects in 50–100 years would likely be observed according to 41%, while 44% thought the effects would be moderate and about 13 percent saw relatively little danger. 5% said they thought human activity did not contribute to greenhouse warming.[117][118][119][120]

Dennis Bray and Hans von Storch conducted a survey in August 2008 of 2058 climate scientists from 34 different countries.[121] A web link with a unique identifier was given to each respondent to eliminate multiple responses. A total of 373 responses were received giving an overall response rate of 18.2%. No paper on climate change consensus based on this survey has been published yet (February 2010), but one on another subject has been published based on the survey.[122]

The survey was composed of 76 questions split into a number of sections. There were sections on the demographics of the respondents, their assessment of the state of climate science, how good the science is, climate change impacts, adaptation

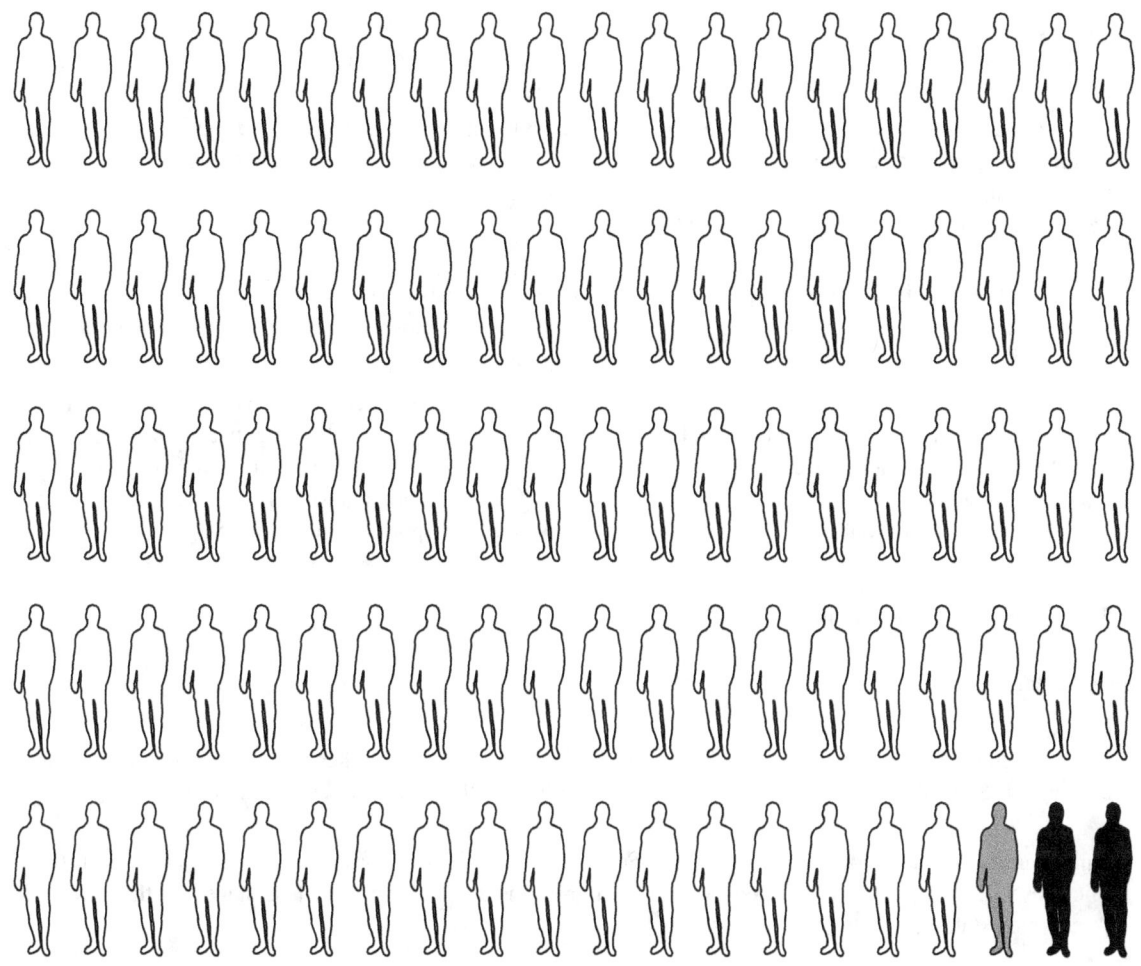

Just over 97% of published climate researchers say humans are causing most global warming.[113][114][115]

and mitigation, their opinion of the IPCC, and how well climate science was being communicated to the public. Most of the answers were on a scale from 1 to 7 from 'not at all' to 'very much'.

To the question "How convinced are you that climate change, whether natural or anthropogenic, is occurring now?", 67.1% said they very much agreed, 26.7% agreed to some large extent, 6.2% said to they agreed to some small extent (2–4), none said they did not agree at all. To the question "How convinced are you that most of recent or near future climate change is, or will be, a result of anthropogenic causes?" the responses were 34.6% very much agree, 48.9% agreeing to a large extent, 15.1% to a small extent, and 1.35% not agreeing at all.

A poll performed by Peter Doran and Maggie Kendall Zimmerman at University of Illinois at Chicago received replies from 3,146 of the 10,257 polled Earth scientists. Results were analyzed globally and by specialization. 76 out of 79 climatologists who "listed climate science as their area of expertise and who also have published more than 50% of their recent peer-reviewed papers on the subject of climate change" believed that mean global temperatures had risen compared to pre-1800s levels. Seventy-five of 77 believed that human activity is a significant factor in changing mean global temperatures. Among all respondents, 90% agreed that temperatures have risen compared to pre-1800 levels, and 82% agreed that humans significantly influence the global temperature. Economic geologists and meteorologists were among the biggest doubters, with only 47 percent and 64 percent, respectively, believing in significant human involvement. The authors summarised the findings:

> It seems that the debate on the authenticity of global warming and the role played by human activity is largely nonexistent among those who understand the nuances and scientific basis of long-term climate processes.[123]

A 2010 paper in the *Proceedings of the National Academy of Sciences of the United States* (PNAS) reviewed publication and citation data for 1,372 climate researchers and drew the following two conclusions:

> (i) 97–98% of the climate researchers most actively publishing in the field support the tenets of ACC (Anthropogenic Climate Change) outlined by the Intergovernmental Panel on Climate Change, and (ii) the relative climate expertise and scientific prominence of the researchers unconvinced of ACC are substantially below that of the convinced researchers.[124]

A 2013 paper in Environmental Research Letters reviewed 11,944 abstracts of scientific papers matching "global warming" or "global climate change". They found 4,014 which discussed the cause of recent global warming, and of these 97.1% endorsed the consensus position.[125]

James L. Powell, a former member of the National Science Board and current executive director of the National Physical Science Consortium, analyzed published research on global warming and climate change between 1991 and 2012 and found that of the 13,950 articles in peer-reviewed journals, only 24 rejected anthropogenic global warming.[126] A follow-up analysis looking at 2,258 peer-reviewed climate articles with 9,136 authors published between November 2012 and December 2013 revealed that only one of the 9,136 authors rejected anthropogenic global warming.[127]

2.17.5 Scientific consensus

See also: Scientific consensus

A question that frequently arises in popular discussion of climate change is whether there is a scientific consensus on climate change.[1] Several scientific organizations have explicitly used the term "consensus" in their statements:

- American Association for the Advancement of Science, 2006: "The conclusions in this statement reflect the scientific consensus represented by, for example, the Intergovernmental Panel on Climate Change, and the Joint National Academies' statement."[38]

- US National Academy of Sciences: "In the judgment of most climate scientists, Earth's warming in recent decades has been caused primarily by human activities that have increased the amount of greenhouse gases in the atmosphere. ... On climate change, [the National Academies' reports] have assessed consensus findings on the science..."[128]

- Joint Science Academies' statement, 2005: "We recognise the international scientific consensus of the Intergovernmental Panel on Climate Change (IPCC)."[129]

- Joint Science Academies' statement, 2001: "The work of the Intergovernmental Panel on Climate Change (IPCC) represents the consensus of the international scientific community on climate change science. We recognise IPCC as the world's most reliable source of information on climate change and its causes, and we endorse its method of achieving this consensus."[32]

- American Meteorological Society, 2003: "The nature of science is such that there is rarely total agreement among scientists. Individual scientific statements and papers—the validity of some of which has yet to be assessed adequately—can be exploited in the policy debate and can leave the impression that the scientific community is sharply divided on issues where there is, in reality, a strong scientific consensus.... IPCC assessment reports are prepared at approximately five-year intervals by a large international group of experts who represent the broad range of expertise and perspectives relevant to the issues. The reports strive to reflect a consensus evaluation of the results of the full body of peer-reviewed research.... They provide an analysis of what is known and not known, the degree of consensus, and some indication of the degree of confidence that can be placed on the various statements and conclusions."[130]

- Network of African Science Academies: "A consensus, based on current evidence, now exists within the global scientific community that human activities are the main source of climate change and that the burning of fossil fuels is largely responsible for driving this change."[35]

- International Union for Quaternary Research, 2008: "INQUA recognizes the international scientific consensus of the Intergovernmental Panel on Climate Change (IPCC)."[78]

- Australian Coral Reef Society,[131] 2006: "There is almost total consensus among experts that the earth's climate is changing as a result of the build-up of greenhouse gases.... There is broad scientific consensus that coral reefs are heavily affected by the activities of man and there are significant global influences that can make reefs more vulnerable such as global warming...."[132]

2.17.6 See also

- 4 Degrees and Beyond International Climate Conference

- Climate change denial

- Economics of global warming

- Effects of global warming

- Global warming controversy

- History of climate change science

- International Year of Planet Earth

- List of authors of *Climate Change 2007: The Physical Science Basis*

- List of climate scientists

- List of scientists opposing the mainstream scientific assessment of global warming

- National Registry of Environmental Professionals#Climate change survey survey on climate change

- Public opinion on climate change

2.17.7 References

[1] Oreskes, Naomi (2007). "The Scientific Consensus on Climate Change: How Do We Know We're Not Wrong?". In DiMento, Joseph F. C.; Doughman, Pamela M. *Climate Change: What It Means for Us, Our Children, and Our Grandchildren*. MIT Press. pp. 65–66. ISBN 978-0-262-54193-0.

[2] "Warming of the climate system is unequivocal, as is now evident from observations of increases in global average air and ocean temperatures, widespread melting of snow and ice and rising global average sea level." IPCC, Synthesis Report, Section 1.1: Observations of climate change, in IPCC AR4 SYR 2007.

[3] IPCC, "Summary for Policymakers", *Detection and Attribution of Climate Change*, «It is extremely likely that human influence has been the dominant cause of the observed warming since the mid-20th century» (page 15) and «In this Summary for Policy-makers, the following terms have been used to indicate the assessed likelihood of an outcome or a result: (...) extremely likely: 95–100%» (page 2)., in IPCC AR5 WG1 2013.

[4] IPCC, Synthesis Report, Section 2.4: Attribution of climate change, in IPCC AR4 SYR 2007."It is likely that increases in GHG concentrations alone would have caused more warming than observed because volcanic and anthropogenic aerosols have offset some warming that would otherwise have taken place."

[5] [Notes-SciPanel] America's Climate Choices: Panel on Advancing the Science of Climate Change; National Research Council (2010). *Advancing the Science of Climate Change*. Washington, D.C.: The National Academies Press. ISBN 0-309-14588-0. (p1) ... there is a strong, credible body of evidence, based on multiple lines of research, documenting that climate is changing and that these changes are in large part caused by human activities. While much remains to be learned, the core phenomenon, scientific questions, and hypotheses have been examined thoroughly and have stood firm in the face of serious scientific debate and careful evaluation of alternative explanations. * * * (p21-22) Some scientific conclusions or theories have been so thoroughly examined and tested, and supported by so many independent observations and results, that their likelihood of subsequently being found to be wrong is vanishingly small. Such conclusions and theories are then regarded as settled facts. This is the case for the conclusions that the Earth system is warming and that much of this warming is very likely due to human activities.

[6] "Summary for Policymakers", *1. Observed changes in climate and their effects*, in IPCC AR4 SYR 2007

[7] "Summary for Policymakers", *2. Causes of change*, in IPCC AR4 SYR 2007

[8] Parry, M.L.; et al., "Technical summary", *Industry, settlement and society, in: Box TS.5. The main projected impacts for systems and sectors*, in IPCC AR4 WG2 2007

[9] IPCC, "Summary for Policymakers", *Magnitudes of impact*, in IPCC AR4 WG2 2007

[10] "Synthesis report", *Ecosystems, in: Sec 3.3.1 Impacts on systems and sectors*, in IPCC AR4 SYR 2007

[11] "Question 1", *1.1*, in IPCC TAR SYR 2001, p. 38

[12] Summary, in US NRC 2001, p. 4

[13] Julie Brigham-Grette; et al. (September 2006). "Petroleum Geologists' Award to Novelist Crichton Is Inappropriate" (PDF). *Eos* **87** (36): 364. Bibcode:2006EOSTr..87..364B. doi:10.1029/2006EO360008. Retrieved 2007-01-23. The AAPG stands alone among scientific societies in its denial of human-induced effects on global warming.

[14] AAPG *Climate Change* June 2007

[15] Oreskes 2007, p. 68

[16] Ogden, Aynslie and Cohen, Stewart (2002). *"Integration and Synthesis: Assessing Climate Change Impacts in Northern Canada"* (PDF). Retrieved 2009-04-12.

[17] "Climate Change 2014 Synthesis Report Summary for Policymakers" (PDF). IPCC. Retrieved 1 August 2015.

[18] Nuccitelli, Dana (31 March 2014). "IPCC report warns of future climate change risks, but is spun by contrarians". The Guardian. Retrieved 1 August 2015.

[19] "U.N. Climate Panel Endorses Ceiling on Global Emissions". The New York Times. 27 September 2013. Retrieved 1 August 2015.

[20] "Warming 'very likely' human-made". *BBC News* (BBC). 2007-02-01. Retrieved 2007-02-01.

[21] Rosenthal, Elisabeth; Revkin, Andrew C. (2007-02-03). "Science Panel Calls Global Warming 'Unequivocal'". *New York Times*. Retrieved 2010-08-28. the leading international network of climate scientists has concluded for the first time that global warming is 'unequivocal' and that human activity is the main driver, 'very likely' causing most of the rise in temperatures since 1950

[22] Stevens, William K. (2007-02-06). "On the Climate Change Beat, Doubt Gives Way to Certainty". *New York Times*. Retrieved 2010-08-28. The Intergovernmental Panel on Climate Change said the likelihood was 90 percent to 99 percent that emissions of heat-trapping greenhouse gases like carbon dioxide, spewed from tailpipes and smokestacks, were the dominant cause of the observed warming of the last 50 years. In the panel's parlance, this level of certainty is labeled "very likely." Only rarely does scientific odds-making provide a more definite answer than that, at least in this branch of science, and it describes the endpoint, so far, of a progression.

[23] "U.N. Report: Global Warming Man-Made, Basically Unstoppable". Fox News. Retrieved 2012-07-30.

[24] Downloads.globalchange.gov

[25] "Impacts of a Warming Arctic: Arctic Climate Impact Assessment New Scientific Consensus: Arctic Is Warming Rapidly". UNEP/GRID-Arendal. 2004-11-08. Retrieved 2010-01-20.

[26] "ACIA Display". Amap.no. Retrieved 2012-07-30.

[27] The literature has been assessed by the IPCC, e.g., see:

- Adger, W.N.; et al., *Ch 17: Assessment of Adaptation Practices, Options, Constraints and Capacity*, in IPCC AR4 WG2 2007

- Barker, T.; et al., *Technical summary*, in IPCC AR4 WG3 2007

[28] 2009 Joint Science Academies' Statement

[29] *Doha Declaration on Climate, Health and Wellbeing.* This statement has been signed by numerous medical organizations, including the World Medical Association.

[30] Arnold, D.G., ed. (March 2011), *The Ethics of Global Climate Change*, Cambridge University Press, ISBN 9781107000698

[31] "Editorial: The Science of Climate Change". *Science* **292** (5520): 1261. May 18, 2001. doi:10.1126/science.292.5520.1261.

[32] The Science of Climate Change, The Royal Society

[33] Joint science academies' statement: Global response to climate change, 2005

[34] 2007 Joint Science Academies' Statement

[35] "Joint statement by the Network of African Science Academies (NASAC) to the G8 on sustainability, energy efficiency and climate change" (PDF). Network of African Science Academies. 2007. Retrieved 2012-08-28.

[36] 2008 Joint Science Academies' Statement

[37] "Stanowisko Zgromadzenia Ogólnego PAN z dnia 13 grudnia 2007 r" (PDF) (in Polish). Polish Academy of Sciences. Retrieved 2009-06-16. *Note*: As of 16 June 2009, PAS has not issued this statement in English, all citations have been translated from Polish.

[38] AAAS Board Statement on Climate Change *www.aaas.org* December 2006

[39] *FASTS Statement on Climate Change* (PDF), 2008 "Global climate change is real and measurable. Since the start of the 20th century, the global mean surface temperature of the Earth has increased by more than 0.7°C and the rate of warming has been largest in the last 30 years. Key vulnerabilities arising from climate change include water resources, food supply, health, coastal settlements, biodiversity and some key ecosystems such as coral reefs and alpine regions. As the atmospheric concentration of greenhouse gases increases, impacts become more severe and widespread. To reduce the global net economic, environmental and social losses in the face of these impacts, the policy objective must remain squarely focused on returning greenhouse gas concentrations to near pre-industrial levels through the reduction of emissions. The spatial and temporal fingerprint of warming can be traced to increasing greenhouse gas concentrations in the atmosphere, which are a direct result of burning fossil fuels, broad-scale deforestation and other human activity."

[40] Committee on the Science of Climate Change, Division on Earth and Life Studies, National Research Council (2001). *Climate Change Science: An Analysis of Some Key Questions*. Washington DC: National Academy Press. ISBN 0-309-07574-2.

[41] Wratt, David; Renwick, James (2008-07-10). "Climate change statement from the Royal Society of New Zealand". The Royal Society of New Zealand. Retrieved 2010-01-20.

[42] Gray, Louise (May 29, 2010). "Royal Society to publish guide on climate change to counter claims of 'exaggeration'". *The Daily Telegraph* (London).

[43] "New guide to science of climate change". The Royal Society. Retrieved 9 June 2010.

[44] Harrabin, Roger (27 May 2010). "Society to review climate message". BBC News. Retrieved 9 June 2010.

[45] Gardner, Dan (8 June 2010). "Some excitable climate-change deniers just don't understand what science is". Montreal Gazette. Archived from the original on 11 June 2010. Retrieved 9 June 2010.

[46] "Joint statement by the Network of African Science Academies (NASAC) to the G8 on sustainability, energy efficiency and climate change". 2007. Retrieved 22 May 2015. A consensus, based on current evidence, now exists within the global scientific community that human activities are the main source of climate change and that the burning of fossil fuels is largely responsible for driving this change... Although we recognize that this nexus poses daunting challenges for the developed world, we firmly believe that these challenges are even more daunting for the most impoverished, science-poor regions of the developing world, especially in Africa.

[47] European Academy of Sciences and Arts *Let's Be Honest*

[48] *European Science Foundation Position Paper* Impacts of Climate Change on the European Marine and Coastal Environment — Ecosystems Approach, 2007, pp. 7–10 "There is now convincing evidence that since the industrial revolution, human activities, resulting in increasing concentrations of greenhouse gases have become a major agent of climate change. These greenhouse gases affect the global climate by retaining heat in the troposphere, thus raising the average temperature of the planet and altering global atmospheric circulation and precipitation patterns. While on-going national and international actions to curtail and reduce greenhouse gas emissions are essential, the levels of greenhouse gases currently in the atmosphere, and their impact, are likely to persist for several decades. On-going and increased efforts to mitigate climate change through reduction in greenhouse gases are therefore crucial."

[49] Panel Urges Global Shift on Sources of Energy

[50] "InterAcademy Council". InterAcademy Council. Retrieved 2012-07-30.

[51] "InterAcademy Council". InterAcademy Council. Retrieved 2012-07-30.

[52] "InterAcademy Council". InterAcademy Council. Retrieved 2012-07-30.

[53] http://www.caets.org/nae/naecaets.nsf/(weblinks)/WSAN-78QL9A?OpenDocument

[54] *American Chemical Society* Global Climte Change "Careful and comprehensive scientific assessments have clearly demonstrated that the Earth's climate system is changing rapidly in response to growing atmospheric burdens of greenhouse gases and absorbing aerosol particles (IPCC, 2007). There is very little room for doubt that observed climate trends are due to human activities. The threats are serious and action is urgently needed to mitigate the risks of climate change. The reality of global warming, its current serious and potentially disastrous impacts on Earth system properties, and the key role emissions from human activities play in driving these phenomena have been recognized by earlier versions of this ACS policy statement (ACS, 2004), by other major scientific societies, including the American Geophysical Union (AGU, 2003), the American Meteorological Society (AMS, 2007) and the American Association for the Advancement of Science (AAAS, 2007), and by the U. S. National Academies and ten other leading national academies of science (NA, 2005)."

[55] *American Institute of Physics Statement supporting AGU statement on human-induced climate change*, 2003 "The Governing Board of the American Institute of Physics has endorsed a position statement on climate change adopted by the American Geophysical Union (AGU) Council in December 2003."

[56] *American Physical Society Climate Change Policy Statement*, November 2007 "Emissions of greenhouse gases from human activities are changing the atmosphere in ways that affect the Earth's climate. Greenhouse gases include carbon dioxide as well as methane, nitrous oxide and other gases. They are emitted from fossil fuel combustion and a range of industrial and agricultural processes. The evidence is incontrovertible: Global warming is occurring. If no mitigating actions are taken, significant disruptions in the Earth's physical and ecological systems, social systems, security and human health are likely to occur. We must reduce emissions of greenhouse gases beginning now. Because the complexity of the climate makes accurate prediction difficult, the APS urges an enhanced effort to understand the effects of human activity on the Earth's climate, and to provide the technological options for meeting the climate challenge in the near and longer terms. The APS also urges governments, universities, national laboratories and its membership to support policies and actions that will reduce the emission of greenhouse gases.

[57] *AIP science policy document.* (PDF), 2005 "Policy: The AIP supports a reduction of the green house gas emissions that are leading to increased global temperatures, and encourages research that works towards this goal. Reason: Research in Australia and overseas shows that an increase in global temperature will adversely affect the Earth's climate patterns. The melting of the polar ice caps, combined with thermal expansion, will lead to rises in sea levels that may impact adversely on our coastal cities. The impact of these changes on biodiversity will fundamentally change the ecology of Earth."

[58] *EPS Position Paper* Energy for the future: The Nuclear Option (PDF), 2007 "The emission of anthropogenic greenhouse gases, among which carbon dioxide is the main contributor, has amplified the natural greenhouse effect and led to global warming. The main contribution stems from burning fossil fuels. A further increase will have decisive effects on life on earth. An energy cycle with the lowest possible CO_2 emission is called for wherever possible to combat climate change."

[59] "AGU Position Statement: Human Impacts on Climate". Agu.org. Retrieved 2012-07-30.

[60] "Human-induced Climate Change Requires Urgent Action". *Position Statement*. American Geophysical Union. Retrieved 14 August 2013.

[61] ASA, CSSA, and SSSA Position Statement on Climate Change

[62] "EFG Website | Home". Eurogeologists.de. 2011-08-10. Retrieved 2012-07-30.

[63] EFG *Carbon Capture and geological Storage*

[64] http://www.egu.eu/statements/position-statement-of-the-divisions-of-atmospheric-and-climate-sciences-7-july-2005.html

[65] http://www.egu.eu/statements/egu-position-statement-on-ocean-acidification.html

[66] "The Geological Society of America - Position Statement on Global Climate Change". Geosociety.org. Retrieved 2012-07-30.

[67] "Geological Society - Climate change: evidence from the geological record". Geolsoc.org.uk. Retrieved 2012-07-30.

[68] IUGG Resolution 6

[69] http://www.nagt.org/index.html

[70] http://nagt.org/nagt/organization/ps-climate.html

[71] "AMS Information Statement on Climate Change". Ametsoc.org. 2012-08-20. Retrieved 2012-08-27.

[72] "Statement". AMOS. Retrieved 2012-07-30.

[73] CFCAS Letter to PM, November 25, 2005

[74] Canadian Meteorological and Oceanographic Society Letter to Stephen Harper (Updated, 2007)

[75] http://www.rmets.org/news/detail.php?ID=332

[76] WMO's Statement at the Twelfth Session of the Conference of the Parties to the U.N. Framework Convention on Climate Change.

[77] AMQUA "Petroleum Geologists' Award to Novelist Crichton Is Inappropriate"

[78] INQUA Statement On Climate Change.

[79] *AAWV* Position Statement on Climate Change, Wildlife Diseases, and Wildlife Health "There is widespread scientific agreement that the world's climate is changing and that the weight of evidence demonstrates that anthropogenic factors have and will continue to contribute significantly to global warming and climate change. It is anticipated that continuing changes to the climate will have serious negative impacts on public, animal and ecosystem health due to extreme weather events, changing disease transmission dynamics, emerging and re-emerging diseases, and alterations to habitat and ecological systems that are essential to wildlife conservation. Furthermore, there is increasing recognition of the inter-relationships of human, domestic animal, wildlife, and ecosystem health as illustrated by the fact the majority of recent emerging diseases have a wildlife origin."

[80] *AIBS Position Statements* "Observations throughout the world make it clear that climate change is occurring, and rigorous scientific research demonstrates that the greenhouse gases emitted by human activities are the primary driver."

[81] Scientific societies warn Senate: climate change is real, Ars Technica, October 22, 2009

[82] *Letter to US Senators* (PDF), October 2009

[83] *Global Environmental Change — Microbial Contributions, Microbial Solutions* (PDF), American Society For Microbiology, May 2006 They recommended "reducing net anthropogenic CO_2 emissions to the atmosphere" and "minimizing anthropogenic disturbances of" atmospheric gases. Carbon dioxide concentrations were relatively stable for the past 10,000 years but then began to increase rapidly about 150 years ago...as a result of fossil fuel consumption and land use change. Of course, changes in atmospheric composition are but one component of global change, which also includes disturbances in the physical and chemical conditions of the oceans and land surface. Although global change has been a natural process throughout Earth's history, humans are responsible for substantially accelerating present-day changes. These changes may adversely affect human health and the biosphere on which we depend. Outbreaks of a number of diseases, including Lyme disease, hantavirus infections, dengue fever, bubonic plague, and cholera, have been linked to climate change."

[84] *Australian Coral Reef Society official letter* (PDF), 2006, archived from the original (PDF) on 22 March 2006 Official communique regarding the Great Barrier Reef and the "world-wide decline in coral reefs through processes such as overfishing, runoff of nutrients from the land, coral bleaching, global climate change, ocean acidification, pollution", etc.: There is almost total consensus among experts that the earth's climate is changing as a result of the build-up of greenhouse gases. The IPCC (involving over 3,000 of the world's experts) has come out with clear conclusions as to the reality of this phenomenon. One does not have to look further than the collective academy of scientists worldwide to see the string (of) statements on this worrying change to the earth's atmosphere. There is broad scientific consensus that coral reefs are heavily affected by the activities of man and there are significant global influences that can make reefs more vulnerable such as global warming....It is highly likely that coral bleaching has been exacerbated by global warming."

[85] *Institute of Biology policy page 'Climate Change'* "there is scientific agreement that the rapid global warming that has occurred in recent years is mostly anthropogenic, *ie* due to human activity." As a consequence of global warming, they warn that a "rise in sea levels due to melting of ice caps is expected to occur. Rises in temperature will have complex and frequently localised effects on weather, but an overall increase in extreme weather conditions and changes in precipitation patterns are probable, resulting in flooding and drought. The spread of tropical diseases is also expected." Subsequently, the Institute of Biology advocates policies to reduce "greenhouse gas emissions, as we feel that the consequences of climate change are likely to be severe."

[86] *SAF* Forest Management and Climate Change (PDF), 2008 "Forests are shaped by climate....Changes in temperature and precipitation regimes therefore have the potential to dramatically affect forests nationwide. There is growing evidence that our climate is changing. The changes in temperature have been associated with increasing concentrations of atmospheric carbon dioxide (CO_2) and other GHGs in the atmosphere."

[87] *SAF* Forest Offset Projects in a Carbon Trading System (PDF), 2008 "Forests play a significant role in offsetting CO_2 emissions, the primary anthropogenic GHG."

[88] *Wildlife Society* Global Climate Change and Wildlife (PDF) "Scientists throughout the world have concluded that climate research conducted in the past two decades definitively shows that rapid worldwide climate change occurred in the 20th century, and will likely continue to occur for decades to come. Although climates have varied dramatically since the Earth was formed, few scientists question the role of humans in exacerbating recent climate change through the emission of greenhouse gases. The critical issue is no longer "if" climate change is occurring, but rather how to address its effects on wildlife and wildlife habitats." The statement goes on to assert that "evidence is accumulating that wildlife and wildlife habitats have been and will continue to be significantly affected by ongoing large-scale rapid climate change." The statement concludes with a call for "reduction in anthropogenic (human-caused) sources of carbon dioxide and other greenhouse gas emissions contributing to global climate change and the conservation of CO_2- consuming photosynthesizers (i.e., plants)."

[89] *AAP* Global Climate Change and Children's Health, 2007 "There is broad scientific consensus that Earth's climate is warming rapidly and at an accelerating rate. Human activities, primarily the burning of fossil fuels, are very likely (>90% probability) to be the main cause of this warming. Climate-sensitive changes in ecosystems are already being observed, and fundamental, potentially irreversible, ecological changes may occur in the coming decades. Conservative environmental estimates of the impact of climate changes that are already in process indicate that they will result in numerous health effects to children. Anticipated direct health consequences of climate change include injury and death from extreme weather events and natural disasters, increases in climate-sensitive infectious diseases, increases in air pollution–related illness, and more heat-related, potentially fatal, illness. Within all of these categories, children have increased vulnerability compared with other groups."

[90] *ACPM Policy Statement* Abrupt Climate Change and Public Health Implications, 2006 "The American College of Preventive Medicine (ACPM) accept the position that global warming and climate change is occurring, that there is potential for abrupt climate change, and that human practices that increase greenhouse gases exacerbate the problem, and that the public health consequences may be severe."

[91] *American Medical Association Policy Statement*, 2008 "Support the findings of the latest Intergovernmental Panel on Climate Change report, which states that the Earth is undergoing adverse global climate change and that these changes will negatively affect public health. Support educating the medical community on the potential adverse public health effects of global climate change, including topics such as population displacement, flooding, infectious and vector-borne diseases, and healthy water supplies."

[92] *American Public Health Association Policy Statement "Addressing the Urgent Threat of Global Climate Change to Public Health and the Environment"*, 2007 "The long-term threat of global climate change to global health is extremely serious and the fourth IPCC report and other scientific literature demonstrate convincingly that anthropogenic GHG emissions are primarily responsible for this threat....US policy makers should immediately take necessary steps to reduce US emissions of GHGs, including carbon dioxide, to avert dangerous climate change."

[93] *AMA* Climate Change and Human Health — 2004, 2004 They recommend policies "to mitigate the possible consequential health effects of climate change through improved energy efficiency, clean energy production and other emission reduction steps."

[94] *AMA* Climate Change and Human Health — *2004. Revised 2008.*, 2008 "The world's climate – our life-support system – is being altered in ways that are likely to pose significant direct and indirect challenges to health. While 'climate change' can be due to natural forces or human activity, there is now substantial evidence to indicate that human activity – and specifically increased greenhouse gas (GHGs) emissions – is a key factor in the pace and extent of global temperature increases. Health impacts of climate change include the direct impacts of extreme events such as storms, floods, heatwaves and fires and the indirect effects of longer-term changes, such as drought, changes to the food and water supply, resource conflicts and population shifts. Increases in average temperatures mean that alterations in the geographic range and seasonality of certain infections and diseases (including vector-borne diseases such as malaria, dengue fever, Ross River virus and food-borne infections such as Salmonellosis) may be among the first detectable impacts of climate change on human health. Human health is ultimately dependent on the health of the planet and its ecosystem. The AMA believes that measures which mitigate climate change will also benefit public health. Reducing GHGs should therefore be seen as a public health priority."

[95] *World Federation of Public Health Associations resolution "Global Climate Change"* (PDF), 2001 "Noting the conclusions of the United Nations' Intergovernmental Panel on Climate Change (IPCC) and other climatologists that anthropogenic greenhouse

gases, which contribute to global climate change, have substantially increased in atmospheric concentration beyond natural processes and have increased by 28 percent since the industrial revolution....Realizing that subsequent health effects from such perturbations in the climate system would likely include an increase in: heat-related mortality and morbidity; vector-borne infectious diseases,... water-borne diseases...(and) malnutrition from threatened agriculture....the World Federation of Public Health Associations...recommends precautionary primary preventive measures to avert climate change, including reduction of greenhouse gas emissions and preservation of greenhouse gas sinks through appropriate energy and land use policies, in view of the scale of potential health impacts...."

[96] *WHO* Protecting health from climate change (PDF), 2008, p. 2, retrieved 2009-04-18

[97] *Statement supporting AGU statement on human-induced climate change*, American Astronomical Society, 2004 "In endorsing the "Human Impacts on Climate" statement [issued by the American Geophysical Union], the AAS recognizes the collective expertise of the AGU in scientific subfields central to assessing and understanding global change, and acknowledges the strength of agreement among our AGU colleagues that the global climate is changing and human activities are contributing to that change."

[98] *ASA Statement on Climate Change*, November 30, 2007 "The ASA endorses the IPCC conclusions.... Over the course of four assessment reports, a small number of statisticians have served as authors or reviewers. Although this involvement is encouraging, it does not represent the full range of statistical expertise available. ASA recommends that more statisticians should become part of the IPCC process. Such participation would be mutually beneficial to the assessment of climate change and its impacts and also to the statistical community."

[99] Lapp, David. "What Is Climate Change". *Canadian Council of Professional Engineers.* Retrieved 18 August 2015.

[100] *Policy Statement, Climate Change and Energy*, February 2007 "Engineers Australia believes that Australia must act swiftly and proactively in line with global expectations to address climate change as an economic, social and environmental risk... We believe that addressing the costs of atmospheric emissions will lead to increasing our competitive advantage by minimising risks and creating new economic opportunities. Engineers Australia believes the Australian Government should ratify the Kyoto Protocol."

[101] *IAGLR Fact Sheet* The Great Lakes at a Crossroads: Preparing for a Changing Climate (PDF), February 2009 "While the Earth's climate has changed many times during the planet's history because of natural factors, including volcanic eruptions and changes in the Earth's orbit, never before have we observed the present rapid rise in temperature and carbon dioxide (CO_2). Human activities resulting from the industrial revolution have changed the chemical composition of the atmosphere....Deforestation is now the second largest contributor to global warming, after the burning of fossil fuels. These human activities have significantly increased the concentration of "greenhouse gases" in the atmosphere. As the Earth's climate warms, we are seeing many changes: stronger, more destructive hurricanes; heavier rainfall; more disastrous flooding; more areas of the world experiencing severe drought; and more heat waves."

[102] *IPENZ Informatory Note,* Climate Change and the greenhouse effect (PDF), October 2001 "Human activities have increased the concentration of these atmospheric greenhouse gases, and although the changes are relatively small, the equilibrium maintained by the atmosphere is delicate, and so the effect of these changes is significant. The world's most important greenhouse gas is carbon dioxide, a by-product of the burning of fossil fuels. Since the time of the Industrial Revolution about 200 years ago, the concentration of carbon dioxide in the atmosphere has increased from about 280 parts per million to 370 parts per million, an increase of around 30%. On the basis of available data, climate scientists are now projecting an average global temperature rise over this century of 2.0 to 4.5°C. This compared with 0.6°C over the previous century – about a 500% increase... This could lead to changing, and for all emissions scenarios more unpredictable, weather patterns around the world, less frost days, more extreme events (droughts and storm or flood disasters), and warmer sea temperatures and melting glaciers causing sea levels to rise. ... Professional engineers commonly deal with risk, and frequently have to make judgments based on incomplete data. The available evidence suggests very strongly that human activities have already begun to make significant changes to the earth's climate, and that the long-term risk of delaying action is greater than the cost of avoiding/minimising the risk."

[103] AAPG Position Statement: Climate Change from dpa.aapg.org

[104] "Climate :03:2007 EXPLORER". Aapg.org. Retrieved 2012-07-30.

[105] *Sunsetting the Global Climate Change Committee*, *The Professional Geologist*, March/April 2010, p. 28

[106] "American Geological Institute Climate Statement". 12 Feb 1999. Archived from the original on July 2012. Retrieved July 2012.

[107] AIPG Climate Change Letters sent to U.S. Government Officials

[108] "AIPG Climate Change and Domestic Energy Statement", *The Professional Geologist*, January/February 2010, p. 42

[109] "The Professional Geologist publications". Archived from the original on July 2012. Retrieved July 2012.

[110] "Climate Change and Society Governance", *The Professional Geologist*, March/April 2010, p. 33

[111] billobrien.coml. "Canadian Federation of Earth Sciences (CFES)". Geoscience.ca. Retrieved 2012-07-30.

[112] Graham Lloyd (June 4, 2014). "Earth scientists split on climate change statement". *The Australian*. Retrieved June 4, 2014.(subscription required)

[113] Anderegg, William R L; Prall, James W.; Harold, Jacob; Schneider, Stephen H. (2010). "Expert credibility in climate change". *Proc. Natl. Acad. Sci. U.S.A.* **107** (27): 12107–9. Bibcode:2010PNAS..10712107A. doi:10.1073/pnas.1003187107. PMC 2901439. PMID 20566872. Retrieved 22 August 2011.

[114] Doran consensus article 2009

[115] John Cook, Dana Nuccitelli, Sarah A Green, Mark Richardson, Bärbel Winkler, Rob Painting, Robert Way, Peter Jacobs. Andrew Skuce (15 May 2013). "Expert credibility in climate change". *Environ. Res. Lett.* **8** (2): 024024. Bibcode:2013ERL.....8b4024C. doi:10.1088/1748-9326/8/2/024024.

[116] Naomi Oreskes (December 3, 2004). "Beyond the Ivory Tower: The Scientific Consensus on Climate Change" (PDF). *Science* **306** (5702): 1686. doi:10.1126/science.1103618. PMID 15576594. (see also for an exchange of letters to Science)

[117] Lavelle, Marianne (2008-04-23). "Survey Tracks Scientists' Growing Climate Concern". U.S. News & World Report. Retrieved 2010-01-20.

[118] Lichter, S. Robert (2008-04-24). "Climate Scientists Agree on Warming, Disagree on Dangers, and Don't Trust the Media's Coverage of Climate Change". Statistical Assessment Service, George Mason University. Retrieved 2010-01-20.

[119] ""Structure of Scientific Opinion on Climate Change" at Journalist's Resource.org".

[120] Stephen J. Farnsworth, S. Robert Lichter (October 27, 2011). "The Structure of Scientific Opinion on Climate Change". International Journal of Public Opinion Research. Retrieved December 2, 2011.

[121] Bray, Dennis; von Storch, Hans (2009). "A Survey of the Perspectives of Climate Scientists Concerning Climate Science and Climate Change" (PDF).

[122] Bray, D.; von Storch H. (2009). "Prediction' or 'Projection; The nomenclature of climate science". *Science Communication* **30** (4): 534–543. doi:10.1177/1075547009333698.

[123] Doran, Peter T.; Maggie Kendall Zimmerman (January 20, 2009). "Examining the Scientific Consensus on Climate Change" (PDF). *EOS* **90** (3): 22–23. Bibcode:2009EOSTr..90...22D. doi:10.1029/2009EO030002.

[124] Anderegg, William R L; Prall, James W.; Harold, Jacob; Schneider, Stephen H. (2010). "Expert credibility in climate change" (PDF). *Proc. Natl. Acad. Sci. U.S.A.* **107** (27): 12107–9. Bibcode:2010PNAS..10712107A. doi:10.1073/pnas.1003187107. PMC 2901439. PMID 20566872.

[125] Cook, J.; Nuccitelli, D.; Green, S.A.; Richardson, M.; Winkler, B.; Painting, R.; Way, R.; Jacobs, P.; Skuc, A. (2013). "Quantifying the consensus on anthropogenic global warming in the scientific literature". *Environ. Res. Lett.* **8** (2): 024024. Bibcode:2013ERL.....8b4024C. doi:10.1088/1748-9326/8/2/024024.

[126] Plait, P. (11 December 2012). "Why Climate Change Denial Is Just Hot Air". *Slate*. Retrieved 14 February 2014.

[127] Plait, P. (14 January 2014). "The Very, Very Thin Wedge of Denial". *Slate*. Retrieved 14 February 2014.

[128] US NRC (2008). *Understanding and Responding to Climate Change. A brochure prepared by the US National Research Council (US NRC)* (PDF). Washington DC, USA: US National Academy of Sciences.

[129] Joint Science Academies' Statement

[130] "Climate Change Research: Issues for the Atmospheric and Related Sciences Adopted by the AMS Council 9 February 2003". Ametsoc.org. 2003-02-09. Retrieved 2012-07-30.

[131] "Australian Coral Reef Society". Australian Coral Reef Society. Retrieved 2012-07-30.

[132] Australian Coral Reef Society official letter, June 16, 2006

- IPCC TAR SYR (2001), Watson, R. T.; and the Core Writing Team, ed., *Climate Change 2001: Synthesis Report*, Contribution of Working Groups I, II, and III to the Third Assessment Report of the Intergovernmental Panel on Climate Change, Cambridge University Press, ISBN 0-521-80770-0 (pb: 0-521-01507-3).

- IPCC AR4 WG2 (2007), Parry, M.L.; Canziani, O.F.; Palutikof, J.P.; van der Linden, P.J.; and Hanson, C.E., ed., *Climate Change 2007: Impacts, Adaptation and Vulnerability*, Contribution of Working Group II to the Fourth Assessment Report of the Intergovernmental Panel on Climate Change, Cambridge University Press, ISBN 978-0-521-88010-7 (pb: 978-0-521-70597-4).

- IPCC AR4 WG3 (2007), Metz, B.; Davidson, O.R.; Bosch, P.R.; Dave, R.; and Meyer, L.A., ed., *Climate Change 2007: Mitigation of Climate Change*, Contribution of Working Group III to the Fourth Assessment Report of the Intergovernmental Panel on Climate Change, Cambridge University Press, ISBN 978-0-521-88011-4 (pb: 978-0-521-70598-1).

- IPCC AR4 SYR (2007), Core Writing Team; Pachauri, R.K; and Reisinger, A., ed., *Climate Change 2007: Synthesis Report (SYR)*, Contribution of Working Groups I, II and III to the Fourth Assessment Report (AR4) of the Intergovernmental Panel on Climate Change, Geneva, Switzerland: IPCC, ISBN 92-9169-122-4.

- US NRC (2001), *Climate Change Science: An Analysis of Some Key Questions. A report by the Committee on the Science of Climate Change, US National Research Council (NRC)*, Washington, D.C., USA: National Academy Press, ISBN 0-309-07574-2, archived from the original on 5 June 2011

2.17.8 External links

- The Radical Emission Reduction Conference

- Robin Lloyd (23 February 2011). "Why Are Americans So Ill-Informed about Climate Change?". Scientific American. Retrieved 31 March 2011.

2.18 Semmelweis reflex

The **Semmelweis reflex** or "**Semmelweis effect**" is a metaphor for the reflex-like tendency to reject new evidence or new knowledge because it contradicts established norms, beliefs or paradigms.

The term originated from the story of Ignaz Semmelweis, who discovered that childbed fever mortality rates reduced ten-fold when doctors washed their hands with a chlorine solution between patients and, most in particular, after an autopsy (the institution where Semmelweis worked, a university hospital, performed autopsies on every deceased patient). Semmelweis's decision stopped the ongoing contamination of patients—mostly pregnant women—with "cadaverous particles".[1] His hand-washing suggestions were rejected by his contemporaries, often for non-medical reasons. For instance, some doctors refused to believe that a gentleman's hands could transmit disease (see Contemporary reaction to Ignaz Semmelweis).

While there is some uncertainty regarding the origin and generally accepted use of the expression, the expression Semmelweis Reflex has been documented and at least used by the author Robert Anton Wilson.[2] In his book *The Game of Life*, Timothy Leary provided the following polemical definition of the Semmelweis reflex: "Mob behavior found among primates and larval hominids on undeveloped planets, in which a discovery of important scientific fact is punished". The expression has found way into philosophy and religious studies as "unmitigated Humean skepticism concerning causality".[3]

2.18.1 See also

- Confirmation bias

- Conservatism (belief revision)

- Cognitive dissonance

- Paradigm shift

- Schopenhauer

2.18.2 References

[1] Levitt, Steven D (2009). "4". *Super Freakanomics*. William Morrow. ISBN 0-06-088957-8.

[2] Wilson, Robert Anton (1991). *The Game of Life*. New Falcon Publications. ISBN 1561840505.

[3] http://www.ctr4process.org/publications/ProcessStudies/PSS/2006-9-ScarfeA-On_Determinations_of_Causal_Connection.pdf

2.19   State of Fear

State of Fear is a 2004 techno-thriller novel by Michael Crichton in which eco-terrorists plot mass murder to publicize the danger of global warming. Despite being a work of fiction, the book contains many graphs and footnotes, two appendices, and a twenty-page bibliography in support of Crichton's beliefs about global warming. Climate scientists, science journalists, environmental groups and science advocacy organisations dispute Crichton's views on the science as being error-filled and distorted,[1][2][3][4][5][6] and it was described as "pure porn for global warming deniers" by journalist Chris Mooney.[7]

The novel had an initial print run of 1.5 million copies and reached the #1 bestseller position at Amazon.com and #2 on the New York Times Best Seller list for one week in January 2005. The novel itself has garnered mixed reviews, with some literary reviewers stating that the book's presentation of facts and stance on the global warming debate detracted from the book's plot.

2.19.1   Overview

State of Fear is, like many of Crichton's books, a fictional work that uses a mix of speculation and real world data, plus technological innovations as fundamental storyline devices. The debate over global warming serves as the backdrop for the book. Crichton supplies a personal afterword and two appendices that link the fictional part of the book with real examples of his thesis.

The main villains in the plot are environmental extremists. Crichton does place blame on "industry" in both the plot line and the appendices. Various assertions appear in the book, for example:

- The science behind global warming is speculative and incomplete, meaning no concrete conclusions can be drawn regarding human involvement in climate change.

- Elites in various fields use either real or artificial crises to maintain the existing social order, misusing the "science" behind global warming.

- As a result of potential conflicts of interest, the scientists conducting research on topics related to global warming may subtly change their findings to bring them in line with their funding sources. Since climatology can not incorporate double-blind studies, as are routine in other sciences, and climate scientists set experiment parameters, perform experiments within the parameters they have set, and analyze the resulting data, a phenomenon known as "bias" is offered as the most benign reason for climate science being so inaccurate.

- A key concept, delivered from the eccentric Professor Hoffman, suggests, in Hoffman's words, the existence of a "politico-legal-media" complex, comparable to the "military industrial complex," of the Cold War era. Hoffman insists climate science began using more extreme, fear-inducing terms such as "crisis," "catastrophe," and, "disaster," shortly after the fall of The Berlin Wall, in order to maintain a level of fear in citizens, for the purpose of social control, since the specter of Soviet Communism was gone. This "state of fear" gives the book its title.

Numerous charts and quotations from real world data, including footnoted charts which strongly suggest mean global temperature is, in this era, lowering. Where local temperatures show a general rise in mean temperature, mostly in major world cities, Crichton's characters infer it is due to urban sprawl and deforestation, not carbon emissions.

Crichton argues for removing politics from science and uses global warming and real-life historical examples in the appendices to make this argument. In a 2003 speech at the California Institute of Technology he expressed his concern about what he considered the "emerging crisis in the whole enterprise of science—namely the increasingly uneasy relationship between hard science and public policy."[8]

2.19.2   Plot summary

Peter Evans is lawyer for a millionaire philanthropist, George Morton. Evans' main duties are managing the legal affairs surrounding Morton's contributions to an environmentalist organization, the National Environmental Resource Fund (NERF) (modeled after the Natural Resources Defense Council [NRDC]).[9]

Morton becomes suspicious of NERF's director, Nicholas Drake, after he discovers that Drake has misused some of the funds Morton had donated to the group. Soon after, Morton is visited by two men, John Kenner and Sanjong Thapa, who appear on the surface to be researchers at MIT, but, in fact, are international law enforcement agents on the trail of an eco-terrorist group, the Environmental Liberation Front (ELF) (modeled on the Earth Liberation Front). The ELF is attempting to create "natural" disasters to convince the public of the dangers of global warming. All these events are timed to happen during a NERF-sponsored climate conference that will highlight the "catastrophe" of global warming. The eco-terrorists have no qualms about how many people are killed in their manufactured "natural" disasters and ruthlessly assassinate anyone who gets in their way (their preferred methods being ones few would recognize as murder; the venom of a rare Australian blue-ringed octopus which causes paralysis, and "lightning attractors" which cause their victims to get electrocuted in electrical storms). Kenner and Thapa suspect Drake of involvement with the ELF to further his own ends (garnering more donations to NERF from the environmentally-minded public).

Evans joins Kenner, Thapa, and Morton's assistant, Sarah Jones on a globe-spanning series of adventures to thwart various ELF-manufactured disasters before these disasters kill thousands of people. Kenner's niece, Jennifer Haynes, joins the group for the final leg as they travel to a remote island in the Solomons to stop the ELF's "piece de resistance", a tsunami that will inundate the coastline of California just as Drake is winding up the international conference on the "catastrophe" of global warming. Along the way they battle man-eating crocodiles and cannibalistic tribesmen (who feast on Ted Bradley, an environmentalist TV actor whom Drake had sent to spy on Kenner and his team). The rest of the group are rescued in the nick of time by Morton who had previously faked his own death to throw Drake off the trail so that he could keep watch on the ELF's activities on the island while he waited for Kenner and his team to arrive. The group has a final confrontation with the elite ELF team on the island during which Haynes is almost killed and Evans kills one of the terrorists who had previously tried to kill both him and Jones in Antarctica. The rest of the ELF team is killed by the backwash from their own tsunami, which Kenner and his team have sabotaged just enough to prevent it from becoming a full-size tsunami and reaching California. Morton, Evans and Jones return to Los Angeles. Evans quits the firm to work for Morton's new (unnamed) organization, which will practice environmental activism as a business, free from potential conflicts of interest. He hopes Evans and Jones will take his place in the new organization after his death.

2.19.3   Allegorical characters

Several critics have suggested that Crichton uses the major characters as proxies for differing viewpoints on the topic of global warming in order to allow the reader to clearly follow the various positions portrayed in the book.

- Joseph Romm suggests that Kenner is a stand-in for Crichton himself.[10]

- David Roberts suggests that Evans is the stand-in for the reader (who Crichton presumes accepts most of the tenets of global warming without any detailed study of it, but not unquestioningly) and also that Ted Bradley is a stand-in for people who accept the "environmentalist" party line without question.[11]

- Ronald Bailey suggests that Drake is a stand-in for the environmental movement "professional" activist.[12]

- Bruce Barcott suggests that Sarah,[13] and Michael B. McElroy and Daniel P. Schrag suggest that Jennifer[14] are stand-ins for the academic community (intelligent enough to follow the debate but undecided until the evidence is presented) with Sarah being the portion of the community likely to believe in global warming on less than undeniable evidence (they will accept "Likely, but not proven" as sufficient proof) and Jennifer representing the part of the community that accepts undeniable evidence only.

- Michael B. McElroy and Daniel P. Schrag also suggest that Jennifer is simultaneously a stand-in for conflicts of interest created by how the research is funded (i.e. her "official" story changes based on who is paying the bills but in private she makes her true feelings known).[14]

- Gregory Mone suggests that Sanjong is a stand-in for the local university library/reputable Internet source verification, etc.[15]

2.19.4 Author's afterword/appendices

Crichton included a statement of his views on global climate change as an afterword. In the "Author's message", Crichton states that the cause, extent, and threat of climate change is largely unknown and unknowable. He finishes by endorsing the management of wilderness and the continuation of research into all aspects of the Earth's environment.

In Appendix I, Crichton warns both sides of the global warming debate against the politicization of science. Here he provides two examples of the disastrous combination of pseudo-science and politics: the early 20th-century ideas of eugenics (which he directly cites as one of the theories that allowed for the Holocaust) and Lysenkoism.

This appendix is followed by a bibliography of 172 books and journal articles that Crichton presents "...to assist those readers who would like to review my thinking and arrive at their own conclusions." (State of Fear, pp, 583).

2.19.5 Reception

Literary reviews

The novel has received mixed reviews from professional literary reviewers.[16]

The Wall Street Journal's Ronald Bailey gave a favorable review, calling it "a lightning-paced technopolitical thriller" and the "novelization of a speech that Mr. Crichton delivered in September 2003 at San Francisco's Commonwealth Club."[17] Entertainment Weekly's Gregory Kirschling gave a favorable A- review and said it was "one of Crichton's best because it's as hard to pigeonhole as greenhouse gas but certainly heats up the room."[18]

In The New Republic, Sacha Zimmerman gave a mixed review. Zimmerman criticized Crichton's presentation of data as condescending to the reader but concluded that the book was a "globe-trotting thriller that pits man against nature in brutal spectacles while serving up just the right amount of international conspiracy and taking digs at fair-weather environmentalists."[19]

Much criticism was given to Crichton's presentation of global warming data and the book's portrayal of the global warming debate as a whole. In the Sydney Morning Herald, John Birmingham criticized the book's usage of real world research and said it was "boring after the first lecture, but mostly in the plotting... It's bad writing and it lets the reader ignore the larger point Crichton is trying to make."[20] In The Guardian, Peter Guttridge wrote that the charts and research in the book got "in the way of the thriller elements" and stated the bibliography was more interesting than the plot.[21] In The New York Times, Bruce Barcott criticized the novel's portrayal of the global warming debate heavily, stating that it only presented one side of the argument.[22]

In the Pittsburgh Post-Gazette, Allan Walton gave a mostly favorable review and offered some praise for Crichton's work. Walton wrote that Crichton's books, "as meticulously researched as they are, have an amusement park feel. It's as if the author channels one of his own creations, "Jurassic Park's" John Hammond, and spares no expense when it comes to adventure, suspense and, ultimately, satisfaction."[23]

Criticism from scientific community

This novel received criticism from climate scientists,[1][6][24] science journalists[7][25] and environmental groups[26][27] for inaccuracies and misleading information. Sixteen of 18 U.S. climate scientists interviewed by Knight Ridder said the author was bending scientific data and distorting research.[6]

Several scientists whose research had been referenced in the novel stated that Crichton had distorted it in the novel. Peter Doran, leading author of the *Nature* paper,[28] wrote in the *New York Times* stating that

> "... our results have been misused as 'evidence' against global warming by Michael Crichton in his novel 'State of Fear'"[24]

Myles Allen, Head of the Climate Dynamics Group, Department of Physics, University of Oxford, wrote in *Nature* in 2005:

> "Michael Crichton's latest blockbuster, State of Fear, is also on the theme of global warming and is likely to mislead the unwary. . . Although this is a work of fiction, Crichton's use of footnotes and appendices is clearly intended to give an impression of scientific authority."[1]

The American Geophysical Union, consisting of over 50,000 members from over 135 countries, states in their newspaper *Eos* in 2006:

> "We have seen from encounters with the public how the political use of State of Fear has changed public perception of scientists, especially researchers in global warming, toward suspicion and hostility."[29]

James E. Hansen, former head of the NASA Goddard Institute for Space Studies at the time, wrote "He (Michael Crichton) doesn't seem to have the foggiest notion about the science that he writes about."[4] Jeffrey Masters, Chief meteorologist for Weather Underground, writes: "Crichton presents an error-filled and distorted version of the Global Warming science, favoring views of the handful of contrarians that attack the consensus science of the IPCC."[2]

The Union of Concerned Scientists devote a section of their website to what they describe as misconceptions readers may take away from the book.[27]

2.19.6   Recognition

U.S. Congress

Despite being a work of fiction, the book has found use by global warming skeptics. For example, United States Senator Jim Inhofe, who once pronounced global warming "the greatest hoax ever perpetrated on the American people",[30][31] made *State of Fear* "required reading"[32] for the Senate Committee on Environment and Public Works, which he chaired from 2003–2007, and to which he called Crichton to testify before in September 2005.[32]

Al Gore said on March 21, 2007 before a US House committee: "The planet has a fever. If your baby has a fever, you go to the doctor [...] if your doctor tells you you need to intervene here, you don't say 'Well, I read a science fiction novel that tells me it's not a problem'". Several commentators interpreted this as a reference to *State of Fear*.[33][34][35][36]

AAPG 2006 Journalism Award

The novel received the American Association of Petroleum Geologists (AAPG) 2006 Journalism Award. AAPG Communications director Larry Nation told the *New York Times*, "It is fiction, but it has the absolute ring of truth". The presentation of this award has been criticized as a promotion of the politics of the oil industry and for blurring the lines between fiction and journalism.[29][37] After some controversy within the organization, AAPG has since renamed the award the "Geosciences in the Media" Award.[38]

Daniel P. Schrag, Director of the Center for the Environment at Harvard University, called the award "a total embarrassment" that he said "reflects the politics of the oil industry and a lack of professionalism" on the association's part. As for the book, he added "I think it is unfortunate when somebody who has the audience that Crichton has shows such profound ignorance".[37]

2.19.7 References

[1] Myles Allen (2005-01-20). "A novel view of global warming — Book Reviewed: State of Fear" (PDF). *Nature* **433** (7023): 198. Bibcode:2005Natur.433..198A. doi:10.1038/433198a. PDF version from climateprediction.net site

[2] Review of Michael Crichton's State of Fear : Weather Underground

[3] Doran, Peter (July 27, 2006). "Cold, Hard Facts". *Opinion* (New York Times).

[4] *Michael Crichton's "Scientific Method"* James E. Hansen

[5] Union of Concerned Scientists *Crichton's Thriller State of Fear: Separating Fact from Fiction*

[6] Borenstein, Seth (2005-02-10). "Novel on global warming gets some scientists burned up". The Seattle Times. Retrieved 2008-08-17.

[7] Mooney, Chris (2005-01-18). "Bad Science, Bad Fiction". Committee for Skeptical Inquiry. Retrieved 2008-08-20.

[8] Aliens Cause Global Warming a speech give at the California Institute of Technology January 17, 2003

[9] Bruce Barcott (2005-01-30). "'State of Fear': Not So Hot". The New York Times. Retrieved 2010-09-15.

[10] Romm, Joseph (May 2005). "Greenhouse Gas
Michael Crichton's new novel fingers the wrong villains in global warming". Technology Review.

[11] http://www.grist.org/advice/books/2005/02/01/roberts-fear/

[12] Wall Street Journal Review

[13] Barcott, Bruce (January 30, 2005). "'State of Fear': Not So Hot". *Sunday Book Review* (New York Times).

[14] McElroy MB, Schrag DP (March–April 2005). "Overheated Rhetoric
A new novel misrepresents global warming and distorts science". Harvard Magazine.

[15] Popular Science's Review

[16] "State of Fear. What the Critics Said". Metacritic.com. Archived from the original on 2008-06-29. Retrieved 2008-09-13.

[17] Ronald Bailey (2004-12-10). "A Chilling Tale. Michael Crichton's "State of Fear"". The Wall Street Journal. Retrieved 2008-08-13.

[18] Gregory Kirschling (2004-12-13). "Book Review State of Fear (2004)". Entertainment Weekly. Retrieved 2008-09-13.

[19] Sacha Zimmerman (2005-01-20). "Review-A-Day: State of Fear. Weather Man". The New Republic. Retrieved 2008-09-13.

[20] John Birminghamn (2005-01-15). "Books: State of Fear". Sydney Morning Herald. Retrieved 2008-09-13.

[21] Peter Guttridge (2005-01-16). "Well, the bibliography sings". London: The Guardian. Retrieved 2008-09-13.

[22] Bruce Barcott (2005-01-30). "'State of Fear': Not So Hot". The New York Times. Retrieved 2008-09-13.

[23] Allan Walton (2004-12-24). "'State of Fear': 'State of Fear' by Michael Crichton". Pittsburgh Post-Gazette. Retrieved 2009-11-22.

[24] Peter Doran (2006-07-27). "Cold, Hard Facts". The New York Times. Retrieved 2008-08-14.

[25] Evans, Harold (2005-10-07). "Crichton's conspiracy theory". BBC NEWS. Retrieved 2008-08-17.

[26] "Michael Crichton's State of Fear: They Don't Call It Science Fiction for Nothing". Natural Resources Defense Council. 2004-12-16. Archived from the original on 2008-07-18. Retrieved 2008-08-17.

[27] "Crichton's Thriller State of Fear". Union of Concerned Scientists. 2005-06-27. Retrieved 2008-08-17.

[28] Doran PT, Priscu JC, Lyons WB, et al. (2002-01-13). "Antarctic climate cooling and terrestrial ecosystem response". *Nature* **415** (6871): 517–520. doi:10.1038/nature710. PMID 11793010. as PDF

[29] <Please add first missing authors to populate metadata.> (5 September 2006). "American Geophysical Union, Petroleum Geologists' *Award to Novelist Crichton Is Inappropriate*" (PDF). *Eos* **87** (36).

[30] Mooney, Chris (2005-01-11). "Warmed Over — American Prospect: Sen. James Inhofe's Science Abuse". CBS News. Retrieved 2008-08-15.

[31] Coile, Zachary (2006-10-11). "Senator says warming by humans just a hoax". San Francisco Chronicle. Retrieved 2008-08-15.

[32] Janofsky, Michael (2005-09-29). "Michael Crichton, Novelist, Becomes Senate Witness". The New York Times. Retrieved 2008-08-15.

[33] Glenn, Joshua (1 April 2007). "Climate of fear". The Boston Globe.

[34] "More from 'Inconvenient Gore'". Alaska Report. 22 March 2007.

[35] "What Al Gore Really Wants". FOX News. 25 March 2007.

[36] Ansible 237, April 2007

[37] Dean, Cornelia (2006-02-09). "Truth? Fiction? Journalism? Award Goes to ...". New York Times. Retrieved 2008-08-17.

[38] President 06:2006 EXPLORER

2.19.8   External links

- *State of Fear* page on Michael Crichton's official web site

- "Why Politicized Science is Dangerous" (Appendix I in *State of Fear*, excerpted on Crichton's official web site)

2.20   The Real Global Warming Disaster

The Real Global Warming Disaster (*Is the Obsession with 'Climate Change' Turning Out to Be the Most Costly Scientific Blunder in History?*) is a 2009 book by English journalist and author Christopher Booker in which he asserts that global warming can not be attributed to humans, and then alleges how the scientific opinion on climate change was formulated.

From a standpoint of environmental scepticism, Booker seeks to combine an analysis of the science of global warming with the consequences of political decisions to reduce CO
2 emissions and claims that, as governments prepare to make radical changes in energy policies, the scientific evidence for global warming is becoming increasingly challenged. He asserts that global warming is not supported by a significant number of climate scientists, and criticises how the UN's Intergovernmental Panel on Climate Change (IPCC) presents evidence and data, in particular citing its reliance on potentially inaccurate global climate models to make temperature projections. Booker concludes, "it begins to look very possible that the nightmare vision of our planet being doomed"[1] may be imaginary, and that, if so, "it will turn out to be one of the most expensive, destructive, and foolish mistakes the human race has ever made".[1]

The book's claims were strongly criticised by science writer Philip Ball,[2] but the book was praised by several columnists. The book opens with an erroneous quotation,[3] which Booker subsequently acknowledged and promised to correct in future editions.[4]

The book was Amazon UK's fourth bestselling environment book of the decade 2000–10.[5]

2.20.1 Synopsis

The book consists of three parts and an epilogue.

Booker sums up the book's contents in a long epilogue, which quotes Theseus in *A Midsummer's Night Dream*:

> In the night, imagining some fear, How easy is a bush supposed a bear

Booker contends that in this quote Shakespeare is identifying that "when we are not presented with enough information for our minds to resolve something into certainty, they may be teased into exaggerating it into something quite different from what it really is".[6]

2.20.2 Reception

The book received a mixed reception in the media.

In his review in *The Observer*, Philip Ball wrote that the book was "the definitive climate sceptics' manual" in that it makes an uncritical presentation of "just about every criticism ever made of the majority scientific view" on global warming. Though expressing "a queer kind of admiration for the skill and energy with which Booker has assembled his polemic", Ball called the claims made by the author "bunk". Ball also criticised Booker's tactic of introducing global warming sceptics "with a little eulogy to their credentials, while their opponents receive only a perfunctory, if not disparaging, preamble".[2]

In *The Spectator*, Rodney Leach wrote that "the shelf of sceptical books keeps filling and Booker's belongs there with the best", remarking that Booker "narrates this story with the journalist's pace and eye for telling detail and the historian's forensic thoroughness which have made him a formidable opponent of humbug".[7] Columnist James Delingpole described the book as "another of those classics which any even vaguely intelligent person who wants to know what's really going on needs to read".[8]

Writing in *The Herald*, Brian Morton was largely sympathetic to the position taken by Booker in the book: "The question isn't whether climate is changing, but what is to blame. A crippling tithe of international political effort and social action is directed to the assumption that we are", and "the climate change debate—or enforced consensus—concerns the way science is done and perceived. As Booker says, 'consensus' is not a term in science but in politics".[9]

A positive review by Henry Kelly in *The Irish Times*, referring to the book as "meticulously researched, provocative and challenging",[10] was criticised by Irish environmental campaigner and *climatechange.ie* website founder John Gibbons, who said that the decision by *The Irish Times* to allow Kelly to review *The Real Global Warming Disaster* was part of a recent trend of "the media giving too much coverage to 'anti-science' climate change deniers and failing to convey the gravity of the threat, making readers and viewers apathetic".[11]

In *The Scotsman*, writer and environmentalist Sir John Lister-Kaye chose *The Real Global Warming Disaster* as one of his books of the year, writing that "though barely credible in places" this was an "important, brave book making and explaining many valid points".[12]

2.20.3 Houghton misquotation

The book opens with an incorrect quotation which wrongly attributes to John T. Houghton the words "Unless we announce disasters, no one will listen".[3] The publishers apologised for this misquotation, confirmed that it would not be repeated, and agreed to place a corrigendum in any further copies of the book. In an article which appeared in *The Sunday Telegraph* on 20 February 2010, Booker wrote "we shall all in due course take steps to correct the record, as I shall do in the next edition of my book".[4] Houghton felt that Booker continued to misstate his position regarding the role of disasters in policy making, and he referred the matter to the Press Complaints Commission (PCC Reference 101959), following whose involvement *The Sunday Telegraph* published on 15 August a letter of correction by Houghton stating his actual position, that adverse events shock people and thereby bring about change.[13] An article supportive of Houghton appeared in the *New Scientist* magazine.[14]

2.20.4 See also

- Global warming controversy

- Public opinion on climate change

- The Hockey Stick Illusion

- The Deniers

- The Great Global Warming Swindle

2.20.5 Bibliography

- Booker, Christopher (2009). *The Real Global Warming Disaster*. Continuum International Publishing Group Ltd. ISBN 1-4411-1052-6.

2.20.6 Notes

[1] Booker 2009, p. 342

[2] Ball, Phillip (15 November 2009). "The Real Global Warming Disaster by Christopher Booker". *The Observer*. Retrieved 4 February 2010.

[3] Connor, Steve (10 February 2010). "Fabricated quote used to discredit climate scientist". *The Independent*. Retrieved 10 February 2010. The quotation has since become the iconic smoking gun of the climate sceptic community. The words are the very first to appear in the "manual" of climate denialism written by the journalist and arch-sceptic Christopher Booker. They get more than a 100,000 hits on Google, and are wheeled out almost every time a climate sceptic has a point to make [...]

[4] Booker, Christopher (20 February 2010). "What the weatherman never said". *The Sunday Telegraph*. Retrieved 21 February 2010.

[5] Carrington, Damian (17 December 2010). "Bestselling green books of the decade". *The Guardian*. Retrieved 12 July 2011.

[6] Booker 2009, p. 341

[7] Leach, Rodney (4 November 2009). "A wild goose chase". *The Spectator*. Retrieved 2 February 2010.

[8] Delingpole, James (28 October 2009). "You Know It Makes Sense". *The Spectator*. Retrieved 5 April 2010.

[9] Morton, Brian (3 November 2009). "Is a climate-change sceptic more like a flat-earther or a Holocaust denier, merely out of touch or mendacious and evil?". *The Herald*. Retrieved 3 April 2010.

[10] Kelly, Henry (19 November 2009). "Myths of global warming skilfully debunked". *The Irish Times*. Retrieved 12 April 2010.

[11] Monaghan, Gabrielle (4 April 2010). "A little warming under the collar". *The Times*. Retrieved 12 April 2010.

[12] Lister-Kaye, John (5 December 2009). "Books of the year: Writers' choice". *The Scotsman*. Retrieved 5 February 2010.

[13] "Letters". *The Daily Telegraph*. 15 August 2010. Retrieved 28 September 2010.

[14] Giles, Jim (15 May 2010). "Giving life to a lie". *New Scientist*. Retrieved 24 June 2010.

2.20.7 Further reading

- Booker, Christopher; North, Richard (2007). *Scared To Death: From BSE To Global Warming, Why Scares Are Costing Us The Earth*. Continuum International Publishing Group Ltd. ISBN 0-8264-8614-2.

- Carter, Robert; Spooner, John (2013). *Taxing air: Facts and fallacies of climate change*. Kelpie Press. ISBN 9781742983189.

- Montford, Andrew (2010). *The Hockey Stick Illusion; Climategate and the Corruption of Science*. Stacey International. p. 482. ISBN 1-906768-35-8.

2.20.8   External links

- Booker's synopsis of the book in *The Daily Mail*

2.21   The Republican War on Science

The Republican War on Science is a 2005 book by Chris C. Mooney, an American journalist who focuses on the politics of science policy. In the book, Mooney discusses the Republican Party leadership's stance on science, and in particular that of the George W. Bush administration, with regard to issues such as global warming, the creation–evolution controversy, bioethics, alternative medicine, pollution, separation of church and state, and the government funding of education, research, and environmental protection. The book argues that the administration regularly distorted and/or suppressed scientific research to further its own political aims.

The book was reviewed in *Science* and *Nature Medicine* as well as the popular press. It was featured on the cover of *The New York Times Book Review* and selected as an "Editors' Choice" by *The New York Times* which described it as, "A frankly polemical survey of scientific findings and procedures in collision with political operations."[1]

Filmmaker Morgan Spurlock (of *Super Size Me* fame) optioned the rights for the book to create a documentary film,[2] but in 2008 announced that he had released the option.[3]

2.21.1   Reviews

In scientific journals

A review in *Science* by Naomi Oreskes states the author recounts the 20-year campaign by "influential Republicans—initially in Congress and now also in the White House—in concert with determined allies in private industry and fundamentalist Christian organizations" to systematically deny, disparage and misrepresent scientific information related to public policy. She gave the following list of topics, "acid rain, global warming, the efficacy of condoms in preventing the spread of sexually transmitted diseases, the health impacts of excess dietary sugar and fat, the alleged link between abortion and breast cancer, the status of endangered species, the efficacy of abstinence-only sex education programs, the therapeutic potential of adult stem cells, and more." Oreskes goes on to detail the tactics used in the attempt to mislead the both the public and politicians, "misrepresenting real debates, exaggerating uncertainty, interfering with the activities of expert agencies, trumpeting the views of outlier scientists whose interpretations are rarely to be found in the refereed literature, and attacking the integrity of genuine experts." She states that Mooney points out that multiple misinformation campaigns have involved the same individuals and groups. Oreskes concludes, "Mooney's book makes it clear that when sensible people stand on the sidelines, a great deal of nonsense can be spread."[4]

Michael Stebbins wrote in *Nature Medicine* that "This book should serve as a harsh wake-up call to the scientific community and the American public." He stated that Mooney "painstakingly documents the roots of the efforts to undercut the influence of science on national policy and the relentless politicization of US science policy by conservatives working on behalf of the Republican Party." He notes that the author clearly documents the "Bush administrations' attacks on the integrity of science information" listing examples that include some of those mentioned by Oreskes and "undercutting the Clean Air Act and the Endangered Species Act, and stacking agencies and advisory committees with unqualified ideologues." Stebbins credits the book with providing context by detailing the tactics employed by "conservative Republicans" and establishing the roots of these undercutting techniques with examples from the last 40 years. He goes on to state, "Mooney's documentation of the willful manipulation of science on the part of conservatives to suit an agenda is well supported and nauseating." Stebbins addresses two criticisms of the book, the first, that it doesn't explain the science involved, which he explains is not the purpose of the book, and the second, that it doesn't detail the misuse of science by democrats and liberals, which he dismisses as untrue. He finds issue with the last chapter, which proposes solutions, stating, "His suggestions are sound and well thought out, but seem more of an afterthought than a real goal of the book."[5]

Daniel Sarewitz panned the book in a review in *Issues in Science and Technology* describing it as a, "tiresome polemic masquerading as a defense of scientific purity."[6]

In the popular press

In a positive review in *Scientific American* Boyce Rensenberger described the book as, "a well-researched, closely argued and amply referenced indictment of the right wing's assault on science and scientists."[7] Lisa Margonelli of *The New York Times Book Review* wrote that Mooney, "juggled extensive research and sharp arguments [...] with precision and a showman's wink that made his unpromising subject fun."[8]

Keay Davidson wrote in the *Washington Post* that "Mooney's political heart is in the right place" but says "Mooney is like a judge who interprets a law one way to convict his enemies and another way to acquit his friends."[9]

Writing in *The New York Times*, John Horgan said of the book that the prose was "often clunky and clichéd", but explains that Mooney "addresses a vitally important topic and gets it basically right." Horgan defends the book against another reviewer's criticism, saying "the journalist Keay Davidson faults Mooney for not acknowledging how hard it can be to distinguish good science from bad... But in many of the cases that [Mooney] examines, demarcation is easy, because one side has an a priori commitment to something other than the truth — God or money, to put it bluntly."[10]

Stuart Derbyshire, a senior lecturer at the University of Birmingham School of Psychology, praises Mooney and notes that he explained how Republicans had manipulated the uncertainty in science "to ensure that Congress rarely hears any consensus opinion that may damage a Bush policy." Derbyshire agrees with Mooney's claim that there is Republican "flagrant twisting" of research findings and that it "violates the integrity of science."[11]

2.21.2   Media

Mooney was interviewed about the book on *Science Friday*.[12]

2.21.3   Publishing information

Mooney, Chris (2005). *The Republican War on Science*. Basic Books. ISBN 9780465046751.

2.21.4   See also

- Antiscience

- Agnotology

- Climate change policy of the United States

- List of books about the politics of science

- *Merchants of Doubt*

- Politicization of science

- William R. Steiger

2.21.5   References

[1] "Editors' Choice". Browsing Books. *The New York Times*. 25 December 2005. Retrieved 2014-04-25.

[2] "'Science' Scooped Up". *Washington Post*. December 17, 2005.

[3] "See Morgan? That's Why You Should Have Made The Republican War on Science". *ScienceBlogs*.

[4] Oreskes, Naomi (7 October 2005). "Anti-realism in government". *Science* **310** (5745): 56. doi:10.1126/science.1115765.

[5] Stebbins, Michael (April 2006). "The wake-up call". *Nature Medicine* **12** (4): 381. doi:10.1038/nm0406-381.

[6] Sarewitz, Daniel (2006). "Scientizing politics". *Issues in Science and Technology* **22** (2): 91–3.

[7] Rensenberger, Boyce (24 September 2005). "Science abuse". *Scientific American* (book review). Retrieved 2014-04-24.

[8] Margonelli, Lisa (1 July 2007). "Wild is the wind". *The New York Times Book Review*. Retrieved 2014-04-23.

[9] "Research and the Right", The republican war on science reviewed by Keay Davidson, The Washington Post September 18, 2005

[10] Horgan, John (18 December 2005). "Political Science". *The New York Times*. Retrieved 25 May 2010.

[11] "Bush isn't the only one who's anti-science". *Spiked*. 28 November 2005.

[12] Flatow, Ira (11 November 2005). "Interview: Chris Mooney discusses 'The Republican War on Science'". *Talk of the Nation: Science Friday* (NPR). Retrieved 2014-04-24 – via HighBeam Research. (subscription required (help)).

2.21.6 External links

- Official website

- Chris C. Mooney, personal website

- Chris Mooney: *The Republican War on Science* on YouTube

2.22 United States Chamber of Commerce

This article is about the American lobbying group. For trade organisations globally, see Chamber of commerce.

The **United States Chamber of Commerce** (**USCC**) is a business-oriented American lobbying group. It is not an agency of the United States government.

Politically, the Chamber is generally considered to be a conservative organization. It usually supports Republican political candidates, though it has occasionally supported conservative Democrats.[1][2] The Chamber is one of the largest lobbying groups in the U.S., spending more money than any other lobbying organization on a yearly basis.[3][4]

2.22.1 History

The U.S. Chamber of Commerce's own history of itself describes it as originating from an April 22, 1912, meeting of delegates.[5] The Chamber was created by President Taft as a counterbalance to the labor movement of the time.[2]

In 1993, the Chamber lost several members over its support for Clinton's healthcare reform efforts. The Chamber had chosen to support healthcare reform at that time due to the spiraling healthcare costs experienced by its members. However, House Republicans retaliated by urging boycotts of the organization. The Chamber operated its own cable television station, Biz-Net until 1997 in order to promote its policies. The Chamber shifted somewhat more to the right when Tom Donohue became head of the organization in 1997. By the time health care reform became a major issue again in 2010-2012, the organization opposed such efforts.[2] The Washington, D.C., headquarters of the U.S. Chamber of Commerce occupies land that was formerly the home of Daniel Webster.[6]

Although all chambers can work with all levels of government, they tend to concentrate their efforts on specific levels: Local chambers of commerce tend to focus on local issues, state chambers on state issues, and the U.S. Chamber of Commerce focuses on national issues at the federal government level.[7] They also work closely with a number of other youth organizations in the country about the value and role of business in our society today. [8]

In late 2011 it was revealed that the Chamber's computer system was breached from November 2009 to May 2010 by Chinese hackers. The purpose of the breach appeared to be gain information related to the Chamber's lobbying regarding Asian trade policy.[9]

United States Chamber of Commerce building at 1615 H Street, NW, in Washington, D.C. The building is listed on the National Register of Historic Places.

Since a 1971 internal memo by Lewis Powell advocating a more active role in cases before United States Supreme Court, the Chamber has found increasing success in litigation. Under the Burger and Rehnquist Courts the Chamber was on the prevailing side 43% and 56% of the time, respectively, but under the Roberts Court, the Chamber's success rate rose to 68% as of 21 June 2012.[10]

2.22.2   Positions taken

Legislation

- Actively lobbies against anti-tobacco policies implemented in other countries.[11][12] In particular, it opposes attempts to carve out tobacco from the Investor-state dispute settlement mechanism negotiated under the Trans-Pacific Partnership agreement.[13]

- Opposed to using the government shutdown and debt ceiling limit as negotiating tactics.[14]

- Supported the Stop Online Piracy Act (SOPA).[15]

- Support for business globalization, free trade, and offshoring.[16]

- Qualified opposition to financial regulation.[2]

    - The Chamber views some reform as necessary, but opposed the Dodd/Frank legislation that was passed, asserting that it would damage loan availability.[2]

- Campaigned against portions of the Sarbanes–Oxley Act.[17]

- Opposes the DISCLOSE Act, which aims to limit foreign influence on U.S. elections.[18]

- Opposed the Affordable Health Care for America Act.[2]

- Opposes action on climate change. The Chamber believes that climate change is real, but disputes the scientific consensus that warming mostly results from human activity and questions whether anything can be done to reduce climate change. The Chamber emphasizes the economic impact of climate decisions.[19][20]

- Supported the American Recovery and Reinvestment Act of 2009.[2]

Court Cases

- Argued against mandatory immigration status checks by employers in Arizona including in a Supreme Court case.[21]

- Filed an amicus brief to the U.S. Supreme Court in Citizens United v FEC. Its position is opposed by some advocates for independent businesses.[22]

2.22.3 Lobbying expenditures

The Chamber has emerged as the largest lobbying organization in America. The Chamber's lobbying expenditures in 2013 were almost twice as high as the next highest spender: National Assn of Realtors, at $38.5 million.

2.22.4 International network

As of October 2010, the Chamber had a worldwide network of 115 American Chamber of Commerce affiliates located in 108 countries.[24] The US Chamber says that a relative handful of the Chamber's 300,000 members are "non-U.S.-based (foreign) companies." It claims that, "No foreign money is used to fund political activities." A US Chamber executive has said that the organization has had "foreign multinationals" (foreign companies) as members for "over a century, many for decades."[25] The US Chamber states that it receives approximately $100,000 annually in membership dues from its foreign affiliates, out of an annual budget of $200 million.[25][26]

AmCham China, with members comprise more than 2,600 individuals from over 1,200 companies,[27] is said to be the largest affiliate outside of United States.

2.22.5 Electoral activities

In the 2008 election cycle, aggressive ads paid for by the USCC attacked a number of Democratic congressional candidates (such as Minnesota's DFL Senate candidate Al Franken) and supported a number of Republican candidates including John Sununu, Gordon Smith, Roger Wicker, Saxby Chambliss and Elizabeth Dole.

During the 2010 campaign cycle, the Chamber spent $32 million, 93 percent of which was to help Republican candidates.[28] The Chamber's spending out of its general funds was criticized as illegal under campaign finance laws.[29][30][31][32] In a front-page article titled "Large Donations Aid U.S. Chamber in Election Drive", *The New York Times* reported that the Chamber used contributions in campaigns without separating foreign and domestic contributions, which if true would appear to contravene prohibitions on lobbying by foreign nations and groups. In question was the Chamber's international branches, "AmChams", whose funds are unaccounted for and perhaps mix into the general collection.[30][33][34][35] All branches, corporations, and members of the Chamber pay dues; the question is how they divide the money for expenses in national campaigns.

The truth of these allegations is unknown, as neither the Chamber nor its detractors can provide any concrete evidence to support or refute the allegations.[36] In reference to the matter, Tom Donohue wrote his council and members on October

12, 2010. He stated, "Let me be clear. The Chamber does not use any foreign money to fund voter education activities—period. We have strict financial controls in place to ensure this. The funds we receive from American Chambers of Commerce abroad, bilateral business councils, and non-U.S. based global companies represent a small fraction of our more than $200 million annual revenues. Under our accounting system, these revenues are never used to support any political activities. We are in full compliance with all laws and regulations."[37][38][39] Organizations Moveon.org, Think Progress, and People for the American Way rallied against the Chamber at the Justice Department to start an injunction for a criminal investigation.[40][41] The Chamber is not required to produce fundraising records.[42]

President Barack Obama and other legislators asked the IRS and Federal Elections Commission to ensure that the foreign funds that the Chamber receives are not used for political activities.[43][44] Obama criticized the Chamber for not disclosing its contributors.[45] The Chamber has responded that "No foreign money is used to fund political activities." [25] After the election, the Chamber reiterated the nature of Obama's policy dictated action from the Chamber, however the conflict would not be made "personal".[46][47]

In addition to the expenditures from the Chamber's own funds, in 2010 its political action committee gave $29,000 (89 percent) to Republican candidates and $3,500 (11 percent) to Democratic candidates.[48] The Chamber's PAC received a total of 76 donations from individual donors ($200 or more donation) totaling $79,852 in 2007-2008, or an average of $1050 per donation, and three donations per month.[49]

In 2011, the Chamber hosted a "GOP Holiday Party" honoring the Republican National Committee.[50]

Despite more than $33 million spent supporting candidates in the 2012 Congressional races, Chamber-backed candidates lost 36 out of the 50 elections in which the Chamber participated.[51]

In late 2013 the Chamber announced it would distribute campaign contributions in "10s" of Republican primary elections to oppose the Tea Party movement and create a "more governable Republican party."[52] In early 2014 Tom Donohue clarified that the push would be to elect "pro-business" members of Congress "who favor trade, energy development and immigration reform".[53]

2.22.6   Current Leadership[54]

- Tom J. Donohue - President and CEO

- Shannon DiBari - Chief Operating Officer, Senior Vice President, and Chief Administrative Officer

- Myron Brillant - Executive Vice President and Head of International

- Lily Fu Claffee - Senior Vice President, Chief Legal Officer, and General Counsel; Executive Vice President, U.S. Chamber Litigation Center

- Suzanne Clark - Executive Vice President, U.S. Chamber of Commerce

- Thomas Collamore - Senior Vice President, Communications and Strategy and Counselor to the President

- Rob Engstrom - Senior Vice President, Political Affairs & Federation Relations and National Political Director

- Amanda Engstrom - Senior Vice President and Chief of Staff; Senior Vice President, U.S. Chamber Center for Capital Markets Competitiveness; Acting President, Center for Advanced Technology and Innovation

2.22.7   Controversies

In April 2009, the Chamber began an ad campaign against the proposed Employee Free Choice Act.[55] Critics such as the National Association of Manufacturers have contended that additional use of card check elections will lead to overt coercion on the part of union organizers. Opponents of the Employee Free Choice Act also claim, referring to perceived lack of access to a secret ballot, that the measure would not protect employee privacy. For this reason the Chamber argued the act would reduce workers' rights.[56]

In November 2009, the Chamber was reported to be seeking to spend $50,000 to hire a "respected economist" to produce a study that could be used to portray health-care legislation as a job killer and threat to the nation's economy.[57]

In December 2009, activist group Velvet Revolution, under the name StopTheChamber, posted a $200,000 reward for "information leading to the arrest and conviction of Chamber of Commerce CEO Tom Donahue."[58]

Some in the business community have criticized the Chamber's approach to public issues as overly aggressive. Hilary Rosen, former CEO of the Recording Industry Association of America added, "Their aggressive ways are out of step with a new generation of business leadership who are looking for more cooperative relationship with Washington."[59]

Climate Change

The climate campaign organisation 350.org estimates that 94% of US Chamber of Commerce electoral contributions went to candidates denying the scientific consensus on climate change.[60]

The Chamber threatened to sue the Environmental Protection Agency in order to have what the Chamber termed "the Scopes monkey trial of the 21st century" on climate science before any federal climate regulation is passed in October 2009.[19] In response to this position, several companies quit the Chamber, including Exelon Corp, PG&E Corp, PNM Resources, and Apple Inc.[20] Nike, Inc resigned from their board of directors position, but continued their membership. Nike stated that they believe they can better influence the policy by being part of the conversation.[61] Peter Darbee, CEO of former chamber member PG&E (a natural gas and electric utility company in California), said, "We find it dismaying that the Chamber neglects the indisputable fact that a decisive majority of experts have said the data on global warming are compelling... In our view, an intellectually honest argument over the best policy response to the challenges of climate change is one thing; disingenuous attempts to diminish or distort the reality of these challenges are quite another."[62] In response to an online campaign of Prius owners organized by Moveon.org, Toyota stated that it would not leave the Chamber.[63] The Aspen Chamber Resort Association of Aspen, Colorado left the U.S. Chamber because of its views on climate change, in light of how climate change could hurt Aspen's winter tourism industry.[64]

Immigration Reform

The Chamber of Commerce has also come under attack by conservatives and others for its support of amnesty for illegal immigrants.[65][66] In 2014, Tom Donohue stated the Chamber will "pull out all stops" for the passage of immigration reform in Congress.[67] According to the Washington Post, Donohue did not offer specifics with regard to provisions or bills on the matter, speaking generally about the impact immigration would have on the U.S. economy.

2.22.8   Opposition to Chamber

Several organization have attacked the Chamber for its advocacy, including Chamber Watch (a campaign of Public Citizen). Advocates for independent business, like the American Independent Business Alliance (AMIBA) and "green businesses," like the American Sustainable Business Council, have fought the Chamber on multiple issues. Among major divisions between the Chamber and these business advocates is allowing corporations to engage in electioneering.[68] Oliver E. Diaz says one example of this was when the Chamber spent $1,000,000+ to elect judicial candidate Keith Starrett and fund negative campaign ads against judicial candidate Oliver E. Diaz. [69]

2.22.9   Affiliate organizations

- Americans for Transportation Mobility

- Business Civic Leadership Center

- Center for Capital Markets Competitiveness

- Center for International Private Enterprise

- Essential Worker Immigration Coalition

- Institute for 21st Century Energy

- Institute for a Competitive Workforce

- Institute for Legal Reform

- National Chamber Litigation Center

- TradeRoots

- U.S. Chamber of Commerce Foundation (previously the National Chamber Foundation)

2.22.10   See also

- Advocacy group

- American Green Chamber of Commerce

- Global Intellectual Property Center

- Lobbying in the United States

- National Federation of Independent Business

- U.S. Women's Chamber of Commerce

- United States Hispanic Chamber of Commerce

2.22.11   References

[1] Gold, Matea; Geiger, Kim (8 October 2010). "Republican-leaning U.S. Chamber of Commerce buys ads supporting Democrats". *Los Angeles Times.*

[2] Verini, James (1912-04-12). "Show Him the Money by James Verini (July, 2010)". Washington Monthly. Retrieved 2010-11-05.

[3] "Top lobbyists in the US". Retrieved 7.8.2009. Check date values in: |access-date= (help)

[4] Allen, Jonathan (2009-10-20). "U.S. Chamber: $34.7 million in lobbying". Politico.

[5] "U.S. Chamber of Commerce website, "History"". Uschamber.com. Retrieved 2010-11-05.

[6] "U.S. Chamber of Commerce website, "History of the building"". Uschamber.com. Retrieved 2010-11-05.

[7] "Frequently Asked Questions". Retrieved 4 August 2014.

[8] American Business BSA Merit Badge Guide, 22 Jun 2015.

[9] Gorman, Siobahn (21 December 2011). "Chinese Hackers Hit U.S. Chamber - WSJ.com". *The Wall Street Journal.* Retrieved 21 December 2011.

[10] Sacks, Mike (21 June 2012). "Supreme Court: U.S. Chamber Of Commerce Undefeated This Term". *Huffington Post.* Retrieved 12 March 2013.

[11] "Making Tobacco's Case". *The New York Times.* 2015-06-30. Retrieved 2015-07-03.

[12] Hakim, Danny (2015-06-30). "U.S. Chamber of Commerce Works Globally to Fight Antismoking Measures". *The New York Times.* Retrieved 2015-07-03.

[13] "No Exclusions! Why Carveouts Would Weaken the Trans-Pacific Partnership". 2014-04-24. Retrieved 2015-07-03.

[14]

[15] "Cautious Optimism Follows SOPA (2011)".

[16] "Tom Donohue, U.S. Chamber of Commerce on outsourcing and offshoring (2007) Interview conducted and hosted by". Ventureoutsource.com. Retrieved 2010-11-05.

[17] Henry, David (17 January 2005). "Death, Taxes, & Sarbanes-Oxley?". *Business Week*.

[18] Montopoli, Brian (26 July 2010). "Obama Slams GOP For Opposing DISCLOSE Act Meant to Expose "Shadow Groups" Behind Ads". *CBS News*.

[19] Tankersley, Jim (25 August 2009). "U.S. Chamber of Commerce seeks trial on global warming". *The Los Angeles Times*. Retrieved 13 December 2010.

[20] Gardner, Timothy (5 October 2009). "Apple, citing climate, tells U.S. Chamber iQuit". *Reuters*. Retrieved 13 December 2010.

[21] Mears, Bill (26 May 2011). "Supreme Court backs Arizona immigration law that punishes businesses". CNN. Retrieved 26 May 2011.

[22] Independent Business Advocates Condemn U.S. Supreme Court's Decision in Citizens United.

[23] Opensecrets.org ranking page for 2013

[24] "President Barack Obama says foreign funds received by the US Chamber may be helping to fund attack ads". Politifact, St. Petersburg Times. October 11, 2010.

[25] "Chamber of Commerce - The White House Wants Our Donor List". *ABC News*.

[26] Abdullah, Halimah (12 October 2010). "Democratic Partisans Up in Arms Against US Chamber Donations". *Kansas City Star*. Retrieved 26 October 2010.

[27] About AmCham China

[28] Murray, Matthew (November 12, 2010). "Chamber Watch: Business Group 'Central' to GOP Gains". *Roll Call*. Retrieved 2011-02-05.

[29] "Chamber of Commerce under fire for foreign cash". Politico.com.

[30] Waters, Clay (2010-10-22). "Lead Story Goes After Fundraising of Obama Foe, the Chamber of Commerce | Media Research Center". Mrc.org. Retrieved 2012-04-05.

[31] Fang, Lee (2010-10-13). "Exclusive: Chamber Receives At Least $885,000 From Over 80 Foreign Companies In Disclosed Donations Alone". ThinkProgress. Retrieved 2012-04-05.

[32] "Vote 2010: Is Foreign Money Behind U.S. Chamber of Commerce Ads? - ABC News". Abcnews.go.com. 2010-10-11. Retrieved 2012-04-05.

[33] Eggen, Dan (8 October 2010). "Chamber and Democrats battle over the midterms and election spending". *The Washington Post*.

[34] "News Headlines". Cnbc.com. 2010-10-22. Retrieved 2012-04-05.

[35] Jacob Sullum (2010-10-22). "NYT Shocker: Chamber of Commerce Promotes Business Interests - Hit & Run : Reason Magazine". Reason.com. Retrieved 2012-04-05.

[36] "The Chamber and Foreign Contributors". *Factcheck.org*. Retrieved 13 December 2010.

[37] Shear, Michael D. (2010-10-12). "Chamber of Commerce Vows to 'Ramp Up' Political Activity". *The New York Times*.

[38] Lipton, Eric. "Large corporate donations fund controversial US Chamber of Commerce campaign of election attack ads". Ocala.com. Retrieved 2012-04-05.

[39] Smith, Ben (2010-10-05). "Chamber: 'We have a system' - Ben Smith". Politico.Com. Retrieved 2012-04-05.

[40] Fang, Lee (2010-10-05). "Exclusive: Foreign-Funded 'U.S.' Chamber Of Commerce Running Partisan Attack Ads". ThinkProgress. Retrieved 2012-04-05.

[41] Graves, Lucia (7 October 2010). "Watchdog Groups Rally Outside Chamber Of Commerce, As Calls For A Justice Department Investigation Mount". *Huffington Post*.

[42] Lipton, Eric; McIntire, Mike; NATTA Jr, DON VAN (21 October 2010). "Top Corporations Aid U.S. Chamber of Commerce Campaign". *The New York Times.*

[43] Shear, Michael D. (2009-10-20). "Rift between Obama and Chamber of Commerce widening". *The Wall Street Journal.*

[44] "Obama's Risky Fight Against the Chamber of Commerce". *Time.*

[45] Calmes, Jackie (2010-12-11). "Obama to Meet With Executives". *New York Times.*

[46] "Donohue: US Chamber won't seek Obama's defeat". Real Clear Politics.

[47] Stein, Sam (17 November 2010). "The White House, Chamber Of Commerce Attempt Rapprochement". *Huffington Post.*

[48] "Center for Responsive Politics, US Chamber of Commerce summary". Opensecrets.org. Retrieved 2010-11-05.

[49] "Center for Responsive Politics, 31Oct 2009". Opensecrets.org. Retrieved 2010-11-05.

[50] Keyes, Scott (December 8, 2011). "U.s. chamber of commerce to host rnc holiday party". ThinkProgress. Retrieved 2011-12-09.

[51] "Chamber of Commerce $33 Million Lost Most Races: BGOV Barometer". *Bloomberg.*

[52] Needham, Vicki (September 13, 2013). "Top business groups vow more involvement in primaries". *The Hill.*

[53] Wingfield, Brian; Bykowicz, Julie (8 January 2014). "Big Business Doubles Down on GOP Civil War With Tea Party". *www. bloomberg.com* (Bloomberg L.P.). Retrieved 8 January 2014.

[54] "Leadership U.S. Chamber of Commerce". *About Us.* U.S. Chamber of Commerce. March 9, 2015. Retrieved March 9, 2015.

[55] "chambergrassroots.com". chambergrassroots.com. Retrieved 2012-04-05.

[56] "Issue Alert: CARD_CHECK". Bipac.net. Retrieved 2010-11-05.

[57] Shear, Michael D. (16 November 2009). "Opponents of health-care effort look to fund a critical economic study". *The Washington Post.* Retrieved 26 May 2010.

[58] "Activist Group Puts Bounty on Chamber of Commerce CEO". *Fox News.* 2009-12-07.

[59] Stier, Ken (31 October 2009). "Is the Chamber of Commerce Its Own Worst Enemy?". *Time.*

[60] Who's holding us back? Full report Greenpeace November 23, 2011

[61] "Nike US Chamber Statement" (PDF). September 30, 2009. Retrieved October 6, 2009.

[62] "U.S. Chamber of Commerce in climate rift". MSNBC. 2009-09-25. Retrieved 2012-04-05.

[63] Goldenberg, Suzanne (26 October 2009). "Toyota: We're staying in US chamber of commerce". *The Guardian* (London). Retrieved 26 October 2010.

[64] Salvail, Andre (24 April 2012). "Aspen chamber to cut ties with national organization". *The Aspen Times.* Retrieved 29 December 2012.

[65] Kasperowicz, Pete. "GOP lawmaker returns award to protest Chamber of Commerce's immigration policy". *The Blaze.*

[66] O'Connor, Patrick (2014-12-26). "U.S. Chamber of Commerce Pushes Priorities in Congress". *The Wall Street Journal.*

[67] Ho, Catherine (2014-01-08). "U.S. Chamber to 'pull out all stops' to pass immigration reform". *Washington Post.* Retrieved 7 May 2015.

[68] "Granting Corporations Bill of Rights Protections Is Not "Pro-business". AMIBA. Retrieved 2015-06-30.

[69] Saladoff, S. (Director). (2011). Hot Coffee [Motion picture]. Docurama Films

2.22.12 Further reading

- Davis, Cory, "The Political Economy of Commercial Associations: Building the National Board of Trade, 1840-1868," *Business History Review,* 88 (Winter 2014), 761-83.

- Heald, Morrell. "Business thought in the twenties: Social responsibility." *American Quarterly* (1961): 126-139. in JSTOR

- Werking, Richard Hume. "Bureaucrats, businessmen, and foreign trade: the origins of the United States Chamber of Commerce." *Business History Review* 52#03 (1978) pp: 321-341.

2.22.13 External links

- Official website

- Real Clear Politics Portal

- Guide to Chamber of Commerce of the United States of America. Publications. 5332. Kheel Center for Labor-Management Documentation and Archives, Martin P. Catherwood Library, Cornell University.

- Chamber of Commerce of the United States of America. Communications Development Division. Videotape collection, 1988-1992. Schlesinger Library, Radcliffe Institute, Harvard University.

Chapter 3

Text and image sources, contributors, and licenses

3.1 Text

- **Climate change denial** *Source:* https://en.wikipedia.org/wiki/Climate_change_denial?oldid=687810387 *Contributors:* Ed Poor, Fxmastermind, Edward, Fred Bauder, Mac, Ronz, William M. Connolley, Evercat, Dragons flight, Raul654, Rossnixon, Chrism, PBS, Stephan Schulz, TMLutas, Auric, Sunray, Alan Liefting, Cool Hand Luke, FeloniousMonk, Theblog, StuartH, BozMo, Mozzie, Lumidek, Anirvan, Now3d, Adambondy, Brevity, Rich Farmbrough, KillerChihuahua, Sladen, Vsmith, ArnoldReinhold, Dave souza, Bender235, Neilrieck, Guettarda, Nigelj, Viriditas, I9Q79oL78KiL0QTFHgyc, Nsaa, Orangemarlin, Rd232, Plumbago, John Quiggin, Tony Sidaway, Count Iblis, A D Monroe III, Stemonitis, Woohookitty, Benhocking, Drbogdan, Rjwilmsi, Nightscream, Ground Zero, Jrtayloriv, Jfraatz, Theo Pardilla, Sharkface217, Bgwhite, Manscher, Wavelength, Bobby1011, Markhoney, Arado, Splette, Jaymax, Brian A Schmidt, Morphh, Rick Norwood, Długosz, GHcool, Number 57, Froth, 2over0, Zzuuzz, Arthur Rubin, DGaw, CWenger, Smurfy, Ricka0, NeilN, SmackBot, BenBurch, Meatpack, Anastrophe, Cla68, Portillo, Squiddy, RDBrown, DroEsperanto, MartinPoulter, Raymond arritt, Tasty monster, Silly rabbit, RayAYang, Annelid, Narco, John Hyams, Threeafterthree, Jquazimodor, Khukri, Ratel, John D. Croft, Topologyrob, Dreadstar, BullRangifer, JzG, AnonEMouse, Khono, Robofish, JoshuaZ, Peterlewis, RMHED, Mbelrose, Iridescent, K, Judgesurreal777, Joseph Solis in Australia, Alexh19740110, RekishiEJ, Markbassett, Cyrusc, THF, Ertdredge, AndrewHowse, Cydebot, Reywas92, Treybien, ST47, Synergy, Teratornis, Alaibot, Ssilvers, SteveMcCluskey, KimDabelsteinPetersen, Rosarinagazo, Itsmejudith, Guy Macon, SmokeyTheCat, Mackan79, J. Langton, Tillman, Harryzilber, AniRaptor2001, Nicholas Tan, Zeeboid, Rothorpe, Coffee2theorems, VoABot II, Ling.Nut, Brusegadi, Benzocane, Gabriel Kielland, DGG, Oren0, Ombudswiki, AstroHurricane001, UBeR, Noahcs, Heyitspeter, Glynth, Juliancolton, Scott Illini, DASonnenfeld, Funandtrvl, Johnfos, HughD, Uyvsdi, Maghnus, Childhoodsend, Spoisp, Aymatth2, IronMaidenRocks, Maracana, Vchimpanzee, Dmcq, Iceage77, Legokid, Lylefor, SieBot, Rlendog, PatronSaintOfEntropy, Ravensfire, Odd nature, Aleding, Jason Patton, Grundle2600, Toddst1, Radon210, Jc-S0CO, SimonTrew, Comingdeer, Zoomwsu, Calatayudboy, Envirocorrector, Capitalismojo, The Four Deuces, Hamiltondaniel, Bowei Huang 2, Sphilbrick, Wahrmund, Trappem, Loren.wilton, ClueBot, Sea.wolf4, Revolutionaryluddite, Franamax, Watti Renew, Monobi, GoRight, Mustang19, Rhododendrites, Arjayay, Hans Adler, N p holmes, VsevolodKrolikov, Redthoreau, Mlaffs, Aprock, Wikinv, Thompsontough, Cshearer19, DumZiBoT, Zenwhat, Fifaworld07, Frenstad, Rankiri, Stickee, Ajcheema, Feenixtim, Jprw, Spoonkymonkey, Kaiwhakahaere, Drolz09, Addbot, Wyatt Stringfellow, CurtisSwain, Cuaxdon, Victim of Changes, Driving and Crying, Thoroughfare, Blazing On To Glory, An American Gigolo, Jaimaster, Verbal, Lightbot, Jarble, Publicly Visible, Luckas-bot, Yobot, Threop, Brougham96, DrFleischman, AnomieBOT, Jim1138, Shock Brigade Harvester Boris, Ulric1313, Mann jess, Ckruschke, Citation bot, BBiiis08, SakaScotii, MonoApe, TracyMcClark, ChildofMidnight, JohnWBarber, GrouchoBot, Off2riorob, Treedel, Moxy, Smallman12q, Shadowjams, Thehelpfulbot, Green Cardamom, FrescoBot, Surv1v4l1st, LucienBOT, Menwith, Adam9389, Airborne84, Citation bot 1, J. Sketter, DigbyDalton, Wikispan, Gaba p, Moonraker, Jandalhandler, Tim1357, Trappist the monk, Drrll, Lotje, Morizbliz, Tbhotch, Kookaburra17, 128.104.truth, RjwilmsiBot, ButOnMethItIs, Meerwind7, DarknessShines2, EmausBot, Santamoly, WikitanvirBot, KurtLC, DuKu, GoingBatty, Djembayz, Solomonfromfinland, Hhhippo, Josve05a, H3llBot, Hyblackeagle22, Elenna-nórë, SporkBot, Seniortrend, Ocaasi, Sbmeirow, Ubikwit, Accotink2, Jess, Sailsbystars, Longshevius, Terra Novus, Teaktl17, LM2000, Tunabex, Helpful Pixie Bot, Bibcode Bot, Plantdrew, BG19bot, NewsAndEventsGuy, Karinsa, ArtifexMayhem, Iselilja, Northamerica1000, ASCIIn2Bme, 86.** IP, Safehaven86, Aisteco, Ammorgan2, Karin Anker, SuperHero2111, Dexbot, SoledadKabocha, TippyGoomba, Andyhowlett, Sarmariemack, Jodosma, Everymorning, Wuerzele, Prokaryotes, MrLinkinPark333, Jora8488, Trayvon1, Sum Christianus, Monkbot, Darreus, Mjmroy, Knife-in-the-drawer, Srednuas Lenoroc, Afterimage33, Stewi101015 and Anonymous: 19

- **Global warming controversy** *Source:* https://en.wikipedia.org/wiki/Global_warming_controversy?oldid=687354131 *Contributors:* Matthew Woodcraft, Slrubenstein, Ed Poor, Grouse, Axon, Graft, Heron, Stevertigo, D, Lexor, MartinHarper, Gabbe, Tannin, Ixfd64, Sheldon Rampton, DavidWBrooks, Williamv1138, Mac, Doom, William M. Connolley, G-Man, Marco Krohn, Ciphergoth, Evercat, Guaka, Timwi, Dcoetzee, Jstanley01, Will, Tpbradbury, Dragons flight, SEWilco, Frihet, Stormie, Raul654, Wetman, Mjmcb1, Lumos3, The lorax, Phil Boswell, ChrisO~enwiki, Stephan Schulz, Naddy, Securiger, Lowellian, Academic Challenger, Gidonb, Andrew Levine, Aetheling, Quadalpha, Drstuey, Phanly, Alan Liefting, Ancheta Wis, Mat-C, ShaneCavanaugh, Webbosoft, Cool Hand Luke, Wwoods, FeloniousMonk, Duncharris, Darrien, MSTCrow, Neilc, Stevietheman, PenguiN42, Gzuckier, Antandrus, BozMo, Beland, Piotrus, Daniel,levine, Llewdor, Anythingyouwant, Goh wz, Spottedowl, Lumidek, Asbestos, Zondor, Mrdarklight, Atlastawake, Spiffy sperry, DanielCD, Discospinster, Rich Farmbrough, Killer-

Chihuahua, Guanabot, Rameses, Vsmith, Tsumetai, Dave souza, GregBenson, TheOuthouseMouse, Bender235, ESkog, Blogjack, Wolfman, Zippedmartin, El C, Clem Powell, Mwanner, Liberatus, Art LaPella, Cacophony, Guettarda, Bobo192, Nigelj, JonGwynne, NetBot, Flying Hamster, Evolauxia, Enric Naval, Viriditas, Gjl, I9Q79oL78KiL0QTFHgyc, DG~enwiki, Sam Korn, Silverback, Crust, Nsaa, Orangemarlin, Schissel, Alansohn, Anthony Appleyard, Polarscribe, Diego Moya, Rd232, Jtalledo, Plumbago, John Quiggin, Hu, MrBudgens, Miltonhowe, Cortonin, Krappie, Peter B., Tony Sidaway, Count Iblis, Frescard, Ianblair23, Axeman89, Dejvid, Stemonitis, Boothy443, Richard Arthur Norton (1958-), Dandv, Merlinme, Yansa, Benhocking, Queerudite, Jeff3000, MONGO, Tedneeman, GregorB, BlaiseFEgan, Doc Richard, Audiovideo, Driftwoodzebulin, Rnt20, Marskell, Minos~enwiki, Kbdank71, Grammarbot, Josh Parris, Drbogdan, Rjwilmsi, Nightscream, Phalseid, Koavf, Eyu100, Tangotango, Dunro, GreetingsEarthling, HappyCamper, Ligulem, Daniel Collins, Czalex, MarnetteD, AySz88, SNIyer12, Titoxd, Isotope23, Gurch, Jrtayloriv, Choess, TeaDrinker, Alphachimp, Christopher Boyd, Theo Pardilla, Bgwhite, Bi, Wavelength, Jsolinsky, Midgley, John Callender, RussBot, PWhittle, Limulus, Kvuo, Jaymax, Brian A Schmidt, Gaius Cornelius, Kimchi.sg, Wimt, Cunado19, NawlinWiki, Ozzykhan, Rick Norwood, EWS23, SEWilcoBot, ChadThomson, Robertvan1, Kvn8907, Długosz, Dilaudid~enwiki, RazorICE, Johantheghost, Nephron, Froth, CrazyLegsKC, RonCram, Cerejota, Nethgirb, Mugwumpjism, Mnyakko, Ke5crz, Nescio, Werdna, Pevos, Saric, American2, Heptazane, Paul Magnussen, 2over0, Likwidshoe, Bayerischermann, Chase me ladies, I'm the Cavalry, Arthur Rubin, NHSavage, Reyk, Petri Krohn, CWenger, Luiscolorado, QmunkE, Dbarefoot, RG2, NeilN, SadaraX, CIreland, That Guy, From That Show!, Hal peridol, Nick R Hill, A bit iffy, SmackBot, Amcbride, Derek Andrews, Unschool, Zazaban, InverseHypercube, The Monster, Spsmiler~enwiki, Jim62sch, Lawrencekhoo, Anastrophe, Delldot, Cla68, Hans Erren, Josephprymak, UrbanTerrorist, DLH, Man with two legs, Markeer, Yamaguchi⬜⬜, UnqstnableTruth, Gilliam, Hmains, Skizzik, Rmosler2100, Squiddy, Chris the speller, Bluebot, Stevenwagner, Johnskrb2, Sumthingweird, Jcc1, Persian Poet Gal, RDBrown, Jprg1966, Raymond arritt, Robocoder, Colonies Chris, Gingi0, Gracenotes, Sentinel75, Royboycrashfan, Dethme0w, Can't sleep, clown will eat me, John Hyams, Jefffire, KaiserbBot, Jdhammer, Pnkrockr, JesseRafe, Threeafterthree, Normxxx, Isonomia, PrometheusX303, Ratel, Nils Simon, Groomson, Professor Chaos, Black Butterfly, Monoape, Nrcprm2026, Itzar, SpiderJon, DMacks, Fredgoat, Wizardman, Filpaul, Kalathalan, Lukebutler, Daniel.Cardenas, Mostlyharmless, Drunken Pirate, Will Beback, The undertow, Nathanael Bar-Aur L., Dave314159, Tanuki-Dori, Gloriamarie, Profg, The great grape ape is straight out of the know, Vgy7ujm, Crazyviolinist, Loodog, Gobonobo, Soumyasch, JH-man, JoshuaZ, Peterlewis, Ghw777, Chris 42, Pflatau, Rm w a vu, Collect, Chrisch, Gordongraff, MarkSutton, Slakr, Boomshadow, Hiiiiiiiiiiiiiiiiiiiiii, Waggers, Winknnudge, Zorxd, Atakdoug, Biglin, Kvng, Bolt Vanderhuge, Freedom Fan, Dean1970, Hu12, SimonD, Fan-1967, K, Colonel Warden, Joseph Solis in Australia, Theone00, Cedrium, Alexh19740110, DreamsReign, IanOfNorwich, Bertport, Tawkerbot2, Harold f, TNeloms, Paul Matthews, Cyrusc, Judyjowers, Vaughan Pratt, CmdrObot, Blouis79, Tobes00, Dnwk, Dycedarg, Van helsing, Nadyes, Pseudo-Richard, Zinjixmaggir, IrishJew, JettaMann, Besidesamiracle, Phase Theory, Green 4 Peace, Jbraun1984, Jlancaster, Cydebot, Abeg92, Kevinp2, Treybien, UncleBubba, Khatru2, Frostlion, Jayen466, Tawkerbot4, DumbBOT, Teratornis, HillChris1234, Ssilvers, Crum375, Www.inquisition.ca, JohnInDC, The machine512, Fat7926, Thijs!bot, Elanthiel, Jaxsonjo, KimDabelsteinPetersen, Daniel, Sagaciousuk, Gralo, Headbomb, Id447, The4ce, Toddlamb, Rosarinagazo, Pacific PanDeist, Dalahäst, West Brom 4ever, Ufwuct, Itsmejudith, Doyley, Sliponshoe, Second Quantization, Chrisdab, Bails247, Ramckay, Sinn, Straussian, Dansphere, Rbrustman, Grand51paul, 00666, Dawnseeker2000, Vsevolod4, LachlanA, Mentifisto, Mgerb, AntiVandalBot, RobotG, Luna Santin, Guy Macon, CPWinter, Tjsynkral, Fyunck(click), Prolog, Prof.Thamm, Sweart1, SadanYagci, Bridgeplayer, Tillman, Wing Nut, Blair Bonnett, Captain canada, C56C, OGGVOB, Narssarssuaq, Sirfrankomac, MER-C, Inks.LWC, Jecates, Dr. JJ, Cary.boyce, Dicksonlaprade, Sln3412, Starflixx, Zeeboid, Jhm15217, MaxPont, Magioladitis, Skyemoor, Bongwarrior, VoABot II, MiguelMunoz, Mast-Cell, Buraianto, Ling.Nut, Poujeaux, Roches, Soulbot, Drollere, Iceberg007, Bernardissimo, Rich257, Froid, Brusegadi, Majestic Lizard, Jdey123, Cgingold, Gabriel Kielland, Bryanpeterson, NimNick, Arthur lastman, Beagel, Demosfoni, N.Nahber, Bobanny, Prester John, Edward321, Jarro 2783, Dan Pangburn, Oren0, Stephenchou0722, MartinBot, Jwbaumann, AussieBoy, Ugajin, Rjwellings, Arjun01, Lesikar, Gelsomina, Roastytoast, Keith D, Coeus, R'n'B, CommonsDelinker, Jarhed, Jascii, Jeffhall318, Exarion, Pharaoh of the Wizards, Nufacion, Smcauliffe, AstroHurricane001, R. Baley, UBeR, Jrsnbarn, Eliz81, RedPoptarts, Xyzt1234, Littlebum2002, Dispenser, Enuja, McSly, Kroush, Patch666, NS Zakeruga, T Steinway, Rossenglish, Plasticup, Aksnitd, Birdbrainscan, SJP, Vanished user 47736712, Jorfer, Student7, Joshua Issac, Curtis Bledsoe, Mleonard85032, Vyn, Saximus, Ratfox, Scott Illini, Brandonromero, Islandman2, Enescot, Duchamps comb, Will henderson, Gav111, Cpt ricard, Fuzzygenius, Johnfos, HughD, Jeff G., Shinju, JoanCaucus, Katydidit, Ymous, Philip Trueman, Childhoodsend, Oshwah, Jogar2, Zamphuor, Staplegunther, Red Act, Peterrhyslewis, Ninarosa, Gwinva, Rob944s2, Qxz, Someguy1221, MrRSMan, Seraphim, Yilloslime, CreateSomeNoise, Wikifan5554, Peekoid, Leafyplant, Rcrookes, PDFbot, Ahm2307, Nodrogp, Mishlai, Foorider, Q Science, Hey jude, don't let me down, Every name is taken12345, OverMyHead, AngryWolf81, Don't lose that number, Sapphic, Joneseypoo, Dmcq, WiseWin, Iceage77, GAPlauche, Legoktm, Sfmammamia, Culliganator, Runewiki777, GirasoleDE, Mikedurtnall, Ymom2, Hi people48, Sundaybrunch, Esthameian, Damorbel, Hertz1888, Thomascartwright, Dawn Bard, Caltas, Odd nature, Jason Patton, LeadSongDog, Mckaulick, Grundle2600, Loliamnickjohnson, Jawshoeaw, Happysailor, Toddst1, Daniloxx, Pburto, MaynardClark, Jc-S0CO, Dominik92, Antonio Lopez, PhilMacD, Godhatesme, Lightmouse, LaidOff, Alex.muller, Jack the Stripper, Belligero, Akarkera, Envirocorrector, Stupidreject09, Capitalismojo, Duae Quartunciae, Latics, Rjc34, Bowei Huang 2, Sphilbrick, Butane Goddess, Req30843, Kreag, Bmedley Sutler, Mrfebruary, Sagredo, Martarius, Peter Neville, Howezhu, ClueBot, Mariordo, Polentario, Binksternet, Foxj, The Thing That Should Not Be, Gft4, WriterListener, Tracys49, Jambla, Lo Scaligero, Pairadox, Sekander94, Richrakh, Vinny Burgoo, Chris Bainbridge, Bobman999, Agustinaldo, Mseall, LuckyPoppa, Niceguyedc, Blanchardb, Phdplayahatadegree, Auntof6, PhGustaf, SamuelTheGhost, Awickert, Dr. B. R. Lang, Bonefishj0e, Nymf, DBlade, Resoru, GoRight, Heartland institue, Gtstricky, Gwguffey, The Founders Intent, Tomthetank94, ZuluPapa5, NuclearWarfare, Svilli, Cenarium, Arjayay, Psinu, VsevolodKrolikov, Redthoreau, Hercule, Aprock, Ozgadgetgirl, Luiang, Bleeding Blue, Mjun88, Belchfire, Bacteriophage, DumZiBoT, Chemical Euphoria, Jean Shaheen, Caldwell matt, Bearsona, Mjharrison, Nathan Johnson, ChyranandChloe, EastTN, FellGleaming, Dwr12, SilvonenBot, Jcosco, Aunt Entropy, Good Olfactory, Lemmey, Parejkoj, Infonation101, Hakuin, Kbdankbot, FlagrantUsername, Tre2, Mistinis, The Squicks, Armored Saint, American Eagle, Wyatt Stringfellow, CurtisSwain, Some jerk on the Internet, DOI bot, Nacho71, Steve1941, Cuspid Groove, Bushcutter, Vanished user oerjio4kdm3, Freezehky08, Victim of Changes, Fothergill Volkensniff IV, Leszek Jańczuk, Greenerdays120606, Eivindbot, Kasmel, CO2 doubter, An American Gigolo, Pakato, Chzz, Dr. Universe, Lucian Sunday, Rodeo90, Patton123, Tassedethe, Jaimaster, Tide rolls, Olsonjs444, Zorrobot, Jarble, J. Johnson, Legobot, Drpickem, Publicly Visible, Yobot, Bunnyhop11, Fraggle81, Shenstar, Jan Arkesteijn, Vanished user sflgjhaerp98q3iv8j3qp8uti, Shawine, Punctilius, AnomieBOT, Climatedragon, 1exec1, Shock Brigade Harvester Boris, Ulric1313, Materialscientist, Ckruschke, Citation bot, Embram, Batpox, Twri, Gemtpm, Gimmethegepgun, MonoApe, Pontificalibus, Tyrol5, Srich32977, NOrbeck, Armbrust, Resident Mario, Gsälzbär, A Quest For Knowledge, Moxy, MerlLinkBot, Miyagawa, David Nemati, Buzz-tardis, FrescoBot, Airship (whoops), Tiramisoo, Charles Edwin Shipp, Airborne84, Citation bot 1, DigbyDalton, Wikispan, Gaba p, Pink Bull, Tom.Reding, Idge62, Moonraker, Jschnur, Trappist the monk, Vrenator, Jeffrd10, Canuckian89, Tbhotch, Danieljaycho, RjwilmsiBot, Marktka, Jlhcpa, NameIsRon, Torontokid2006, Walkinxyz, I love SUV's, DarknessShines2, Desmon Dantes, Dewritech, GoingBatty, Chris Vanderpump, Lynn Dulsok, Wikipelli, K6ka, Solomonfromfinland, Tonnex, Thepisky, Thargor Orlando, Minor4th, Jagr0068, H3llBot, Arduinot, Picklejarr, Makecat, Δ, Rostz, Keulian, Buu206, Jess, Sailsbystars, Yuppyyummy, Wikiuser953, ThePowerofX, Phirmbutzian, Haktl23, AKeenEye, Casey399, Fits and Starts, ClueBot NG, Prioryman, Zytigon, Joffos, Samwell Pinkus, TheBigCatcher, Icantwait, Widr, Helpful Pixie Bot, Guy who reads a lot, Tholme, Wbm1058, Bibcode Bot, Sokavik, BG19bot, NewsAndEventsGuy, ArtifexMayhem, Restreusion, Northamerica1000, Wheiner, Sweetbreads, MusikAnimal, Darkness Shines, Sam Lefferts, Hoktshok, Majjor Payne, Mark Arsten, Maxellus, BenDen1, Harizotoh9, MrBill3, 86.** IP, Chandlej2011, Oliseh87, Otus scops, Glacialfox, Roymcdo, Klilidiplomus, MaytheFarcebewithYou, Tranh Phanh, BattyBot, Rwenonah, Solntsa90, ChrisGualtieri, FUNPOW, Saedon, Timelezz, Sunvox, Dexbot, Webclient101, Leucosticte, TippyGoomba, Stephenrkirby, FonsScientiae, Ruska25, Steve Handersman, Punashay, Dr. R. Rosen, Metina9182, Kindyungsir, Junaji, MrWorshipMe, Cole132132, Hide the Decline II, GelPaks, Proodenhamer, I am One of Many, Wheels of Steal, Melonkelon, Everymorning, Evano1van, Robinlarson, Cescam, DavidLeighEllis, Dellerex, Maw368, Citytownhome, MrLinkinPark333, Acalycine, Icensnow42, IHaveAMastersDegree, Femkemilene, Austrartsua, Gfcan777, A.Todd English122, Graihagh, Monkbot, Ordinary reader, Dave18wheeler, Bammie73, Co2denier, Prconfer, Skoritz, Mediify, Mjmroy, Emilyscarr, Serten II, Prostar190, NetworkOP, K scheik, Jerodlycett, The edit master 123, Scaravich105nj, Pandaninja982, Afterimage33, Revelacious, Robinm1973, Iaepetus, Disprosiam and Anonymous: 807

- **American Petroleum Institute** *Source:* https://en.wikipedia.org/wiki/American_Petroleum_Institute?oldid=688004816 *Contributors:* Gabbe, Lquilter, Typhoon, William M. Connolley, Saltine, Wetman, Ashdurbat, HaeB, Mattflaschen, Geni, BozMo, Histrion, User2004, Jensbn, Sole Soul, Remuel, John Vandenberg, Rd232, Dachannien, JoaoRicardo, Pol098, Ruud Koot, Rogerd, Vegaswikian, Corto, FlaBot, Gurch, YurikBot, Barneygumble, Ampacific, Falcon9x5, BOT-Superzerocool, Arthur Rubin, SmackBot, Gilliam, Mrubx, Fuhghettaboutit, Will Beback, Harryboyles, Gobonobo, Mbeychok, Adam sk, Billy Hathorn, Cyrusc, CmdrObot, Jbgilm, Neelix, Cydebot, Marqueed, Evanclifthorne, MKil, Apostmodernist, ShadowGuy, Teratornis, K001, Uruiamme, PhiLiP, Glrx, Sanjivkumarsharma, Harshalraje, Robone03, HughD, Ziounclesi, McM.bot, S2grand, Plazak, LuigiManiac, HybridBoy, Jrats, Capitalismojo, Treekids, Lebron2k5, Yonskii, Kyslyi, Versus22, Addbot, Willking1979, Akadonnew, Myk60640, Legobot, Yobot, AnomieBOT, Srich32977, Mykjoseph, Off2riorob, RjwilmsiBot, Angrytoast, GoingBatty, Seansimpson, Helpful Pixie Bot, Afimka, Mdann52, Rp8083, Michipedian, Mjmroy and Anonymous: 52

- **Business action on climate change** *Source:* https://en.wikipedia.org/wiki/Business_action_on_climate_change?oldid=676740216 *Contributors:* Edward, SebastianHelm, Timwi, Raul654, Alan Liefting, Vanished user wdjklasdjskla, Rich Farmbrough, NrDg, Vsmith, Xezbeth, Tarndt, Rd232, Tony Sidaway, Woohookitty, Benhocking, Lapsed Pacifist, Mandarax, BD2412, Rjwilmsi, Simesa, Wavelength, RussBot, Splette, Ozzykhan, THB, Arthur Rubin, Chriswaterguy, Fram, Naught101, DocendoDiscimus, SmackBot, Brossow, Hmains, Hectorguinness, Chris the speller, Deli nk, Colonies Chris, Nakon, BullRangifer, CartoonDiablo, Gobonobo, Ckatz, Dean1970, Cydebot, KimDabelstein-Petersen, Gralo, Prolog, Brusegadi, Lesikar, DadaNeem, Alinaboey, 28bytes, Johnfos, Fillinchen, Grocc, Wede~enwiki, Flyer22 Reborn, Lightmouse, Mikimomen, Sphilbrick, Notesfromtheroad, MenoBot, Lawrence Cohen, Jamesboxhead, Dclinton, DumZiBoT, Zenwhat, Addbot, Xp54321, Roystonea, Maestra222, Thomas Yeardly, Led zec, Luckas-bot, Themfromspace, Bathandbristol, Guy1890, Kingpin13, Materialscientist, Climateneutral, FrescoBot, Menwith, Nepomuk 3, Shqipëria për Europën, Skyerise, Scullybuns, Starnsworth, John of Reading, H3llBot, Teaktl17, Jack Greenmaven, Coastwise, Helpful Pixie Bot, Titodutta, Technical 13, NewsAndEventsGuy, Northamerica1000, Matulkar, BattyBot, Hmainsbot1, StressOverStrain and Anonymous: 78

- **Climate Audit** *Source:* https://en.wikipedia.org/wiki/Climate_Audit?oldid=688231971 *Contributors:* Booyabazooka, Lquilter, William M. Connolley, Dragons flight, Johnleemk, Bearcat, Stephan Schulz, Desmay, TMLutas, Lumidek, Spiffy sperry, Vsmith, Dave souza, Guettarda, Evolauxia, Timecop, Crust, Nsaa, PaulHanson, Bacteria, Woohookitty, GregorB, Titoxd, Ground Zero, Nimur, Coold00d, GeeJo, Ms2ger, Arthur Rubin, Chris Chittleborough, SmackBot, Cla68, DLH, John A, Gobonobo, ATren, Colonel Warden, IanOfNorwich, Amalas, Sweet Potato, Themightyquill, Cydebot, DumbBOT, KimDabelsteinPetersen, Prolog, Tillman, RebelRobot, Brusegadi, KConWiki, R'n'B, Katharineamy, VolkovBot, Q Science, Ravensfire, OKBot, Capitalismojo, John Nevard, DumZiBoT, Bearsona, Nathan Johnson, FellGleaming, Ecolabs, Addbot, Arbitrarily0, Poznan, SausageLady, Sionk, A Quest For Knowledge, Wikispan, Tbhotch, DarknessShines2, Thepisky, Jess, Sailsbystars, Kparthka, Tunshayan, Rendahl, Barney780, Tender Nelson, AbiDabie, Helpful Pixie Bot, Soontwob, Particals of Phoof, Haabulu, BattyBot, Malerooster, IHaveAMastersDegree, Monkbot, Sinbit68, Whenzler and Anonymous: 28

- **Climate change policy of the George W. Bush administration** *Source:* https://en.wikipedia.org/wiki/Climate_change_policy_of_the_George_W._Bush_administration?oldid=676127090 *Contributors:* Andrewman327, Alan Liefting, Rjwilmsi, Cydebot, Antony-22, Jarble, Yobot, Solomonfromfinland, AvicBot, Everymorning, ArmbrustBot, Monkbot and Anonymous: 1

- **Environmental skepticism** *Source:* https://en.wikipedia.org/wiki/Environmental_skepticism?oldid=682356365 *Contributors:* Ed Poor, William M. Connolley, Jpatokal, TUF-KAT, Vespristiano, Henrygb, PxT, Wayland, Alan Liefting, Gzuckier, Scott Burley, Spiffy sperry, Rich Farmbrough, Vsmith, FWBOarticle, Dagonet, Nigelj, Viriditas, Cmdrjameson, Pearle, Orangemarlin, Rd232, John Quiggin, Kurieeto, Omphaloscope, Raygirvan, TheEvilBlueberryCouncil, Dangerous Angel, Gctegpipes, Common Man, UnlimitedAccess, Wavelength, Bhny, Dilaudid~enwiki, Tony1, Arthur Rubin, Opiaterein, SmackBot, Nil Einne, Can't sleep, clown will eat me, Nils Simon, Weregerbil, Polonium, Wjejskenewr, CmdrObot, KimDabelsteinPetersen, Tillman, MastCell, AstroHurricane001, Henry Cassini, Nlarcher, Johnfos, QuackGuru, Yilloslime, Dmcq, Wjl2, Swliv, Lightmouse, KathrynLybarger, Capitalismojo, Bowei Huang 2, ImperfectlyInformed, TotesBoats, Mild Bill Hiccup, Alexbot, SpikeToronto, Cshearer19, Vjlenin, DOI bot, Jedes, Jarble, Legobot, Sageo, Punctilius, Mann jess, Strayson, Feraljyce, Belsavis, Wikispan, MeUser42, Jonkerz, RjwilmsiBot, Cvazquez09, Keilandreas, Werieth, ZéroBot, ThePowerofX, Wikiwind, Helpful Pixie Bot, Northamerica1000, 86.** IP, Karin Anker, Sarmariemack, Monkbot, Skrakov and Anonymous: 34

- **Global warming conspiracy theory** *Source:* https://en.wikipedia.org/wiki/Global_warming_conspiracy_theory?oldid=687130982 *Contributors:* Ed Poor, ESnyder2, William M. Connolley, Greenrd, Stephan Schulz, TMLutas, Alan Liefting, Giftlite, Iota, Utcursch, Eep², Vsmith, Dave souza, Bishonen, WegianWarrior, Bobo192, Nigelj, Viriditas, John Quiggin, Staeiou, Tony Sidaway, Woohookitty, Benhocking, WadeSimMiser, Drbogdan, Rjwilmsi, Koavf, Tarc, Gurch, Kolbasz, RussBot, Arjuna909, Matt Fitzpatrick, NawlinWiki, Pagrashtak, Cerejota, Nethgirb, Arthur Rubin, CWenger, Katieh5584, NeilN, SmackBot, KnowledgeOfSelf, Lawrencekhoo, MediaMangler, Gilliam, Squiddy, Afa86, Tasty monster, Roscelese, John Hyams, DHeyward, Kittybrewster, Ratel, BullRangifer, Wizardman, Lukebutler, Will Beback, Loodog, Peterlewis, Ckatz, Domokato, Midnightblueowl, Dean1970, Joseph Solis in Australia, Frankieleelee, RekishiEJ, JForget, Wafulz, Makeemlighter, Kalaong, Phase Theory, Michfan2123, Cydebot, Lonenut2000, AlexanderLevian, Doug Weller, KimDabelsteinPetersen, Id447, Rosarinagazo, Itsmejudith, Peter Gulutzan, Northumbrian, Noroton, Tillman, VoABot II, DGG, SquidSK, Oren0, Agricolae, Huzzlet the bot, Pharaoh of the Wizards, UBeR, ChrisfromHouston, BashBrannigan, McSly, Birdbrainscan, Duchamps comb, James Callahan, QuackGuru, Philip Trueman,

Childhoodsend, Yilloslime, Q Science, Don't lose that number, Falcon8765, @pple, Cwmacdougall, The Devil's Advocate, Dmcq, 19merlin69, GirasoleDE, Madman, MagicErik, Dougrny, Smsarmad, Jc-S0CO, Oxymoron83, Hello71, Polbot, Capitalismojo, Tesi1700, Bowei Huang 2, ClueBot, Andrew Nutter, TribeCalledQuest, Snigbrook, Foxj, NiD.29, Blanchardb, Trivialist, Deselliers, John Nevard, Feline Hymnic, DarkDevotion, Shinkolobwe, VsevolodKrolikov, Thingg, JessicaJames777, DumZiBoT, The Noosphere, XLinkBot, Nathan Johnson, Jprw, Habitaddict, WikiDao, Osarius, Monktons, Addbot, CurtisSwain, Tyw7, Tide rolls, MuZemike, Jarble, J. Johnson, Fraggle81, DisillusionedBitterAndKnackered, Rikanderson, AnomieBOT, A More Perfect Onion, Shock Brigade Harvester Boris, Jaswantium, Materialscientist, Kasaalan, Twri, Basilisk4u, Gimmethegepgun, MonoApe, JohnWBarber, MalcolmMcDonald, Shattered Gnome, Punksta, Abcgeek, RightCowLeftCoast, Dogposter, Charles Edwin Shipp, Thorenn, Markeilz, Airborne84, Wikispan, Gaba p, Pinethicket, Moonraker, Serols, Submissivesquat, Full-date unlinking bot, Servant David, Trappist the monk, Wdcraven, Tbhotch, MarkOfBondi, RjwilmsiBot, Globalposter, TonySeales14, BeachedOne, Burmiester, Georgiedrink, GoingBatty, YosemiteFudd, Dialecticexpert, Ocaasi, L Kensington, Jess, Info2012, ThePowerofX, Hmcst1, SJTH, EdoBot, Ldvhl, ClueBot NG, Jaredtheevill, Jack Greenmaven, Dcd96, Gulf+6, Widr, Helpful Pixie Bot, Guy who reads a lot, Sokavik, BigJim707, NewsAndEventsGuy, ArtifexMayhem, Gise-354x, MusikAnimal, Writ Keeper, MrBill3, 86.** IP, IsraphelMac, Hmainsbot1, Steve Handersman, GranChi, I am One of Many, Jamesmcmahon0, Robinlarson, Zenibus, Prokaryotes, Realist2013, SpecMade, Monkbot, Bosom, Bikerjim500, Fgxghvjhkgh, HylianDomination, Lilbruh123, NetworkOP, Jerodlycett, Stabila711 and Anonymous: 250

- **Information Council on the Environment** *Source:* https://en.wikipedia.org/wiki/Information_Council_on_the_Environment?oldid=666107005 *Contributors:* SimonP, GCarty, SEWilco, Raul654, Vsmith, Pavel Vozenilek, Durban32, Duk, Rd232, RJFJR, Kbdank71, Erachima, SmackBot, George100, Scott Illini, Yilloslime, Alex Middleton, AnomieBOT, Strayson, ClueBot NG, Jakeybean, Everymorning and Anonymous: 6

- **Leipzig Declaration** *Source:* https://en.wikipedia.org/wiki/Leipzig_Declaration?oldid=686621570 *Contributors:* Eloquence, Ed Poor, Lorenzarius, MartinHarper, Sheldon Rampton, Minesweeper, Radicalsubversiv, William M. Connolley, RadicalBender, Securiger, CorpDan, Zigger, Kate, Vsmith, JonGwynne, John Quiggin, Tony Sidaway, YurikBot, Taganov, Welsh, Mnyakko, SmackBot, Chris the speller, Afasmit, Pflatau, Van helsing, Cydebot, KimDabelsteinPetersen, Qwarto, Magioladitis, David Eppstein, Nwbeeson, Michaeldsuarez, Kbdankbot, Addbot, CurtisSwain, Lightbot, Margin1522, Gongshow, AnomieBOT, Shock Brigade Harvester Boris, Full-date unlinking bot, Cnwilliams, Gordyz75, ThePowerofX, Levdr1, Everymorning, Skr15081997 and Anonymous: 15

- **List of scientists opposing the mainstream scientific assessment of global warming** *Source:* https://en.wikipedia.org/wiki/List_of_scientists_opposing_the_mainstream_scientific_assessment_of_global_warming?oldid=686916709 *Contributors:* The Epopt, The Anome, Ed Poor, Leandrod, Michael Hardy, Lquilter, Cyde, Ronz, William M. Connolley, Angela, Andrevan, Kbk, DJ Clayworth, Dragons flight, SEWilco, Floydian, Raul654, Wetman, Lumos3, Rossnixon, Altenmann, Stephan Schulz, Merovingian, Wereon, LetterRip, Phanly, Lupo, Pengo, Terjepetersen, Alan Liefting, Cool Hand Luke, Marcika, Jfdwolff, Ceejayoz, Theblog, Bobblewik, Jmcnamera, BozMo, Thparkth, Lumidek, Gscshoyru, Engleman, Spiffy sperry, Rich Farmbrough, Rameses, Vsmith, Dave souza, Brian0918, RoyBoy, Triona, Guettarda, Causa sui, Nigelj, Longhair, Viriditas, I9Q79oL78KiL0QTFHgyc, Silverback, Crust, DannyMuse, Diego Moya, Hipocrite, John Quiggin, SlimVirgin, Daniel.inform, B3virq3b, Snowolf, Gdavidp, Tony Sidaway, LFaraone, Pytom, Ondrejk, OwenX, Woohookitty, Henrik, Merlinme, Benhocking, Scjessey, The Wordsmith, GregorB, BlaiseFEgan, Betsythedevine, Emerson7, Rjwilmsi, Jivecat, Matt Deres, Ayla, Ggb667, Bgwhite, Manscher, The Rambling Man, Wavelength, Sceptre, Arzel, RussBot, Geologician, Splette, Skydot, Ergzay, Wiki alf, Arker, Tony1, RonCram, Nethgirb, Mnyakko, PyroGamer, Elysianfields, Mütze, Hobit, Sandstein, Zzuuzz, Arthur Rubin, DGaw, CWenger, Kevin, Nealparr, JeffBurdges, Meegs, NeilN, Kyaa the Catlord, TechBear, SmackBot, Stifle, Cla68, DLH, Nil Einne, Hmains, Squiddy, Bluebot, Jcc1, Sduplessie, RDBrown, Thumperward, Raymond arritt, Can't sleep, clown will eat me, John Hyams, DHeyward, Nickcoop, Lantrix, Addshore, Blueboar, Khukri, Nakon, John D. Croft, Nils Simon, DRJ, Monoape, Tomtefarbror, BullRangifer, Glover, Wizardman, Ohconfucius, Mukadderat, Tanuki-Dori, JzG, Scientizzle, SilkTork, Gobonobo, JoshuaZ, Mbeychok, Peterlewis, Pflatau, Collect, Buddyglass, Gordongraff, Ehheh, Ryulong, RMHED, ShakingSpirit, Dean1970, Vanished user, Echofloripa, Colonel Warden, Joseph Solis in Australia, AlexLibman, Alexh19740110, Alan Baskin, David Guest, GDallimore, DavidOaks, Courcelles, IanOfNorwich, Tawkerbot2, Ayanoa, Regress, Zeke pbuh, EABSE, Paul Matthews, CmdrObot, Avanu, BeenAroundAWhile, Jibal, LotR, Óðinn, Neelix, Yopienso, JettaMann, Rotiro, Cydebot, Fl, Jepp, Jayen466, Rracecarr, DumbBOT, Ssilvers, ErrantX, Matt Garnett, Casliber, The machine512, Komdori, Dscott8186, Gaijin42, KimDabelsteinPetersen, O, HappyInGeneral, Martin Hogbin, Dexterbrown, Headbomb, Id447, Jeffwishart, Rosarinagazo, Mtobis, Itsmejudith, Second Quantization, Elhector, Pcbene, 00666, Thomas Paine1776, Blue Tie, Smartse, Tillman, Wing Nut, Spartaz, LegitimateAndEvenCompelling, Dimawik, DuncanHill, Epeefleche, Inks.LWC, Sln3412, Nicholas Tan, Britcom, Zeeboid, J-stan, Tergadare, Coopercmu, Severo, MaxPont, DrLove829, Skyemoor, VoABot II, Appraiser, Nyttend, Brusegadi, Fabrictramp, WhatamIdoing, Theroadislong, NimNick, Ours18, Prester John, Dan Pangburn, Irate velociraptor, Oren0, Freiheitkrieger, Sympa, 1223334444, Jim2345, R. Baley, UBeR, Hodja Nasreddin, Andy0093, Keesiewonder, Xyzt1234, Pyrospirit, Chriswiki, RobinGrant, Birdbrainscan, Nwbeeson, Mrs.dog, TehCell, Prhartcom, Dhaluza, Mrmuk, ACV777, Bdixon, DASonnenfeld, Wilhelm meis, Beckyvolley, Enescot, Duchamps comb, Bpplowman, Black Kite, Johnfos, Denis.g.rancourt, QuackGuru, Childhoodsend, Jogar2, Zamphuor, Mhaag7, Red Act, Gwinva, Feberle, Hunterhogan, CreateSomeNoise, Bentley4, Skookumuk, UnitedStatesian, Prnd3825, Esmehwp, Onore Baka Sama, CoolHandNuke, Ltbx.com, Brittainia, Juanfermin, Q Science, HaroldHolmyard, TheVerum, Y, Falcon8765, Dmcq, Iceage77, Timothyrood, Brissbane, Dpawkng, RedRabbit1983, Showman60, Swliv, Bugguyak, Tescomaturecheese, Jason Patton, AndrewJlockley, DonaldDuck07, Africangenesis, Nopetro, Jc-S0CO, Lightmouse, Macy, Capitalismojo, Sphilbrick, Meltwaternord, ClueBot, Dylan6207, Jannyshoe, Lawrence Cohen, Jambla, Meekywiki, Chris Bainbridge, Ottawahitech, Rockfang, Awickert, GoRight, Shinkolobwe, The Founders Intent, Rhododendrites, Soccergrls rock555, ZuluPapa5, TheRedPenOfDoom, Bonewah, VsevolodKrolikov, Mlaffs, Obsidi, Aprock, Dylan38, Studip101, DumZiBoT, Zenwhat, The Noosphere, Mmarque, Mjharrison, Frenstad, Nathan Johnson, FellGleaming, HappyJake, Jprw, WikiHead, Aunt Entropy, Shoemaker's Holiday, The Squicks, Addbot, CurtisSwain, Simonm223, Thailboat, Kenneth Cooke, Download, Nigel Montcrief, Debresser, Deamon138, Akasofu, Tassedethe, Dyuku, Verbal, Jarble, J. Johnson, Yobot, Cap'nTrade, Raintwoto, Mnation2, Backslash Forwardslash, AnomieBOT, Indulis.b, Shock Brigade Harvester Boris, JackieBot, Alice Lyddel, Bluerasberry, Mann jess, Eumolpo, Rowlenthunder, ChildofMidnight, Doviel, LVAustrian, Armbrust, MalcolmMcDonald, A Quest For Knowledge, MerlLinkBot, Hongsy, Polargeo, JayJay, Middle 8, FrescoBot, Paine Ellsworth, Menwith, 🔲🔲🔲, Ivanelo, Citation bot 1, J. Sketter, Gaba p, Jonesey95, Moonraker, Irbisgreif, Bioextra, Trappist the monk, MrX, Flegelpuss, EvanHarper, Udippuy, BlueSal, RjwilmsiBot, Michaele and Tareq, Everton12, Aircorn, Tesseract2, Highly Unlikely, AQFK, Dominus Vobisdu, Racerx11, Griswaldo, GoingBatty, Lithistman, Solomonfromfinland, Minor4th, SporkBot, Δ, Jess, Ego White Tray, ThePowerofX, AndyTheGrump, Jonathan Lane Studeman, 9Questions, Neil P. Quinn, Terraflorin, MoonLichen, Helpful Pixie Bot, Bibcode Bot, Sokavik, Lowercase sigmabot, BG19bot, NewsAndEventsGuy, Iselilja, Northamerica1000, Darkness Shines, Chris the Paleontologist, Mark Arsten, ASCIIn2Bme, MrBill3, 86.** IP, LilDabL, BattyBot, Mestesugarul, Mollskman, Sundibar, Dexbot, TippyGoomba, IvarianFactor, Pogrump, Jo-Jo Eumerus, Junaji, Randykitty, Jodosma, Everymorning, Froglich, Serten, Sol1, Prokaryotes, Bladesmulti, Sportfan5000, Keith McClary, Monkbot, User2534, DoctorTerrella and Anonymous: 183

- **Media coverage of climate change** *Source:* https://en.wikipedia.org/wiki/Media_coverage_of_climate_change?oldid=687785398 *Contributors:* Edward, William M. Connolley, Julesd, Phil Boswell, Alan Liefting, Vsmith, Dave souza, Bender235, Circeus, Rd232, MONGO, Rjwilmsi, Choess, 10stone5, Arthur Rubin, SmackBot, RDBrown, CartoonDiablo, Will Beback, Ckatz, Makyen, Cydebot, Itsmejudith, JustA-Gal, Prolog, Tillman, DGG, AstroHurricane001, Dmcq, E8, Mrfebruary, Nymf, Belchfire, EdChem, XLinkBot, Dthomsen8, Dawynn, Jarble, Legobot, Yobot, AnomieBOT, Citation bot, FrescoBot, Citation bot 1, RjwilmsiBot, Solomonfromfinland, DavidMCEddy, Thargor Orlando, SporkBot, ThePowerofX, FurrySings, Teaktl17, Lathre, Snotbot, Cco1983, Helpful Pixie Bot, AzureAnt, Bibcode Bot, Northamerica1000, CitationCleanerBot, The Almightey Drill, 86.** IP, Tomlangridge, BattyBot, Arcandam, Dexbot, Creator011, Harlem Baker Hughes, Knovak07, Paeronet, Monkbot, 664ql.384.612, Stewi101015 and Anonymous: 77

- **Merchants of Doubt** *Source:* https://en.wikipedia.org/wiki/Merchants_of_Doubt?oldid=685943921 *Contributors:* Ed Poor, William M. Connolley, Desertphile, AnonMoos, Tsavage, Alan Liefting, Vsmith, Nigelj, Viriditas, Wavelength, Arado, Arthur Rubin, NeilN, Squiddy, RDBrown, Snori, Frap, Fuhghettaboutit, AB, Alanmaher, RekishiEJ, Cydebot, Ssilvers, JohnInDC, JamesAM, Tillman, Adrian J. Hunter, DadaNeem, Prhartcom, Johnfos, BoogaLouie, Yilloslime, Larklight, Dmcq, Capitalismojo, Nymf, Sun Creator, Favonian, Tassedethe, Yobot, AnomieBOT, Senor Freebie, Jim1138, Wikispan, Slavickp, JournalScholar, WikitanvirBot, GA bot, Dewritech, GoingBatty, Solomonfromfinland, Psw808, ThePowerofX, Wsjacobs, Jaydee000, MRDXII, Oddbodz, Helpful Pixie Bot, KLBot2, Northamerica1000, Peaceandlonglife, Funez Remiaw, Adam Trox, Serten, DangerousJXD, Stabila711 and Anonymous: 98

- **National Association of Manufacturers** *Source:* https://en.wikipedia.org/wiki/National_Association_of_Manufacturers?oldid=684511957 *Contributors:* Bryan Derksen, Alan Liefting, Richard Myers, Wmahan, Mike R, DragonflySixtyseven, Rich Farmbrough, Clawed, LeeHunter, Closeapple, Hooperbloob, Bookandcoffee, Kbdank71, Tim!, Lockley, Malcolma, Avraham, Pegship, SmackBot, Frap, Richardjames444, IdioT.SavanT.i4, Jockgill, Adam sk, Jac16888, Cydebot, Bellerophon5685, JohnnE, Tewapack, Burt57, Escarbot, Darklilac, MECU, Magioladitis, BateRaiko, PubliusPresent, Ugajin, Jayden54, Funandtrvl, SieBot, Busy Stubber, NoVaGOPer, DumZiBoT, Addbot, Zarcadia, Jimiskelly, Yobot, Johnzoeller, Carrite, SD5, FrescoBot, Namwiki09, Chaiswside, Tim1357, Rahvusooper, Ardmore56, ClueBot NG, Groupuscule, Cberedo, Jadererth, HistoricMN44, Magnunath, Michipedian and Anonymous: 32

- **Oregon Petition** *Source:* https://en.wikipedia.org/wiki/Oregon_Petition?oldid=685045534 *Contributors:* Ezra Wax, William M. Connolley, Dragons flight, Raul654, Securiger, TMLutas, Alan Liefting, Isidore, Gunnar Larsson, Tothebarricades.tk, Lumidek, Neutrality, Darkroom, Spiffy sperry, Rich Farmbrough, Vsmith, Guettarda, Causa sui, Viriditas, Rd232, John Quiggin, Arvedui, Tony Sidaway, Count Iblis, Ondrejk, BrianGormanly, NCdave, Rjwilmsi, Tim!, Common Man, YurikBot, RussBot, Gaius Cornelius, Dysmorodrepanis~enwiki, Velibos, JPMcGrath, Mnyakko, Crumley, 2over0, Arthur Rubin, Wsiegmund, John Broughton, SmackBot, Jim62sch, DLH, Squiddy, RDBrown, Raymond arritt, Colonies Chris, VJDocherty, Cybercobra, Monoape, BullRangifer, Peteforsyth, Valfontis, Pflatau, Cbrown1023, Bertport, Aristotle1990, Zinjixmaggir, Cydebot, Vanished user 2340rujowierfj08234irjwfw4, Reywas92, Garik, Thijs!bot, KimDabelsteinPetersen, Qwarto, Second Quantization, AntiVandalBot, Blue Tie, Prolog, Zeeboid, Absolon, R'n'B, AgarwalSumeet, Trusilver, Xyzt1234, McSly, Aboutmovies, Rwbrick, Birdbrainscan, Andyvphil, Criterion99, TXiKiBoT, Java7837, Knowsetfree, Feberle, Almostacowboy77, Yilloslime, CreateSomeNoise, Jimmi Hugh, GirasoleDE, Vampromero, Araignee, 621PWC, Sbowers3, Contontos, Lightmouse, Spartan-James, Dust Filter, Petzl, Sagredo, ClueBot, Callonjim, Artrobinson, Ftsnorf, Nymf, The Founders Intent, DumZiBoT, Kbdankbot, Addbot, CurtisSwain, Balthazar132, Deamon138, Keepcalmandcarryon, SpBot, PropagandaPolice, Olsonjs444, Idislikeusernames, Ben45750, Yngvadottir, AnomieBOT, Shock Brigade Harvester Boris, Citation bot, Twri, Apples grow on pines, WPisgreat, FrescoBot, Wikispan, CONAMERICAN, Bioextra, Trappist the monk, JournalScholar, Fresno Area Rapid Transit, Efb18, DarknessShines2, KurtR, Carol Whit, Tucci78, BobbieCharlton, Solomonfromfinland, Thepisky, H3llBot, Hmelman, 78Mustang, ThePowerofX, AndyTheGrump, Zytigon, BG19bot, Northamerica1000, Mogism, Everymorning, Monkbot and Anonymous: 58

- **Pattern Recognition in Physics** *Source:* https://en.wikipedia.org/wiki/Pattern_Recognition_in_Physics?oldid=669163144 *Contributors:* Kku, William M. Connolley, Vsmith, Ohnoitsjamie, Cydebot, DumbBOT, MichaelMaggs, Seaphoto, Edokter, Scott Illini, TheRedPenOfDoom, Yobot, AnomieBOT, Ouadfeul, ClueBot NG, Mark Arsten, Dreambeaver, BattyBot, Jeremy112233, Randykitty, Everymorning, Monkbot, Josef albert, Intuitive2000, Brother2100 and Anonymous: 4

- **Politics of global warming** *Source:* https://en.wikipedia.org/wiki/Politics_of_global_warming?oldid=687770212 *Contributors:* Ed Poor, Michael Hardy, Mac, William M. Connolley, Tpbradbury, SEWilco, Indefatigable, Raul654, Pstudier, Denelson83, The lorax, Stephan Schulz, Mattflaschen, Pengo, Alan Liefting, MSGJ, Cool Hand Luke, Spiffy sperry, Rich Farmbrough, Vsmith, LindsayH, Bobo192, Nigelj, Nsaa, Rd232, John Quiggin, Wdfarmer, Woohookitty, Benhocking, ESMtll, Queerudite, Tabletop, Rnt20, BD2412, Kbdank71, Drbogdan, Jorunn, Rjwilmsi, Ground Zero, Ahunt, Grafen, Kvn8907, Ashwinr, Nethgirb, Pym98, Closedmouth, Arthur Rubin, NHSavage, Johnpseudo, Shubi, Wizofaus, DocendoDiscimus, SmackBot, Tobias Schmidbauer, Chris the speller, BrendelSignature, Colonies Chris, John D. Croft, Richard001, Nrcprm2026, Will Beback, Gobonobo, Woer$, NYCJosh, Chris 42, Pflatau, WedgeMan, Dean1970, Hu12, Iridescent, Joseph Solis in Australia, Morgan Wick, Cyrusc, Pseudo-Richard, Zinjixmaggir, Cydebot, Dr.enh, Ssilvers, SteveMcCluskey, Sweikart, Dubc0724, KimDabelsteinPetersen, Gralo, Id447, Ninten, BenB4, Geniac, Skyemoor, Tedickey, Twotablets, Brusegadi, Cgingold, Gabriel Kielland, Demosfoni, IwantCleanAir, Oren0, Pauly04, MartinBot, Sm8900, Mschel, PrestonH, AstroHurricane001, R. Baley, UBeR, InspectorTiger, Antony-22, Birdbrainscan, Misha bb, Woood, Atama, VolkovBot, Johnfos, Kjell.kuehne, Dfarrar, Don't lose that number, Ruanua, Jason Patton, Gmb92, Smaug123, Decoratrix, Calatayudboy, Vice regent, FifeOpp08, Cyberfan, Joernscherzer, Mrfebruary, ClueBot, Polentario, Kennvido, Watti Renew, NovaDog, Niceguyedc, MssngrDeath, GoRight, Rhododendrites, DumZiBoT, Zenwhat, The Noosphere, ChyranandChloe, Addbot, JackFeschuk, Driving and Crying, Jarble, Yobot, Legobot II, AnomieBOT, Nepomuk 3, Fifth Fish Finger, Plucas58, Googlemeister, Fulldate unlinking bot, Hessamnia, Vanished user eijw98u34oi23ihf, RjwilmsiBot, Valentin Zahrnt, Torontokid2006, John of Reading, Rami radwan, Dewritech, GoingBatty, Solomonfromfinland, Vanished 1850, ThePowerofX, Coastwise, Widr, Helpful Pixie Bot, Curb Chain, Gob Lofa, NewsAndEventsGuy, Northamerica1000, 1292simon, Justanonymous, BattyBot, TheJJJunk, Dobie80, Koopatrev, Serten, Monkbot, Xylocode, Emilyscarr, Stewi101015 and Anonymous: 49

- **Public opinion on climate change** *Source:* https://en.wikipedia.org/wiki/Public_opinion_on_climate_change?oldid=679430779 *Contributors:* Fred Bauder, William M. Connolley, ChrisO~enwiki, Alan Liefting, Varlaam, Vsmith, Dave souza, Art LaPella, Nigelj, Viriditas, Rd232,

Tony Sidaway, Russil Wvong, Jrtayloriv, Arthur Rubin, SmackBot, InverseHypercube, Squiddy, Vgy7ujm, Peterlewis, ATren, IanOfNorwich, Noha307, Cydebot, DumbBOT, JohnInDC, KimDabelsteinPetersen, Peter Gulutzan, Rbrustman, Jheiv, MastCell, Poujeaux, Oren0, R'n'B, Ian.thomson, Johnfos, HughD, Jeff G., StAnselm, Cirt, Awickert, GoRight, ZuluPapa5, NuclearWarfare, SchreiberBike, Chyranand-Chloe, CurtisSwain, Pcap, AnomieBOT, Jim1138, Shock Brigade Harvester Boris, TracyMcClark, Armbrust, MerlLinkBOT, Airborne84, Gaba p, Moonraker, Brian Everlasting, Jonkerz, Miracle Pen, RjwilmsiBot, NameIsRon, John of Reading, GoingBatty, Winner 42, K6ka, Solomonfromfinland, Fick's First, Styp68, Slikmo, Dougmcdonell, Helpful Pixie Bot, NewsAndEventsGuy, Iselilja, PhnomPencil, Frze, Darkness Shines, ChrisGualtieri, TheJJJunk, Dexbot, Cwobeel, I am One of Many, Dustin V. S., Serten, Prokaryotes, Femkemilene, Monkbot and Anonymous: 146

- **Scientific opinion on climate change** *Source:* https://en.wikipedia.org/wiki/Scientific_opinion_on_climate_change?oldid=688094870 *Contributors:* The Cunctator, Eloquence, Ed Poor, Axon, Gabbe, Sheldon Rampton, Mac, Ronz, William M. Connolley, LittleDan, Marco Krohn, Cyan, Evercat, ShaunOfTheLive, Dragons flight, SEWilco, J D, Raul654, Wetman, Chrisjj, Shantavira, Stephan Schulz, Seglea, Dumky, Ace-Myth, Rursus, TMLutas, Drstuey, Phanly, Pengo, Terjepetersen, Alan Liefting, Giftlite, MSGJ, Cool Hand Luke, NeoJustin, Duncharris, Gracefool, Matt Crypto, Bobblewik, Erich gasboy, Pcarbonn, Piotrus, Llewdor, Brian Jackson, Anirvan, Canterbury Tail, Atlastawake, Spiffy sperry, Rich Farmbrough, KillerChihuahua, Vsmith, Calion, Dave souza, GregBenson, TheOuthouseMouse, Bender235, Jonathanischoice, El C, Phoenix Hacker, Shanes, Guettarda, Causa sui, Bobo192, Nigelj, JonGwynne, Viriditas, I9Q79oL78KiL0QTFHgyc, Osbojos, Silverback, Crust, Alansohn, Diego Moya, Rd232, Hipocrite, John Quiggin, Achernar~enwiki, Cortonin, Tony Sidaway, Count Iblis, Bsadowski1, Pytom, Alai, Dryman, Dan100, Simetrical, Woohookitty, Mindmatrix, Merlinme, Jersyko, Benhocking, BlaiseFEgan, Macaddct1984, U$er, Gimboid13, Karbinski, Palica, Pmj, Drbogdan, Rjwilmsi, XP1, Josiah Rowe, Seraphimblade, Mentality, Denis Diderot, ElKevbo, Yamamoto Ichiro, Ground Zero, Jrtayloriv, TeaDrinker, Tedder, Gareth E Kegg, Bgwhite, Wavelength, Arzel, Splette, Jaymax, Brian A Schmidt, Nawlin-Wiki, EWS23, SEWilcoBot, Grafen, Georgesdelatour, Slarson, RonCram, Nethgirb, Lcl~enwiki, Rktect, Wknight94, Twelvethirteen, 2over0, Closedmouth, Arthur Rubin, Pb30, NHSavage, DGaw, Fram, Naught101, John Broughton, Hal peridol, SmackBot, Amcbride, InverseHypercube, KnowledgeOfSelf, Bigbluefish, Gnangarra, Davewild, Josephprymak, DLH, Fentonrobb, Gaff, Gilliam, Squiddy, Chris the speller, Stevenwagner, Jcc1, RDBrown, Raymond arritt, Cormagh, Lexlex, Can't sleep, clown will eat me, John Hyams, DHeyward, Cregox, Rrburke, Khukri, Ratel, Earthsky, Nils Simon, Monoape, BullRangifer, Plaasjapie, Edbanky, John, Scientizzle, Gobonobo, Evildictaitor, Jaganath, Tktktk, AstroChemist, Mbeychok, Peterlewis, Ghw777, Chris 42, Pflatau, Ckatz, BoyliciousDarian, Beetstra, ATren, Optakeover, Fangfufu, Nehrams2020, Iridescent, K, Frankieleelee, Adambiswanger1, Courcelles, IanOfNorwich, CalebNoble, Croctotheface, Cyrusc, CRGreathouse, CmdrObot, Wafulz, Van helsing, Mystylplx, Zinjixmaggir, Smoove Z, Yopienso, CompRhetoric, Phase Theory, Cydebot, HawkShark, Foofish, Tawkerbot4, DumbBOT, Ssilvers, Michael Johnson, The machine512, Mattisse, Thijs!bot, Epbr123, Pstanton, InSpace~enwiki, KimDabelsteinPetersen, Dfm25, N5iln, Hnchan01, Gralo, Id447, Rosarinagazo, Marek69, Itsmejudith, Second Quantization, Peter Gulutzan, Elhector, Leshalfhill, Porqin, AntiVandalBot, Prolog, Jayron32, Tillman, Wing Nut, Db099221, Nicholas Tan, Zeeboid, Nmcclana, Magioladitis, Skyemoor, VoABot II, MastCell, The Enlightened, Iceberg007, Brusegadi, Gabriel Kielland, NimNick, Spellmaster, JohnDziak, Mrathel, Seba5618, DGG, Oren0, MartinBot, GimliDotNet, Jim.henderson, Alsee, LedgendGamer, Twisted Bunny, GoodSamaritan, J.delanoy, R. Baley, Supernedved, UBeR, Andy0093, When Muffins Attack, McSly, Fairness And Accuracy For All, AntiSpamBot, RobinGrant, Birdbrainscan, Vanished user 47736712, Jorfer, Jaw123, Juliancolton, GregJackP, Foofighter20x, Merzul, Enescot, Steel1943, VolkovBot, Johnfos, Quack-Guru, Suprcel, Albabe, Childhoodsend, JohnMashey, Rasotis, Knowsetfree, Anna Lincoln, IronMaidenRocks, Voiceofreason01, Esmehwp, Gavin.collins, Saintbrendan, Alex.rosenheim, Tankred6, Theclubhq, Mackabean, Dmcq, Iceage77, Oliepedia, Snowman frosty, DonCoyote51, Brissbane, Esthameian, Freesoul111, Dawn Bard, Jfendrick, Ravensfire, Jason Patton, Araignee, Andrewjlockley, Diafygi, Flyer22 Reborn, Special4k, Gmb92, Tezp, PhilMacD, Lacessere, Zoomwsu, LonelyMarble, Vice regent, TypoBot, Artman772000, Sjoffutt, Babakathy, Mrfebruary, Martarius, ClueBot, Ropata, NickCT, The Thing That Should Not Be, Callonjim, Ralree, Ewawer, Richrakh, Der Golem, Nursebhayes, Niceguyedc, Jmccgod, Oriolpont, Awickert, Director Re, Northernhenge, GoRight, Telekenesis, Wndl42, Readin, The Founders Intent, Rhododendrites, ZuluPapa5, Cenarium, Promethean, SpudHawg948, Hans Adler, N p holmes, Obsidi, Aprock, Tired time, DumZiBoT, Zenwhat, Semitransgenic, Caldwell malt, Forbes72, Nathan Johnson, TaalVerbeteraar, Mitch Ames, Jprw, Skarebo, Jcosco, Vianello, Aunt Entropy, Renegade570829, ESO Fan, Hermoine Gingold, Addbot, CurtisSwain, Half The Way Valley, Tcncv, Edgy01, Short Brigade Harvester Boris (original), Riding on the Wind, Nigel Montcrief, SaunderM, Sirwells, Squandermania, Phactotum, Tide rolls, Verbal, Lightbot, OlEnglish, WuBot, Medallion of Phat, Jarble, J. Johnson, Yllie, Yobot, Threop, Adi, Membre, QueenCake, Neilperth, AnomieBOT, Andrewrp, Archon 2488, Jim1138, Shock Brigade Harvester Boris, Piano non troppo, Marianolu, Inthend9, Citation bot, JohnnyB256, LilHelpa, Xqbot, MonoApe, Mononomic, Unilli, ChildofMidnight, Srich32977, MikeR613, MalcolmMcDonald, Treedel, Ehermann2223, AungKhinOo, A Quest For Knowledge, Moxy, E0steven, ZevArnold, FrescoBot, Menwith, Riventree, PinkTentacle, Leightonwalter, BFJCRICKLEWOOD, MrKimber, Airborne84, Utrechtse, Citation bot 1, Wikispan, Gaba p, Edderso, Debrajon, Algoreisamonkey, Istranix, Full-date unlinking bot, Femfacal, Xfastor, SkyMachine, Lightowemon, Trappist the monk, Lotje, Dapa22, Yearston, Jeffrd10, Vanished user aoiowaiuyr894isdik43, Brauntonian, Ebb and Flo, Tbhotch, WVBluefield, Forest001, Mean as custard, RjwilmsiBot, NameIsRon, SanAntonioPete, Levelpanictwiceplus, TonySeales14, Noommos, DHooke1973, World Lever, Dusty14, John of Reading, Bobfreshwater, RA0808, Tommy2010, Geotype, Solomonfromfinland, ZéroBot, Azza1995, Sanford123445, Other Choices, H3llBot, AndrewOne, SporkBot, Ocaasi, Kean Thomas, Δ, Paulduffill, Jess, Maktesh, ThePowerofX, Sciencenews, 9Questions, Funkysurfdude, EthicsEdinburgh, Rendahl, ClueBot NG, Meat Sweats, Coastwise, Snotbot, Icantwait, Kelvin Modest, Tmeste, Helpful Pixie Bot, Bibcode Bot, NewsAndEventsGuy, ArtifexMayhem, Northamerica1000, Frze, Jbrunswick, Paulaskins, Harizotoh9, Jakebennett, Alarbus, Razzat99, Emperor Zhark, BattyBot, Besetho2, Koshvanhorn, Enonamous, Benutzer~enwiki, Padenton, EagerToddler39, Jockzain, Cwobeel, Webclient101, Mogism, Cerabot~enwiki, TippyGoomba, CuriousMind01, Truthistrue, ComfyKem, Jo-Jo Eumerus, RotlinkBot, Ruby Murray, Cybersaur, Everymorning, A Bluenose., Dustin V. S., Robertbeets, Serten, Kharkiv07, Philipmessing, Prokaryotes, Mandruss, db, Rrhthehe, Monkbot, Trackteur, Greying Wizard, In passion of time, MissPiggysBoyfriend, Leevarns, Skrakov, Qv22edwardsr, DTHerrin, Scaravich105nj, Dr. Strange1001, Parker Case, WillieMusk, Stewi101015, Nzmcraft and Anonymous: 426

- **Semmelweis reflex** *Source:* https://en.wikipedia.org/wiki/Semmelweis_reflex?oldid=677448680 *Contributors:* William Avery, Auric, Adam78, DragonflySixtyseven, Florian Blaschke, Bender235, Bdamokos, SmackBot, InverseHypercube, Alaibot, Boffob, Misarxist, Doranchak, Gracoo2, Seb az86556, Tapalmer99, Flyer22 Reborn, DumZiBoT, MystBot, Addbot, Power.corrupts, Margaret9mary, Yobot, Plasticbot, Citation bot, LilHelpa, Wireless Keyboard, Citation bot 1, Seren-dipper, Zellskelington, Solomonfromfinland, AguC, Polisher of Cobwebs, LordIlford, Frietjes, BattyBot and Anonymous: 13

- **State of Fear** *Source:* https://en.wikipedia.org/wiki/State_of_Fear?oldid=681250469 *Contributors:* Ed Poor, SimonP, Edward, TimShell,

William M. Connolley, Furrykef, K1Bond007, Raul654, Owen, Dale Arnett, Stephan Schulz, Lowellian, Sunray, Wally, Alan Liefting, Nat Krause, Inter, SWAdair, Dvavasour, Gzuckier, Fangz, GeoGreg, Lumidek, Spec, TimLambert, AliveFreeHappy, Discospinster, Rhobite, Kdammers, Vsmith, Xezbeth, Pavel Vozenilek, Goochelaar, CanisRufus, Remember, Sietse Snel, RoyBoy, Korivak, Arcadian, Gary, Eleland, Sheehan, Jtalledo, DrBat, Ninebelow, John Quiggin, Splat, Mac Davis, Batmanand, VladimirKorablin, Ronark, Cburnett, Tony Sidaway, Woohookitty, Fred J, Tabletop, Stancollins, Isnow, Ashmoo, Magister Mathematicae, Rjwilmsi, Miros~enwiki, Chwyatt, MattWright, Phantomsteve, Splette, Davemck, Historymike, 2over0, Arthur Rubin, Rlove, A bit iffy, SmackBot, Tobias Schmidbauer, Grey Shadow, Cla68, Evanreyes, Alexan es, Kevinalewis, Squiddy, RDBrown, Julian Morrison, Afasmit, Somewildthingsgo, Schrodinger82, Pennydreadful, DHeyward, Juancnuno, Shunpiker, Madman2001, Iam4Lost, Metamagician3000, Vina-iwbot~enwiki, Gobonobo, Jaganath, JoshuaZ, BillFlis, Optakeover, Cbrown1023, Octane, Tawkerbot2, CmdrObot, Gilroy0, Sidewinder468, Kalaong, JettaMann, Cydebot, Reywas92, DumbBOT, Ssilvers, Query~enwiki, Thijs!bot, KimDabelsteinPetersen, Horologium, Peter Gulutzan, Catsmoke, Seaphoto, Prolog, Tjmayerinsf, JAnDbot, Pejorative.majeure, Albany NY, Curious Violet, Barney Gumble, .anacondabot, RS12, Meeples, Magioladitis, JamesBWatson, Brusegadi, Diotime, DGG, The velociraptor, Q Original, R'n'B, RJBurkhart3, Jkane181, Eowbotm1, Yeti Hunter, Alex2706, Jaundre100, Plasticup, Wavemaster447, DASonnenfeld, Enescot, GrahamHardy, Managerpants, BoogaLouie, DirkLangeveld, Kyle the bot, Blahaccountblah, Snowbot, Billinghurst, Araken Starway, GirasoleDE, SieBot, Rwos, Jmj713, Sagredo, ClueBot, Mariordo, Iserra, GoRight, John Nevard, Wprlh, Zhokuai, Thingg, Chaparral2J, Vanished User 1004, DumZiBoT, Tsagoy, WikHead, Rndmthght, Addbot, Hda3ku, Donkament2K, Captain Obvious and his crime-fighting dog, David0811, Jarble, Ride4Ruin, Yobot, Sageo, Punctilius, Vrinan, MassimoAr, AnomieBOT, Aryeh M. Friedman, Bubblessoc, Shock Brigade Harvester Boris, Vextration, Materialscientist, Citation bot, A Cut Above Ye, LilHelpa, YAG490, Xqbot, Ystil, Unilli, Sam Yi, J04n, Chasethesky, LivingBot, Chaheel Riens, Strang Butz, Plot Spoiler, Menwith, Citation bot 1, Tedeshi, Rell Karpish, Thad Riley, GutturalJeffo, Full-date unlinking bot, Kut or Bait Fish, Trappist the monk, WesUGAdawg, A Kut Above You, Tarminagaia, CobraBot, EvanHarper, Realplex, Qstars, RjwilmsiBot, Drakester6969, Slon02, DarknessShines2, Gouldani, Tucci78, Strangebuddix, Afeard, Wham Bam Rock II, MithrandirAgain, Stympkin, H3llBot, JackHeslop91, Seniortrend, Sailsbystars, Pinktus, Vesmit, CCRidre, ClueBot NG, Theaitetos, Hargrove89, GeorgPl, Axxas, Helpful Pixie Bot, Blake Burba, Maxellus, 86.** IP, 23haveblue, Luke 19 Verse 27, Comatmebro, Hmainsbot1, Mogism, BreakfastJr, CensoredScribe, IHaveAMastersDegree, Monkbot, Spanachan, Transcendentalist01, Revelacious, Iaepetus and Anonymous: 248

- **The Real Global Warming Disaster** *Source:* https://en.wikipedia.org/wiki/The_Real_Global_Warming_Disaster?oldid=685071280 *Contributors:* Gabbe, William M. Connolley, ChrisO~enwiki, Stephan Schulz, Vsmith, Dave souza, Guettarda, Nigelj, Viriditas, I9Q79oL78KiL0QTFHgyc, Nsaa, Hipocrite, Tony Sidaway, Tim!, Wavelength, RadioFan, Pyrotec, 2over0, SmackBot, Cla68, DroEsperanto, Tasty monster, John, Peterlewis, Collect, Alanmaher, Cydebot, Michael C Price, DumbBOT, KimDabelsteinPetersen, JustAGal, Prolog, Harryzilber, Magioladitis, Hamiltonstone, GregJackP, Scott Illini, DASonnenfeld, GrahamHardy, Hammersoft, Johnfos, Majoreditor, Geometry guy, Jack Merridew, Sphilbrick, EoGuy, Drmies, ZuluPapa5, Frederico1234, Nathan Johnson, Jprw, Tassedethe, Jarble, Yobot, AnomieBOT, LilHelpa, A Quest For Knowledge, Moxy, Green Cardamom, Lynn Dulsok, Gimmetoo, Bunfutzian, Xtzou, Fleurdalis, Kenoshay, Plush93, JustisK, Bfd5789, Fred G. Hall, Ὁ οἶστρος, VLB Pocketspup, BillMarrsx, BryantLee, Caring Butz, The Last Methane Bender, Pinktus, SJTH, JJrar7, Zytigon, Helpful Pixie Bot, NewsAndEventsGuy, Northamerica1000, Coldnorthwind and Anonymous: 5

- **The Republican War on Science** *Source:* https://en.wikipedia.org/wiki/The_Republican_War_on_Science?oldid=682159034 *Contributors:* Mrand, AnonMoos, FeloniousMonk, Loremaster, Vsmith, Viriditas, Awk~enwiki, Rjwilmsi, Gadget850, Pegship, Arthur Rubin, JQF, Ratarsed, Aelfthrytha, Bduke, PullTheWires, Cydebot, Teratornis, RobotG, Egpetersen, David Shankbone, CycloneChibi, Cgingold, Yyy555, Johnfos, Digwuren, KPH2293, Randy Kryn, Twinsday, ClueBot, Fadesga, Baegis, Sesquihypercerebral, CurtisSwain, Chimeric Glider, El Gato Gordo, Citation bot, Lionelt, RjwilmsiBot, Tesseract2, Solomonfromfinland, Anaconda88775, Wingman4l7, Helpful Pixie Bot, MrBill3, Zackmann08, Stabila711 and Anonymous: 44

- **United States Chamber of Commerce** *Source:* https://en.wikipedia.org/wiki/United_States_Chamber_of_Commerce?oldid=686444701 *Contributors:* SimonP, Andrewman327, Wetman, David Gerard, JamesMLane, Bhuck, Grunners, Rich Farmbrough, Chadlupkes, Viriditas, Pschemp, Grutness, Arthena, Rd232, Mailer diablo, Snowmanmelting, Woohookitty, Zaorish, Tabletop, SDC, Betsythedevine, Plau, Tim!, Gryffindor, Vegaswikian, Ian Pitchford, Ground Zero, Crazycomputers, JYOuyang, Common Man, Wavelength, Arzel, RussBot, Matt Fitzpatrick, Keithonearth, Korny O'Near, Rjensen, Moe Epsilon, Doncram, Djdaedalus, Bdell555, Delirium of disorder, Ageekgal, True Pagan Warrior, SmackBot, Rrius, Gilliam, Colonies Chris, Senatorpjt, Threeafterthree, Dan.omaley, VictorFRodriguez~enwiki, DDima, Kuru, Gobonobo, JHunterJ, Astuishin, Mets501, Levineps, WilliamJE, JHP, Eastlaw, Cydebot, DumbBOT, JamesAM, Thijs!bot, IIcobb, Fatid iot1234, Opertinicy, Roleplayer, SiobhanHansa, Magioladitis, VoABot II, MastCell, Raftensim, Tedickey, Schurkey, Pvosta, R'n'B, Ancapistan, Elkost, Moglex~enwiki, White 720, Sigmundur, Steel1943, Philip Trueman, TXiKiBoT, Flyte35, Jenseits~enwiki, Samiharris, Bclc uscc, Temporaluser, S.Örvarr.S, Barrympls, Aspects, Onopearls, Randy Kryn, Explicit, Blpdc, Watti Renew, Trivialist, Amcham, John Nevard, Aitias, Indiejade, Fcdc, AgnosticPreachersKid, WikHead, Addbot, National Chamber Foundation, Slref, Cuaxdon, Tuscumbia, Zarcadia, Inetdog, Myk60640, Lightbot, JEN9841, Luckas-bot, Yobot, DrFleischman, Penguino35, AnomieBOT, Mierk, Shock Brigade Harvester Boris, Bluerasberry, Xqbot, TheAMmollusc, MakeBelieveMonster, Blucruz, Kernel.package, Alvin Seville, Naseredin, A.amitkumar, Captain00dle, FrescoBot, Tapas123, Aleister Wilson, BenzolBot, Jun Nijo, Msmith122, Tinton5, Lars Washington, Croatia79, Rasaddin, Larry Dunn of Bakersfield, Kgrad, Lotje, Lucobrat, RjwilmsiBot, SafeScience, Richard LaBorde, Érico Júnior Wouters, Napkin65, Daveburstein, H3llBot, Sailsbystars, Amchampoland, NeutralityPersonified, Druluv75, JosephMichaelChic, Megwd, ClueBot NG, Wikitheskyline, AveVeritas, NewsAndEventsGuy, Kukja33, Njdemocrat, FiveColourMap, Greatstufft, Cberedo, Lieutenant of Melkor, BattyBot, Ziggypowe, Redtidal, Listroiderbob, HistoricMN44, Ydecreux, SomeFreakOnTheInternet, EcoNews08, Matthias7490, HoundPowder, SailsbyGPS, Jurisdicta, Coriantumr15 and Anonymous: 128

3.2 Images

- **File:2000_Year_Temperature_Comparison.png** *Source:* https://upload.wikimedia.org/wikipedia/commons/c/c1/2000_Year_Temperature_Comparison.png *License:* CC-BY-SA-3.0 *Contributors:* ? *Original artist:* ?

- **File:2007_Arctic_Sea_Ice.jpg** *Source:* https://upload.wikimedia.org/wikipedia/commons/f/f0/2007_Arctic_Sea_Ice.jpg *License:* Public domain *Contributors:* NASA, http://earthobservatory.nasa.gov/Newsroom/NewImages/images.php3?img_id=17800 en:NASA Earth Observatory *Original artist:* NASA image created by Jesse Allen, using AMSR-E data courtesy of the National Snow and Ice Data (NSIDC), and sea ice extent contours courtesy of Terry Haran and Matt Savoie, NSIDC, based on Special Sensor Microwave Imager (SSM/I) data.

- **File:2010_Updated_NAM_Logo_from_Rebranding.jpg** *Source:* https://upload.wikimedia.org/wikipedia/commons/9/9b/2010_Updated_ NAM_Logo_from_Rebranding.jpg *License:* CC BY-SA 3.0 *Contributors:* Template:National Association of Manufacturers *Original artist:* National Association of Manufacturers

- **File:97%_of_Climate_Scientists_Confirm_Anthroprogenic_Global_Warming.svg** *Source:* https://upload.wikimedia.org/wikipedia/commons/ 4/41/97%25_of_Climate_Scientists_Confirm_Anthroprogenic_Global_Warming.svg *License:* CC BY-SA 3.0 *Contributors:* Own work. Original source: "Skeptical Science" blog, 11 May 2011 by John Cook *Original artist:* Sagredo

- **File:A_coloured_voting_box.svg** *Source:* https://upload.wikimedia.org/wikipedia/en/0/01/A_coloured_voting_box.svg *License:* Cc-by-sa-3.0 *Contributors:* ? *Original artist:* ?

- **File:Aegopodium_podagraria1_ies.jpg** *Source:* https://upload.wikimedia.org/wikipedia/commons/b/bf/Aegopodium_podagraria1_ies.jpg *License:* CC-BY-SA-3.0 *Contributors:* Own work *Original artist:* Frank Vincentz

- **File:Ambox_current_red.svg** *Source:* https://upload.wikimedia.org/wikipedia/commons/9/98/Ambox_current_red.svg *License:* CC0 *Contributors:* self-made, inspired by Gnome globe current event.svg, using Information icon3.svg and Earth clip art.svg *Original artist:* Vipersnake151, penubag, Tkgd2007 (clock)

- **File:Ambox_important.svg** *Source:* https://upload.wikimedia.org/wikipedia/commons/b/b4/Ambox_important.svg *License:* Public domain *Contributors:* Own work, based off of Image:Ambox scales.svg *Original artist:* Dsmurat (talk · contribs)

- **File:Antarctic_Temperature_Trend_1981-2007.jpg** *Source:* https://upload.wikimedia.org/wikipedia/commons/9/99/Antarctic_Temperature_ Trend_1981-2007.jpg *License:* Public domain *Contributors:* en:Internet Archive - https://web.archive.org/web/20070823123915/http://earthobservatory. nasa.gov/Newsroom/NewImages/images.php3?img_id=17838 (originally http://earthobservatory.nasa.gov/Newsroom/NewImages/images.php3? img_id=17838 NASA Earth Observatory) *Original artist:* Robert Simmon

- **File:Arctic_September_sea_ice_decline.png** *Source:* https://upload.wikimedia.org/wikipedia/en/2/2d/Arctic_September_sea_ice_decline. png *License:* Public domain *Contributors:*

  http://nsidc.org/arcticseaicenews/files/2014/10/monthly_ice_NH_09.png *Original artist:*

  NSIDC

- **File:Brain.png** *Source:* https://upload.wikimedia.org/wikipedia/commons/7/73/Nicolas_P._Rougier%27s_rendering_of_the_human_brain. png *License:* GPL *Contributors:* http://www.loria.fr/~{}rougier *Original artist:* Nicolas Rougier

- **File:Climate_Change_Attribution.png** *Source:* https://upload.wikimedia.org/wikipedia/commons/a/a2/Climate_Change_Attribution.png *License:* CC-BY-SA-3.0 *Contributors:* This figure was created by Robert A. Rohde from published data *Original artist:* Robert A. Rohde

- **File:Climate_change_awareness_by_country_2008-2009.png** *Source:* https://upload.wikimedia.org/wikipedia/commons/4/47/Climate_change_ awareness_by_country_2008-2009.png *License:* GFDL *Contributors:* Pelham, Brett (22 Apr 2009). Awareness, Opinions About Global Warming Vary Worldwide. Gallup. Retrieved on 22 Dec 2009. *Original artist:* GunnMap

- **File:Climate_change_concern_by_country_2008-2009.png** *Source:* https://upload.wikimedia.org/wikipedia/commons/0/0e/Climate_change_ concern_by_country_2008-2009.png *License:* GFDL *Contributors:* Top-Emitting Countries Differ on Climate Change Threat. Gallup (7 Dec 2009). Retrieved on 22 Dec 2009. *Original artist:* GunnMap

- **File:Climate_change_opinion_cause_is_human_by_country_2008-2009.png** *Source:* https://upload.wikimedia.org/wikipedia/commons/ 6/6d/Climate_change_opinion_cause_is_human_by_country_2008-2009.png *License:* GFDL *Contributors:* Pelham, Brett (22 Apr 2009). Awareness, Opinions About Global Warming Vary Worldwide. Gallup. Retrieved on 22 Dec 2009. *Original artist:* GunnMap

- **File:Climate_science_opinion2.png** *Source:* https://upload.wikimedia.org/wikipedia/commons/a/a7/Climate_science_opinion2.png *License:* CC BY 3.0 *Contributors:* ? *Original artist:* ?

- **File:Commons-logo.svg** *Source:* https://upload.wikimedia.org/wikipedia/en/4/4a/Commons-logo.svg *License:* ? *Contributors:* ? *Original artist:* ?

- **File:Crystal_energy.svg** *Source:* https://upload.wikimedia.org/wikipedia/commons/1/14/Crystal_energy.svg *License:* LGPL *Contributors:* Own work conversion of Image:Crystal_128_energy.png *Original artist:* Dhatfield

- **File:DanielPatrickMoynihan.jpg** *Source:* https://upload.wikimedia.org/wikipedia/commons/f/fd/DanielPatrickMoynihan.jpg *License:* Public domain *Contributors:* See http://bioguide.congress.gov/scripts/biodisplay.pl?index=M001054. *Original artist:* ?

- **File:Decrease_Positive.svg** *Source:* https://upload.wikimedia.org/wikipedia/commons/9/92/Decrease_Positive.svg *License:* Public domain *Contributors:*

- Decrease2.svg *Original artist:* Decrease2.svg: Sarang

- **File:Edit-clear.svg** *Source:* https://upload.wikimedia.org/wikipedia/en/f/f2/Edit-clear.svg *License:* Public domain *Contributors:* The *Tango! Desktop Project*. *Original artist:*

  The people from the Tango! project. And according to the meta-data in the file, specifically: "Andreas Nilsson, and Jakub Steiner (although minimally)."

- **File:Enso-global-temp-anomalies.png** *Source:* https://upload.wikimedia.org/wikipedia/commons/f/f9/Enso-global-temp-anomalies.png *License:* Public domain *Contributors:* http://www.ncdc.noaa.gov/sotc/global/2012/13 *Original artist:* NOAA

- **File:Fourier2.jpg** *Source:* https://upload.wikimedia.org/wikipedia/commons/a/aa/Fourier2.jpg *License:* Public domain *Contributors:* Originally from en.wikipedia; description page is/was here. *Original artist:* Original uploader was User:Bunzil at en.wikipedia

- **File:GISS_temperature_2000-09_lrg.png** *Source:* https://upload.wikimedia.org/wikipedia/commons/b/bc/GISS_temperature_2000-09_lrg.png *License:* Public domain *Contributors:* NASA Earth Observatory Image of the Day: 2009 Ends Warmest Decade on Record http://earthobservatory.nasa.gov/IOTD/view.php?id=42392 *Original artist:* NASA images by Robert Simmon, based on data from the Goddard Institute for Space Studies.

- **File:GISS_temperature_palette.svg** *Source:* https://upload.wikimedia.org/wikipedia/commons/1/10/GISS_temperature_palette.svg *License:* Public domain *Contributors:* Based on File:GISS temperature palette.png, from NASA Earth Observatory Image of the Day: 2009 Ends Warmest Decade on Record http://earthobservatory.nasa.gov/IOTD/view.php?id=42392 *Original artist:* NASA images by Robert Simmon, based on data from the Goddard Institute for Space Studies. Vectorized by User:Dcoetzee.

- **File:GISTEMPvsHansen1988.png** *Source:* https://upload.wikimedia.org/wikipedia/commons/2/21/GISTEMPvsHansen1988.png *License:* CC BY-SA 3.0 *Contributors:* Own work *Original artist:* Sailsbystars

- **File:GISTEMPvsIPCC1990.png** *Source:* https://upload.wikimedia.org/wikipedia/commons/a/af/GISTEMPvsIPCC1990.png *License:* CC BY-SA 3.0 *Contributors:* Own work *Original artist:* Sailsbystars

- **File:Global_Temperature_Anomaly.svg** *Source:* https://upload.wikimedia.org/wikipedia/commons/f/f8/Global_Temperature_Anomaly.svg *License:* Public domain *Contributors:* http://data.giss.nasa.gov/gistemp/graphs/ *Original artist:* NASA Goddard Institute for Space Studies

- **File:Global_Warming_Map.jpg** *Source:* https://upload.wikimedia.org/wikipedia/commons/8/8c/Global_Warming_Map.jpg *License:* CC-BY-SA-3.0 *Contributors:* ? *Original artist:* ?

- **File:Global_Warming_Observed_CO2_Emissions_from_fossil_fuel_burning_vs_IPCC_scenarios.svg** *Source:* https://upload.wikimedia.org/wikipedia/commons/2/2d/Global_Warming_Observed_CO2_Emissions_from_fossil_fuel_burning_vs_IPCC_scenarios.svg *License:* CC BY-SA 3.0 *Contributors:* Based on File:Global Warming Observed CO2 Emissions from fossil fuel burning vs IPCC scenarios.jpg, originally from http://www.skepticalscience.com/graphics.php *Original artist:* Dana Nuccitelli, vectorized by User:Dcoetzee

- **File:Gnome-searchtool.svg** *Source:* https://upload.wikimedia.org/wikipedia/commons/1/1e/Gnome-searchtool.svg *License:* LGPL *Contributors:* http://ftp.gnome.org/pub/GNOME/sources/gnome-themes-extras/0.9/gnome-themes-extras-0.9.0.tar.gz *Original artist:* David Vignoni

- **File:H&WWonWashMon.png** *Source:* https://upload.wikimedia.org/wikipedia/en/7/70/H%26WWonWashMon.png *License:* ? *Contributors:* ? *Original artist:* ?

- **File:Holocene_Temperature_Variations.png** *Source:* https://upload.wikimedia.org/wikipedia/commons/c/ca/Holocene_Temperature_Variations.png *License:* CC-BY-SA-3.0 *Contributors:* ? *Original artist:* ?

- **File:Karl_15_Temps_before_and_after_corrxnx.png** *Source:* https://upload.wikimedia.org/wikipedia/en/9/98/Karl_15_Temps_before_and_after_corrxnx.png *License:* Fair use *Contributors:*
http://www.yaleclimateconnections.org/2015/06/new-noaa-reports-shows-no-recent-warming-slowdown-or-pause/ *Original artist:* ?

- **File:MichaelCrighton_StateOfFear.jpg** *Source:* https://upload.wikimedia.org/wikipedia/en/1/1e/MichaelCrighton_StateOfFear.jpg *License:* Fair use *Contributors:*
It is believed that the cover art can or could be obtained from HarperCollins. *Original artist:* ?

- **File:OECD-non-OECD-GDP-1990-2035-DOEEIA-IEO-2011.png** *Source:* https://upload.wikimedia.org/wikipedia/commons/2/26/OECD-non-OECD-GDP-png *License:* Public domain *Contributors:* http://www.eia.gov/forecasts/ieo/pdf/0484(2011).pdf *Original artist:* US Government

- **File:PRIP_journal_cover.png** *Source:* https://upload.wikimedia.org/wikipedia/en/4/4f/PRIP_journal_cover.png *License:* Fair use *Contributors:* http://wattsupwiththat.files.wordpress.com/2014/01/prp-cover-web.png *Original artist:* Copernicus Publications

- **File:Plato-raphael.jpg** *Source:* https://upload.wikimedia.org/wikipedia/commons/4/4a/Plato-raphael.jpg *License:* Public domain *Contributors:* Unknown *Original artist:* Raphael

- **File:Polarbearonice.jpg** *Source:* https://upload.wikimedia.org/wikipedia/commons/1/19/Polarbearonice.jpg *License:* Public domain *Contributors:* http://alaska.usgs.gov/science/biology/polar_bears/size.html *Original artist:* USGS

- **File:Politics_of_global_warming.jpg** *Source:* https://upload.wikimedia.org/wikipedia/en/1/1e/Politics_of_global_warming.jpg *License:* CC-BY-SA-3.0 *Contributors:*
This item was created in December 2012 using vector graphics tools and GNU v3 licensed art
Previously published: 2012-12-03
Original artist:
Justanonymous

- **File:Psi2.svg** *Source:* https://upload.wikimedia.org/wikipedia/commons/6/6c/Psi2.svg *License:* Public domain *Contributors:* ? *Original artist:* ?

- **File:Question_book-new.svg** *Source:* https://upload.wikimedia.org/wikipedia/en/9/99/Question_book-new.svg *License:* Cc-by-sa-3.0 *Contributors:*
Created from scratch in Adobe Illustrator. Based on Image:Question book.png created by User:Equazcion *Original artist:*
Tkgd2007

- **File:Real_Global_Warming_Disaster_book_cover.jpg** *Source:* https://upload.wikimedia.org/wikipedia/en/4/45/Real_Global_Warming_Disaster_book_cover.jpg *License:* Fair use *Contributors:*
Derived from a digital capture (photo/scan) of the book cover (creator of this digital version is irrelevant as the copyright in all equivalent images is still held by the same party). Copyright held by the publisher or the artist. Claimed as fair use regardless.
Original artist: ?

- **File:Satellite_Temperatures.png** *Source:* https://upload.wikimedia.org/wikipedia/commons/7/7e/Satellite_Temperatures.png *License:* CC-BY-SA-3.0 *Contributors:* http://www.cru.uea.ac.uk/cru/data/temperature/ *Original artist:* Robert A. Rohde

3.3 Content license

www.ingramcontent.com/pod-product-compliance
Lightning Source LLC
Chambersburg PA
CBHW080805180526
45168CB00006B/2333